Alexander Henry

A glossary of scientific terms for general use

Alexander Henry

A glossary of scientific terms for general use

ISBN/EAN: 9783337414269

Printed in Europe, USA, Canada, Australia, Japan

Cover: Foto ©berggeist007 / pixelio.de

More available books at **www.hansebooks.com**

A GLOSSARY

OF

SCIENTIFIC TERMS

FOR GENERAL USE

BY

ALEXANDER HENRY, M.D.

LONDON
JAMES WALTON, 137, GOWER STREET
1861.

PREFACE.

THIS GLOSSARY is intended to assist the student of scientific works, and the general reader, by giving the etymologies and significations of such words as are peculiar to the various sciences, together with those of common use having special meanings in science.

In drawing up the work, the author has collected the definitions, wherever practicable, from the most modern standard treatises on the different sciences. He has also availed himself of the assistance derivable from the "Imperial Dictionary," and the excellent "Expository Lexicon" of Dr. Mayne. In all cases he has endeavoured to give the definitions in as concise and simple a form as is compatible with clearness. The accentuation of the words has been carefully marked; and, for the use of those unacquainted with Greek, the Greek words have been printed in both Greek and Roman characters.

15, George Street, Portman Square, W.
November, 1860.

GLOSSARY.

A.

Ab'acus (Lat. a slab or board). An instrument for calculating, consisting of an oblong frame, across which are stretched wires, each supplied with ten balls; in *architecture*, a table forming the upper part or crowning of a column and its capital.

Abattoir' (Fr. *abattre*, to fell or strike down). A public slaughter-house.

Abdo'men (Lat. *abdo*, I hide). That cavity of the animal body in vertebrates which contains the organs of digestion; in insects, the hinder part of the body, which appears united to the fore part by a thread.

Abdom'inal (Lat. *abdômen*). Belonging to the abdomen: applied to an order of fishes which have the ventral fins attached under the abdomen behind the pectoral fins.

Abdu'cent (Lat. *ab*, from ; *duco*, I lead). Drawing away or separating.

Abduc'tion (Lat. *ab*, from ; *duco*, I lead). A drawing away.

Abduc'tor (Lat. *ab*, from ; *duco*, I lead). A leader or drawer away : applied to certain muscles.

Aber'rant (Lat. *ab*, from ; *erro*, I wander). Deviating from the type of the natural group.

Aberra'tion (Lat. *ab*, from ; *erro*, I wander). A wandering away; in *optics*, spherical aberration is indistinctness in the optical image produced by a convex lens, from the formation of images on the exterior part of the lens ; chromatic aberration, false colouring of an optical image from the decomposition of light by a lens into its primary colours ; in *astronomy*, an apparent motion of the fixed stars, by which they appear at a small distance from their real place ; in *medicine*, insanity.

Ablacta'tion (Lat. *ab*, from ; *lac*, milk). Weaning.

Abla'tion (Lat. *ab*, from; *latus*, carried). A taking away.

Ab'lative (Lat. *ab*, away; *latus*, borne). Taking away ; in *grammar*, applied to a case of nouns, denoting an action of taking away.

Ablu'tion (Lat. *ab*, from ; *lavo*, I wash). A washing.

Abnor'mal (Lat. *ab*, from ; *norma*, a rule). Not according to rule ; unnatural.

Aboma'sum (Lat. *ab*, from ; *omāsum*, the paunch). The fourth stomach of ruminant animals.

Aborig'inal (*Lat. ab*, from ; *origo*, an origin). First ; primitive ; original.

Aborig'ines (Lat. *ab*, from ; *origo*, an origin). The first or primitive inhabitants of a country.

Abor'tion (Lat. *aborto*, I miscarry). The expulsion of a fœtus before the proper term ; a miscarriage ; an incomplete formation.

Abor'tive (Lat. *aborto*, I miscarry). Unfruitful ; incomplete ; having the property of arresting development.

Abran'chiate (Gr. ἀ, *a*, not; βραγχια, *bran'chia*, gills). Without gills.

Abra'sion (Lat. *ab*, from ; *rado*, I shave). A tearing or rubbing off, as of a piece of skin.

B

Abrupt (Lat. *ab*, from ; *rumpo*, I break). Broken off ; in *botany*, applied to leaves and roots which appear as if the extremity had been cut off.

Absoess (Lat. *abscēdo*, I depart). A collection of pus or matter.

Abscis'sa (Lat. *abscin'do*, I cut off). That part of the diameter of a conic section which lies between the vertex or some other fixed point and a semi-ordinate, or the half of a straight line drawn at right angles to the axis.

Absois'sion (Lat. *ab*, away ; *scindo*, I cut). A cutting away, or removal.

Ab'solute (Lat. *ab*, from ; *solvo*, I loosen). Independent ; perfect or complete in itself ; pure.

Absorb'ent (Lat. *absorb'eo*, I sup up). Having the property of sucking or supping up fluids, as a sponge.

Absorp'tion (Lat. *absorb'eo*, I sup up). The act or process of sucking or supping up moisture.

Abster'gent (Lat. *abster'geo*, I wipe clean). Cleansing.

Abstract (Lat. *abs*, from ; *traho*, I draw). Separate ; applied to the ideas of number, properties of matter, &c., considered by themselves without reference to the subject which they qualify ; an outline of a treatise or writing.

Abstraction (Lat. *abs*, away ; *traho*, I draw). Removal ; a taking away ; the consideration of a part or property of an object independently of the rest.

Acale'phæ (Gr. ἀκαληφη, *acalēphē*, a nettle). A class of sea-animals of the radiated division ; so called because some of them, when taken in the hand, sting like nettles.

Acantha'ceous (Gr. ἀκανθα, *acantha*, a spine). Having prickles.

Acanthoceph'ala (Gr. ἀκανθα, *acantha*, a spine ; κεφαλη, *keph'alē*, the head). Intestinal worms having the head armed with spines or hooks.

Acanthopteryg'ii (Gr. ἀκανθα, *acantha*, a spine ; πτερυγιον, *pteru'gion*, a fin). An order of fishes having the first fin supported by bony spiniform rings.

Acar'diac (Gr. ἀ, *a*, not ; καρδια, *kar'dia*, a heart). Without a heart.

Acaules'cent (Gr. ἀ, *a*, not ; Lat. *caulis*, a stem). Having no stem.

Acau'lous (Gr. ἀ, *a*, not ; Lat. *caulis*, a stem). Stemless.

Accel'erate (Lat. *ad*, to ; *celer*, quick). To quicken.

Accelerated motion. In *mechanics*, that motion which constantly receives additional velocity.

Accel'erator (Lat. *ad*, to ; *cel'er*, quick). That which quickens : applied in *anatomy* to certain muscles.

Acces'sory (Lat. *accēdo*, I approach, or am added to). Added to some person or thing in a secondary relation.

Accip'itres (Lat. *ad*, to ; *capio*, I take). An order of birds including the rapacious fowl, as the eagle, vulture, hawk, &c.

Acclima'tion (Lat. *ad*, to ; Gr. κλιμα, *klima*, a region of the earth). The process of becoming accustomed to a climate.

Accliv'ity (Lat. *acclīvus*, ascending). A slope of the earth, considered as ascending.

Accre'tion (Lat. *ad*, to ; *cresco*, I grow). A growing or increase ; a growing together.

Accumula'tion (Lat. *ad*, to ; *cūmulo*, I heap up). A heaping together ; in *mechanics*, accumulation of power is the quantity of motion existing in machines after constant acceleration of the velocity of the moving body.

Aceph'ala (Gr. ἀ, *a*, not ; κεφαλη, *keph'alē*, a head). An order of invertebrate animals without a head ; including oysters, mussels, and other bivalve animals.

Aceph'alocyst (Gr. ἀκεφαλος, *akeph'alos*, headless ; κυστις, *kustis*, a bladder). A parasitic hydatid consisting of a headless cyst or bag.

Aceph'alous (Gr. ἀ, *a*, not ; κεφαλη, *keph'alē*, a head). Without a head.

Ac'erose (Lat. *acus*, chaff). In *botany*, resembling chaff : applied to leaves which are linear and permanent, as in the pine or juniper.

Aces'cent (Lat. *acesco*, I grow sour).

Having a tendency to become sour.

Acetabulif'erous (*Acetab'ulum; fero,* I bear). Having acetabula or sucking-cups.

Acetab'ulum (Lat. a saucer). The round cup-like cavity in the pelvic bone, into which the head of the thigh-bone is received; applied also to the sucking-cups of some invertebrate animals.

Ac'etate (Lat. *acētum,* vinegar). A compound of acetic acid with a base.

Ace'tic (Lat. *acētum,* vinegar). Belonging to vinegar.

Acetom'eter (Lat. *acētum,* vinegar; Gr. μετρον, *metron,* a measure). An instrument for measuring the strength of vinegar.

Ace'tous (Lat. *acētum,* vinegar). Sour: producing vinegar.

Ac'etyl (Lat. *acētum,* vinegar; Gr. ὕλη, *hulē,* material). The supposed base of vinegar and its allies.

Ache'nium (Gr. ἀ, *a,* not; χαινω, *chainō,* I gape). A form of fruit consisting of a single hard pericarp, not splitting, and inclosing a single non-adherent seed.

Achlamyd'eous (Gr. d, *a,* not; χλαμυς, *chlamus,* a garment). A term applied to plants, the flowers of which have neither calyx nor corolla.

Achromat'ic (Gr. d, *a,* not; χρωμα, *chrōma,* colour). Free from colour: applied to optical instruments in which the confusing effect of chromatic aberration, or decomposition of light into colours, is avoided.

Achro'matism (Gr. ἀ, *a,* not; χρωμα, *chrōma,* colour). Freedom from colour: applied to optical instruments which do not decompose light so as to produce colours.

Acic'ular (Lat. *acic'ula,* a little needle). Occurring in needle-like crystals.

Acid (Lat. *aceo,* I am sour). In common meaning, sour; in *chemistry,* applied to all bodies which combine with bases to form salts.

Acidifi'able (Lat. *ac'idus,* acid; *fio,* I become). Capable of being converted into an acid, or made acid.

Acid'ify (Lat. *ac'idus,* acid; *facio,* I make). To make acid, or change into an acid.

Acidim'eter (Lat. *ac'idus,* acid; Gr. μετρον, *metron,* a measure). An instrument for ascertaining the quantity of acid in a fluid.

Acid salt. In *chemistry,* a name given to some salts which have an acid reaction.

Acid'ulate (Lat. *ac'idus,* acid; dim. *ulus*). To make slightly acid.

Acid'ulous (Lat. *ac'idus,* acid; dim. *ulus*). Slightly or mildly acid.

Acinac'iform (Gr. ἀκινακης, *akinăkēs,* a scimitar; Lat. *forma,* shape). Like a scimitar; in *botany,* applied to leaves which are convex and sharp on one side, and straight and thick on the other.

Ac'ini (Lat. *ac'inus,* a grape-stone). The secreting parts of glands, when suspended like grains or small berries to a slender stem.

Acin'iform (Lat. *ac'inus,* a grape-stone; *forma,* shape). In clusters like grapes.

Ac'inose (Lat. *ac'inus,* a grape-stone). Consisting of small granular concretions.

Acme (Gr. ἀκμή, *acmē,* a point). The height or extreme limit.

Acotyle'donous (Gr. d, *a,* not; κοτυληδων, *kotulēdōn,* a cup, or seed-lobe). Having no seed-lobes, or leaves which first appear above ground.

Acous'tic (Gr. ἀκουω, *akouō,* I hear). Relating to sound and hearing.

Acous'tics (Gr. ἀκουω, *akouō,* I hear). The science which describes the phenomena of sound.

Ac'rita (Gr. ἀκριτος, *ak'ritos,* unarranged). A term applied to the lowest animals,.in which the tissues were supposed to be confusedly blended together.

Ac'rodont (Gr. ἀκρος, *akros,* at the summit; ὁδους, *odous,* a tooth). A term applied to fossil scaly saurians, which have the teeth anchylosed to the summit of the alveolar ridge.

Ac'rogen (Gr. ἀκρος, *akros,* high or extreme; γενναω, *gennaō,* I produce). A class of vegetables characterised by growing from the top or point.

Acro'mial (*Acromion*). Belonging to the acromion.

Acro'mion (Gr. ἀκρος, *akros*, high or extreme ; ὠμος, *ōmos*, a shoulder). The projecting or outer point of the shoulder.

Acrop'olis (Gr. ἀκρος, *akros*, highest ; πολις, *polis*, a city). The highest part or citadel of a city ; in particular that of Athens.

Ac'rospire (Gr. ἀκρος, *akros*, a summit ; σπειρα, *speira*, a spire). The shoot or sprout of a seed.

Acrote'rium (Gr. ἀκροτηριον, *akrotērion*). In *architecture*, a small pedestal at the angle or vertex of a pediment.

Actin'ic (Gr. ἀκτιν, *aktin*, a ray). Applied to those rays of the sun by which chemical effects are produced.

Actin'iform (Gr. ἀκτιν, *aktin*, a ray ; Lat. *forma*, form). Having a radiated form.

Ac'tinism (Gr. ἀκτιν, *aktin*, a ray). A property of certain rays of the sunbeam, by which chemical changes are produced.

Actinoc'eros (Gr. ἀκτιν, *aktin*, a ray ; κερας, *keras*, a horn). A term implying the radiated disposition of parts like horns.

Actin'olite (Gr. ἀκτιν, *aktin*, a ray or thorn ; λιθος, *lithos*, a stone). A granitic mineral composed of radiated thorn-like crystals.

Actinom'eter (Gr. ἀκτιν, *aktin*, a ray ; μετρον, *metron*, a measure). An instrument for measuring the heating power of the sun's rays.

Aculea'ta (Lat. *acu'leus*, a sting). A group of hymenopterous or membrane-winged insects, provided with stings, as wasps and bees.

Acu'leate (Lat. *acu'leus*, a prickle). Having prickles or stings.

Acu'minate (Lat. *acu'men*, a sharp point). Having a long projecting point.

Acupunc'ture (Lat. *acus*, a needle ; *pungo*, I prick). The operation of pricking with a needle.

Acute (Lat. *acūtus*, sharp). Sharp, in *geometry*, applied to an angle which is less than a right-angle ;

in *medicine*, applied to diseases which speedily come to an end.

Adaptation (Lat. *ad*, to ; *aptus*, fit). A fitting.

Addu'cent (Lat. *ad*, to ; *duco*, I lead). Leading or bringing towards.

Adduc'tion (Lat. *ad*, to ; *duco*, I lead). The act of bringing towards.

Adduc'tor (Lat. *ad*, to ; *duco*, I lead). A leader or bringer towards.

Ade'niform (Gr. ἀδην, *adēn*, a gland ; Lat. *forma*, shape). Shaped like a gland.

Adeni'tis (Gr. ἀδην, *adēn*, a gland ; *itis*, denoting inflammation). Inflammation of glands.

Ad'enoid (Gr. ἀδην, *adēn*, a gland ; ειδος, *eidos*, form). Like a gland.

Adenol'ogy (Gr. ἀδην, *adēn*, a gland ; λογος, *logos*, a word or discourse). A description of glands.

Adfec'ted (Lat. *ad*, to ; *facio*, I make). Compounded ; containing different powers of the same quantity.

Adhe'sion (Lat. *ad*, to ; *hæreo*, I stick fast). A sticking together.

Adhe'sive (Lat. *ad*, to ; *hæreo*, I stick.) Having the power of adhering ; or promoting this process.

Adipoce're (Lat. *adeps*, fat ; *cera*, wax). A peculiar substance produced in dead animal bodies in moist situations.

Ad'ipose (Lat. *adeps*, fat). Belonging to, or consisting of fat.

Adit (Lat. *adeo*, I go to). A passage or approach to a mine.

Adja'cent (Lat. *ad*, to ; *jaceo*, I lie). Lying near to.

Adjustment (Lat. *ad*, to ; *justus*, just). A fitting ; the means by which an optical instrument is fitted for taking a correct view of an object.

Admixtion (Lat. *ad*, to ; *misceo*, I mix). A mixing of different substances, without change of nature.

Adnascent (Lat. *ad*, to ; *nascor*, I am born). Growing to or on.

Adnate (Lat. *ad*, to ; *nascor*, I am born). Growing together.

Adoles'cence (Lat. *adoles'co*, I grow). The period between childhood and full growth.

Adul'terate (Lat. *ad*, to ; *alter*, the other). To corrupt or make impure by an admixture of materials of inferior quality.

Adus'tion (Lat. *ad*, to ; *uro*, I burn). A burning or heating to dryness.

Adventit'ious (Lat. *ad*, to ; *venio*, I come). Coming accidentally, or out of place.

Adynam'ic(Gr. ἀ, *a*, not; δυναμις, *du'namis*, power). Without power ; applied to invalids, in which there is diminution of the powers of life to resist the disease.

Ægoph'ony (Gr. ἀιξ, *aix*, a goat ; φωνη, *phōnē*, voice). In *medicine*, a peculiar trembling sound of the voice as heard through the chest in some diseased states, resembling the bleating of a goat.

A'ërated (Lat. *aēr*, the air). Charged with air ; applied to waters charged with carbonic acid gas.

Aëra'tion (Lat. *aēr*, the air). The art of charging with air or gas ; or of exposing soils to the action of the air.

Aë'rial (Lat. *aēr*, the air). Belonging to, or consisting of air.

A'ëriform (Lat. *aēr*, the air ; *forma*, shape). Resembling air.

Aërodynam'ics (Gr. ἀηρ, *aēr*, air ; δυναμις, *du'namis*, power). The science of the mechanical effects of air in motion.

A'ërolite (Gr. ἀηρ, *aēr*, air ; λιθος, *lithos*, a stone). A meteoric stone ; a mineral mass which falls through the air.

Aërol'ogy (Gr. ἀηρ, *aēr*, air ; λογος, *logos*, a word or description). A description of the air.

Aërom'eter (Gr. ἀηρ, *aēr*, air ; μετρον, *metron*, a measure). An instrument for ascertaining the weight of air, or the bulk of gases.

Aërom'etry (Gr. ἀηρ, *aēr*, air ; μετρον, *metron*, a measure). The science of measuring air.

A'ëronaut (Gr. ἀηρ, *aēr*, air ; ναυτης, *nautēs*, a sailor). One who sails in the air by means of a balloon.

Aëropho'bia (Gr. ἀηρ, *aēr*, air ; φοβος, *phobos*, fear). A dread of air.

A'ërophyte (Gr. ἀηρ, *aēr*, air ; φυω, *phuō*, I grow). A plant which lives in air.

Aërostat'ics (Gr. ἀηρ, *aēr*, air ; ἱστημι, *histēmi*, I weigh). The science which describes the properties of air at rest.

Æsthet'ics (Gr. αισθανομαι, *aisthan'omai*, I perceive). The science of sensation, or of the cause of mental pleasure and pain derivable from observing the works of nature and art.

Æstivation. *See* Estivation.

Affection (Lat. *ad*, to ; *facio*, I make). A disposition ; used in *medicine* in the same sense as disease.

Af'ferent (Lat. *ad*, to ; *fero*, I bring). Bringing to.

Affin'ity (Lat. *affi'nis*, near to, or bordering on). Relationship ; an agreement in most essential characters ; disposition to unite, so as to form a new substance.

Afflux (Lat. *ad*, to ; *fluo*, I flow). A flowing towards.

Affusion (Lat. *ad*, to ; *fundo*, I pour). A pouring on.

After-damp. A gas emitted in coalmines, very fatal to life ; chokedamp or carbonic acid.

Ag'amous (Gr. ἀ, *a*, not; γαμος, *gamos*, marriage). A term applied to cryptogamic plants, or those which appear to have no distinction of sexes.

Agas'tric (Gr. ἀ, *a*, not; γαστηρ, *gastēr*, a stomach). Without a stomach or intestines.

Agglom'erate (Lat. *ad*, to ; *glomus*, a roll of yarn or thread). To collect together like thread on a ball.

Agglu'tinant (Lat. *ad*, to ; *gluten*, glue). Fastening together like glue.

Agglu'tinate (Lat. *ad*, to ; *gluten*, glue). To fasten together like glue.

Ag'gregate (Lat. *ad*, to ; *grex*, a herd). To collect together into a mass ; collected together.

Aggregation (Lat. *ad*, to ; *grex*, a herd). A collection ; the act of collecting together into a mass.

Agon'ic (Gr. ἀ, *a*, not ; γωνια, *gōnia*, an angle). Without an angle : applied to two lines on the surface of the earth in which there is no declination of the magnetic needle from the meridian.

Agra'rian (Lat. *ager*, a field). Relating to lands.

Agricul'ture (Lat. *ager*, a field ; *colo*, I cultivate). The science of cultivating the ground.

Aiguille (Fr. a needle). In *physical geography*, applied to the sharp needle-like points of lofty mountains.

Air-bladder. A bladder containing air ; generally applied to a bag in the interior of fishes, capable of being filled with air—a rudimentary lung.

Air-cell. A cell or cavity containing air.

Air-pump. An instrument for withdrawing air from a vessel.

Air-sac. A receptacle for holding air.

Ak'era (Gr. ἀ, *a*, not ; κερας, *keras*, a horn). A family of mollusca without horns or feelers.

A'la (Lat.) A wing, or a projection like a wing.

A'lar (Lat. *ala*, a wing). Belonging to a wing.

A'late (Lat. *ala*, a wing). Having wings.

Albi'no (Lat. *albus*, white). A person or animal in whom the natural colouring matter of the skin, hair, and eyes, is absent.

Albugin'ea (Lat. *albūgo*, a white spot in the eye). The white appearance in front of the eye, formed by the expanded tendons of the muscles which move the organ.

Albugin'eous (Lat. *albūgo*, a white spot in the eye). Belonging to or resembling the white of the eye.

Albu'men (Lat. *albus*, white). A substance found in animals and vegetables, of which the white of egg is an example.

Albuminip'arous (Lat. *albumen*, the white of egg ; *pario*, I produce). Producing or secreting albumen.

Albu'minoid (*Albumen* ; Gr. ειδος, *eidos*, form). Resembling albumen.

Albu'minous (Lat. *albumen*, the white of egg). Belonging to or containing albumen.

Albur'num (Lat. *albus*, white). The softer wood or sap-wood, between the bark and the heart-wood.

Al'chemist (Arabic, *al*, the ; *kimia*, secret ; or Gr. χεω, *cheō*, I pour). A person who practises alchemy.

Al'chemy (Arab. *al*, the ; *kimia*, secret ; or Gr. χεω, *cheō*, I pour). The pretended science of changing other metals into gold, &c.

Al'cohol (Arab. *al*, the ; *kohol*). A fluid body produced by distillation from fermented spirits, in which it has been formed from sugar.

Alcoholism (*Alcohol*). A diseased state, arising from the excessive use of alcoholic liquors.

Alcohom'eter (*Alcohol ;* Gr. μετρον, *metron*, a measure). An instrument for determining the strength of spirits by indicating the percentage of pure alcohol.

Alem'bic (Arab. *al*, the ; *ambik*, a chemical vessel). A vessel used in distillation.

Algæ (Lat. *alga*, sea-weed). An order of cryptogamous plants, including sea-weeds.

Al'gebra (Arab. *al*, the ; *gabar* or *chabar*, to reduce parts to a whole). A method of computation in which signs (usually the letters of the alphabet) represent quantities.

Algebra'ical (*Algebra*). Pertaining to or performed by means of Algebra.

Al'gia (Gr. ἀλγος, *algos*, pain). Used as the ending of a word, denotes pain in the part spoken of.

Al'gide (Lat. *al'geo*, I am cold). Accompanied by great coldness ; applied to diseases, such as fevers and cholera.

Aliena'tion (Lat. *aliēnus*, belonging to another ; foreign). A transferring to another ; in *medicine*, insanity.

A'lienist (Fr. *alie'né*, insane). Relating to insanity : applied to physicians who specially study insanity.

A'liform (Lat. *ala*, a wing ; *forma*, shape). Shaped like a wing.

Al'iment (Lat. *alo*, I nourish). Food or nourishment.

Aliment'ary (Lat. *alo*, I nourish). Belonging to food.

Alimenta'tion (Lat. *alo*, I nourish). The act of receiving or imparting food.

Al'iquot (Lat. *aliquot*, some certain). A part which, multiplied by any entire number, exactly makes up a given whole.

Alisphe'noid (Lat. *ala*, a wing; *sphenoid*). A term applied to the part of the skull in fishes which corresponds to the alæ or wings of the sphenoid bone.

Alkales'cent (*Alkali*). Having a tendency to be or to become alkaline.

Al'kali (Arab.). A substance having the property of changing vegetable blues to red, and turmeric and rhubarb to brown, and of neutralising acids.

Alkalig'enous (*Alkali*; Gr. γενναω, *gennaō*, I produce). Producing alkali.

Alkalim'eter (*Alkali*; Gr. μετρον, *metron*, a measure). A graduated measure used by chemists in processes for ascertaining the amount of alkali in any substance.

Alkalim'etry (*Alkali*; Gr. μετρον, *metron*, a measure). The process by which the quantity of alkali in any substance is measured.

Al'kaline (*Alkali*). Having the properties of or containing an alkali.

Alkalin'ity (*Alkali*). The condition produced by an alkali.

Al'kaloid (*Alkali*; Gr. ειδος, *eidos*, form). An organic body consisting of carbon, hydrogen, nitrogen, and oxygen, having the general properties of an alkali.

Allan'toid (*Allantois*). A term applied to the vertebrate animals of which the fœtus is provided with an allantois; including mammals, birds, and reptiles.

Allan'tois (Gr. ἀλλας, *allas*, a sausage; ειδος, *eidos*, form). One of the membranes which invest the fœtus.

Allia'ceous (Lat. *allium*, garlic). Belonging to or resembling garlic.

Alliga'tion (Lat. *ad*, to; *ligo*, I bind). A tying together; a rule in *arithmetic* for finding the average price of a compound of different substances.

Allophyl'ian (Gr. ἀλλος, *allos*, another; φυλη, *phulē*, a tribe). A term applied to the races supposed to have inhabited Europe before the passage into it of the Asian nations.

Allotrop'ic (Gr. ἀλλος, *allos*, another; τρεπω, *trepō*, I turn). Having the property of existing in two or more forms with different physical properties, the composition remaining the same.

Alloy (Lat. *ad*, to; *ligo*, I bind). A compound of two or more metals.

Allu'vial (Lat. *allu'vies*, a muddy stream). Produced by deposit of mud, &c., washed down by water.

Allu'vium (Lat. *ad*, to; *lavo*, I wash). The soil or land formed of matter washed together by the ordinary operations of water.

Alope'cia (Gr. ἀλωπηξ, *alōpex*, a fox). Loss of hair: foxes have been said to be subject to it.

Alt-az'imuth. A term applied to an astronomical instrument for observing both the altitude and azimuth.

Al'terative (Lat. *alter*, another). A medicine which gradually produces a change in the constitution.

Altern'ate (Lat. *alternus*, belonging to one another). Being by turns; in *botany*, applied to branches and leaves which rise on opposite sides alternately; in *geometry*, to the internal angles made by a straight line cutting two parallel lines, and lying on opposite sides of the cutting line.

Alternate generation. A form of reproduction in which the young do not resemble the parent but the grand-parent.

Alt'itude (Lat. *altus*, high). Height; in *astronomy*, applied to the real or apparent height of a heavenly body from the horizon; in *geometry*, the distance of the vertex or summit from the base.

A'lula (Lat. *ala*, a wing). A little wing.

Aluminif'erous (Lat. *alumen*, alum; *fero*, I bear). Producing alum.

Alve'olar (Lat. *alveolus*, a socket).

Belonging to the sockets in which the teeth are fixed; containing cells or pits.

Alve'olus (Lat.) A cell or socket; in *anatomy*, the socket of a tooth; the minute depressions in the mucous membrane of the stomach are also called alveoli.

Al'vine (Lat. *alvus*, the belly). Belonging to the bowels.

Amal'gam (Gr. μαλασσω, *malas'sō*, I soften). A compound of mercury with another metal.

Amalgama'tion. A process by which silver ore is purified by mixture with mercury; a blending.

Amauro'sis (Gr. ἀμαυρος, *amauros*, dark). Blindness from loss of power in the nervous system of the eye to receive or transmit the impression of light.

Amblyg'onous (Gr. ἀμβλυς, *amblus*, obtuse; γωνια, *gōnia*, an angle). Having an obtuse angle.

Amblyo'pia (Gr. ἀμβλυς, *amblus*, dim; ὠψ, *ōps*, the eye). Amaurosis in a milder degree.

Ambula'cra (Lat. *am'bulo*, I walk). The perforated plates in the shell of echinoderms.

Am'bulance (Lat. *am'bulo*, I walk). A moveable hospital attached to an army in the field.

Am'bulatory (Lat. *am'bulo*, I walk). Made for walking.

Amenta'ceous (*Amentum*). Having flowers arranged in amenta or catkins.

Amen'tia (Lat. *a*, from or without; *mens*, the mind). Want of intellect; idiocy.

Amen'tum (Lat., a thong). In *botany*, a form of inflorescence, resembling a spike.

Ammoni'acal (*Ammonia*, the volatile alkali). Pertaining to, or containing ammonia.

Am'monite (*Ammon*, one of the titles of Jupiter, under which his statue was represented with ram's horns). A fossil shell of a cephalopod, of a spiral form.

Am'nion (Gr. ἀμνιον, *amnion*, a bowl). One of the membranes surrounding the foetus; in *botany*, a thin substance in which the embryo of a plant is suspended when it first appears.

Amniot'ic (*Amnion*). Belonging to the amnion.

Amor'phous (Gr. ἀ, *a*, not; μορφη, *morphē*, form). Without regular form; shapeless.

Amorphozo'a (Gr. ἀ, *a*, not; μορφη, *morphē*, form; ζωον, *zōon*, an animal). Animals without definite shape: applied to sponges and their allies.

Amphi (Gr. ἀμφις, *amphis*, on both sides; or, ἀμφω, *amphō*, both). A prefix signifying the co-existence of two things or properties; sometimes signifying around (from ἀμφι, *amphi*, around).

Amphiarthro'sis (Gr. ἀμφις, *amphis*, on both sides; ἀρθρον, *arthron*, a joint). A form of joint which has the properties of two others, named diarthrosis and synarthrosis, and allows slight motion.

Amphibich'nites (*Amphib'ia*, animals living both on land and in water; Gr. ἰχνος, *ichnos*, a footstep). Fossil footprints of amphibious reptiles.

Amphib'ious (Gr. ἀμφις, *amphis*, on both sides; βιος, *bios*, life). Living both on land and in water.

Am'phibrach (Gr. ἀμφις, *amphis*, on both sides; βραχυς, *brachus*, short). In *versification*, a foot consisting of two short syllables with a long one between.

Amphicœ'lia (Gr. ἀμφις, *amphis*, on both sides; κοιλος, *koilos*, hollow). A term applied to a sub-order of crocodiles which have the vertebral bones hollowed at both ends.

Am'phipods (Gr. ἀμφις, *amphis*, on both sides; πους, *pous*, a foot). An order of crustacea having feet for both walking and swimming.

Amphis'cians (Gr. ἀμφις, *amphis*, on both sides; σκια, *skia*, a shadow). The inhabitants of the tropics, whose shadows are thrown to the north in one part of the year and to the south in the other.

Amphit'ropous (Gr. ἀμφις. *amphis*, on both sides; τρεπω, *trepō*, I turn). In *botany*. applied to ovules or seeds which are attached by the middle.

Amphor'ic (Lat. *amphora*, a pitcher). Belonging to a pitcher ; in *medicine*, applied to a sound resembling that produced by speaking into an empty pitcher.

Amplex'icaul (Lat. *amplexor*, I embrace ; *caulis*, a stem). Embracing or surrounding a stem.

Am'plitude (Lat. *amplus*, large). Size, extent.

Ampul'la (Lat. a pitcher). In *botany*, applied to a leaf in which the petiole is dilated and hollowed out in the shape of a hollow vessel, open at the upper end ; in *anatomy*, to the diluted part of the membranous semicircular canals in the ear.

Amputation (Lat. *amputo*, I cut or lop off). A cutting off a limb, or some part of the body.

Amy'elous (Gr. ἀ, a, not ; μυελος, *mu'elos*, marrow). Without a spinal cord.

Amyg'daloid (Lat. *amyg'dala*, an almond ; Gr. εἶδος,*eidos*, form). Like an almond : applied in *geology* to igneous rocks containing small almond-shaped cavities filled with some mineral of a different nature from the mass of the rock.

Amyla'ceous (Lat. *amylum*, starch, from Gr. ἀ, a, not ; μυλη, *mulē*, a mill). Belonging to or containing starch.

Am'yloid (Lat. *amylum*, starch ; Gr. εἶδος, *eidos*, shape). Resembling starch.

Anach'ronism (Gr. ἀνα, ana, implying inversion ; χρονος, *chronos*, time). An error in stating dates.

Anæ'mia (Gr. ἀ, a, not; αἱμα, *haima*, blood). Want of blood.

Anæ'mic (Gr. ἀ, a, not; αἱμα, *haima*, blood). Bloodless ; having a very insufficient quantity of blood.

Anæsthe'sia(Gr. ἀ, a, not ; αἰσθανομαι, *aisthan'omai*, I feel). Loss of feeling or sensation.

Anæsthet'ic (Gr. ἀ, a, not ; αἰσθανομαι, *aisthan'omai*, I feel). Producing loss of feeling or sensation.

A'nal (Lat. *anus*, the excretory orifice). Belonging to or like the anus ; applied to certain fins in fishes, from their position.

Analep'tic (Gr. ἀναλαμβανω, *analam'banō*, I take up or restore). Restoring health and strength.

Anallan'toid (Gr. ἀ, a, not; *allan'tois*). A term applied to the vertebrate animals, of which the fœtus is not provided with an allantois, — including batrachians and fishes.

Anal'ogous (Gr. ἀνα, ana, with ; λογος, *logos*, ratio). Having a degree of similarity, but not identical ; applied to parts which perform a similar function, but are not identical in structure.

An'alogue. That which bears a great resemblance to something else ; a part or organ in an animal which, though anatomically different, has the same function as another part or organ in a different animal.

Anal'ogy (Gr. ἀναλογια). An agreement in some characters, not in all.

An'alyse (Gr. ἀνα, ana, back ; λυω, *luō*, I loosen). To separate anything into the parts or elements of which it is composed.

Anal'ysis (Gr. ἀνα, ana, back ; λυω, *luō*, I loosen). Separation of anything into its component parts or elements.

Analyt'ical (*Analysis*). Pertaining to or performed by analysis.

Anamnes'tic (Gr. ἀνα, ana, back ; μναομαι, *mna'omai*, I remember). Calling to remembrance.

An'apæst (Gr. ανα, ana, back ; παιω, *paiō*, I strike). In *versification*, a foot consisting of three syllables, the first two short, the last long.

Anasar'ca (Gr. ἀνα, ana, through ; σαρξ, *sarx*, flesh). Dropsy of the parts lying beneath the skin.

Anas'tomose (Gr. ἀνα, ana, through ; στομα, *stoma*, a mouth). To unite as if by open mouths, as blood-vessels.

Anastomo'sis (Gr. ἀνα, ana, through ; στομα, *stoma*, a mouth). A communication as if by mouths.

Anat'omy (Gr. ἀνα, ana, apart ; τεμνω, *temnō*, I cut). The science which teaches the structure of animals and plants, as learned by dissection. *Vegetable anatomy* teaches the structure of plants ; *human anatomy*, that of man ; *comparative anatomy*, that of all animals, with the object of comparing them with each other ; *microscopic anatomy* teaches the appearances of structures as seen under the microscope ; *pathological anatomy*, the changes in position and appearance produced by disease ; *surgical anatomy* describes regions of the body in reference to surgical operations.

Anat'ropous (Gr. ἀνα, *ana*, back ; τρεπω, *trepo*, I turn). In *botany*, applied to a seed or ovule which is curved down and grown to the lower half.

Anchylo'sis (more properly *Ancylosis* ; Gr. ἀγκυλεω, *anku'leo*, I bend). An immoveable state of a joint, from union of the surfaces which should move on each other.

Ancone'us (Gr. ἀγκων, *ankon*, the elbow). A name applied to a muscle situated over the elbow.

Anco'noid (Gr. ἀγκων, *ankon*, an elbow ; εἰδος, *eidos*, shape). Like an elbow.

Androg'ynous (Gr. ανηρ, *aner*, a man ; γυνη, *gune*, a female). Having two sexes : applied to plants of which some flowers have stamens only, and others pistils only, on the same plant.

Anelec'trode (Gr. ἀνα, *ana*, up ; *electricity* ; ὁδος, *hodos*, a way). The positive pole of a galvanic battery.

Anella'ta (Lat. *annellus*, a little ring). See Annulata.

Anemog'raphy (Gr. ἀνεμος, *an'emos*, wind ; γραφω, *grapho*, I write). A description of the winds.

Anemol'ogy (Gr. ἀνεμος, *an'emos*, wind ; λογος, *logos*, discourse). The doctrine of winds.

Anemom'eter (Gr. ἀνεμος, *an'emos*, wind ; μετρον, *metron*, a measure). An instrument for measuring the direction and force of wind.

Anem'oscope (Gr. ἀνεμος, *an'emos*, wind ; σκοπεω, *skopeo*, I look). An instrument for showing the direction of the wind.

Anencephal'ic (Gr. ἀ, *a*, not ; ἐγκεφαλον, *enkeph'alon*, the contents of the skull). Without brain.

Anen'terous (Gr. ἀ, *a*, not ; ἐντερον, *en'teron*, an intestine). Without intestines.

An'eroid (Gr. ἀ, *a*, not ; ἀηρ, *aēr*, air ; εἰδος, *eidos*, form). Without air : applied to a peculiar kind of barometer, consisting of a small box from which air is exhausted.

An'eurism (Gr. ἀνα, *ana*, through ; εὐρυνω, *euru'no*, I widen). A diseased state of an artery, in which it is widened at any part (generally from injury) so as to form a pouch or bag.

Aneuris'mal (*Aneurism*). Pertaining to an aneurism.

Anfractuos'ity (*Anfractuous*). A turning or winding ; in *anatomy*, applied to the windings on the surface of the brain.

Anfrac'tuous (Lat. *anfrac'tus*, a winding). Winding ; in *botany*, applied to the lobes of an anther which are folded back on themselves, and doubled and bent, as in the cucumber.

Angien'chyma (Gr. ἀγγειον, *angei'on*, a vessel ; ἐγχυμα, *en'chuma*, any thing poured in). The vascular tissue of plants.

Angi'na (Gr. ἀγχω, *ancho*, I strangle). Quinsey ; a choking.

Angiocar'pous (Gr. ἀγγειον, *angei'on*, a vessel ; καρπος, *karpos*, a fruit). In *botany*, applied to seed-vessels inclosed in a case which does not form part of themselves, as the filbert.

Angiol'ogy (Gr. ἀγγειον, *angei'on*, a vessel ; λογος, *logos*, discourse). A description of blood-vessels.

Angiomonosper'mous (Gr. ἀγγειον, *angei'on*, a vessel ; μονος, *monos*, single ; σπερμα, *sperma*, a seed). Having one seed only in a pod.

Angiosper'mous (Gr. ἀγγειον, *angei'on*, a vessel ; σπερμα, *sperma*, a seed). Applied to plants the seeds of which are enclosed in a vessel.

Angle of contact. The angle which a circle, or other curve, makes with a tangent at the point of contact.

Angle of depression. The angle at

which a straight line drawn from the eye to any object dips below the horizon.

Angle of direction. In *mechanics*, the angle contained by the lines of direction of two forces tending to the same point.

Angle of elevation. In *trigonometry*, the angle formed by two straight lines drawn in the same vertical plane from the observer's eye, one to the top of the object, the other parallel to the horizon.

Angle of incidence. The angle which a body, or a ray of light, forms at the surface on which it falls with a perpendicular to that surface.

Angle of inclination. The mutual approach of two bodies, so as to make an angle where their lines of direction meet.

Angle of polarization. In *optics*, the angle of incidence of a reflecting surface which, added to the corresponding angle of refraction, supposing the ray to enter the medium, would make up a right angle, or 90 degrees.

Angle of position. In *astronomy*, the angle contained by two great circles passing through the earth, one perpendicular to the plane of the ecliptic, the other to that of the equator.

Angle of reflection. The angle which a body or a ray of light rebounding from a surface makes with a perpendicular to that surface.

Angle of refraction. In *optics*, the angle which a ray of light passing from one medium to another makes with a perpendicular drawn through the line of incidence.

Angle, solid. An angle made by more than two plane angles meeting in a point, and not lying in the same plane.

Angle, spherical. An angle on the surface of a sphere, contained within the arcs of two intersecting circles.

Angle, visual. In *optics*, the angle formed in the centre of the eye by lines drawn from the extremities of an object.

An'gular (Lat. *an'gulus*, a corner). Having or relating to angles.

An'gulate (Lat. *an'gulus*, an angle). Having an angular shape.

Anhelation (Lat. *anhēlo*, I breathe short). Short breathing; panting.

Anhy'drous (Gr. ἀ, *a*, not; ὕδωρ, *hudōr*, water). Free from water; without water of crystallization.

Animal (Lat. *anima*, life, breath). A body having life, sensation, and voluntary motion.

Animal'cule (Lat. *animal*, an animal; *ule*, signifying smallness). An animal of very small size.

Animal heat. The warmth which animals possess in themselves.

Animalisa'tion (Lat. *animal*, an animal). The art of imparting the properties belonging to an animal, or to animal structures; a peopling with animals.

An'ion (Gr. ἀνα, up; ἰων, *iōn*, going). That substance which passes to the anode in electrolysis.

Anneal (Saxon *on*, on; *ælan*, to burn). To heat glass, &c., for the purpose of rendering it less brittle.

Annual (Lat. *annus*, a year). Occurring every year.

An'nelids (Lat. *annel'lus*, a little ring; Gr. εἶδος, *eidos*, form). A class of invertebrate animals, so called because apparently composed of rings, including earth-worms and leeches.

An'nular (Lat. *annulus*, a ring). Shaped like a ring.

Annula'ta (Lat. *annulus*, a ring). Having rings: applied to a division of the animal kingdom, including invertebrates having the body arranged in rings.

An'ode (Gr. ἀνα, *ana*, up; ὁδος, *hod'os*, a way). The way by which electricity enters substances.

An'odyne (Gr. ἀ, *a*, not; ὀδυνη, *odu'nē*, pain). Relieving pain.

Anom'alous (Gr. ἀ, *a*, not; ὁμαλος, *hom'alos*, level, or equal). Departing from a general rule; irregular.

Anom'aly (Gr. ἀ, *a*, not; ὁμαλος, *hom'alos*, level or equal). Irregularity ; deviation from an ordinary law or type ; in *astronomy*, the angle formed by a line drawn from the sun to the place of a planet, with the greater axis of the planet's orbit.

Anomodon'tia (Gr. ἄνομος, *an'omos*, irregular ; ὀδούς, *odous*, a tooth). An extinct order of reptiles, with teeth wanting, or in various irregular forms.

Anomou'ra (Gr. ἄνομος, *an'omos*, irregular ; οὐρά, *oura*, a tail). A section of decapodous or ten-footed crustaceans, having tails of intermediate length between the long-tailed and short-tailed, as the hermit crab.

Anoplothe'rium (Gr. δ, *a*, not ; ὅπλον, *hoplon*, a weapon ; θηρίον, *thērion*, a beast). A fossil pachydermatous animal, having no evident organs of defence.

Anoplu'res (Gr. δ, *a*, not ; ὅπλον, *hoplon*, a weapon ; οὐρά, *oura*, a tail). An order of wingless and stingless insects, living as parasites on other animals.

Anorex'ia (Gr. δ, *a*, not ; ὄρεξις, *orexis*, desire). Loss of appetite for food.

Anor'mal (Lat. *a* from ; *norma*, a rule). *See* Abnormal.

Anou'rous (Gr. δ, *a*, not ; οὐρά, *oura*, a tail). Without a tail.

An'serine (Lat. *anser*, a goose). Belonging to or resembling a goose.

Antac'id (Gr. ἀντί, *anti*, against ; *acid*). Opposed to acids ; counteracting their effects.

Antæ. In *architecture*, the pier-formed ends of the side-walls of temples, when they are prolonged beyond the face of the walls ; pilasters standing opposite a column.

Antag'onism (Gr. ἀντί, *anti*, against ; ἀγωνίζομαι, *agōni'zomai*, I contend). Active opposition.

Antag'onistic (Gr. ἀντί, *anti*, against ; ἀγωνίζομαι, *agōni'zomai*, I contend). In direct or active opposition to.

Antarc'tic (Gr. ἀντί, *anti*, against or opposite ; ἄρκτος, *arktos*, the north pole). Relating to the south pole.

Ante. A Latin preposition used in composition, signifying before.

Antece'dent (Lat. *ante*, before ; *cedo*, I go). Going before.

Ante'cian (Gr. ἀντί, *anti*, opposite ; οἰκέω, *oikeō*, I dwell). In *geography*, applied to the inhabitants of the earth, under the same meridian of longitude, but at equal distances on opposite sides of the equator.

Antefix'æ (Lat. *ante*, before ; *fingo*, I fix). In *architecture*, upright ornamental blocks placed at intervals on the cornice along the sides of a roof ; also heads of animals as water-spouts below the eaves of temples.

Anteflex'ion (Lat. *ante*, before ; *flecto*, I bend). A bending forwards.

Antemu'ral (Lat. *ante*, before ; *murus*, a wall). In *architecture*, the outward wall of a castle.

Anten'næ (Lat. *anten'na*, a sail-yard). Filaments, apparently organs of touch, projecting from the heads of insects and crustacea.

Antepenult' (Lat. *ante*, before ; *pene*, almost ; *ul'timus*, last). The last syllable but two.

Antever'sion (Lat. *ante*, before ; *verto*, I turn). A turning forwards.

Anthe'lion (Gr. ἀντί, *anti*, opposite ; ἥλιος, *hēlios*, the sun). A mock-sun.

Anth'elix (Gr. ἀντί, *anti*, opposite ; ἕλιξ, *helix*, a spiral). A part of the external human ear, before or rather within the helix.

Anthelmin'tic (Gr. ἀντί, *anti*, against ; ἕλμινς, *helmins*, a worm). Capable of destroying or removing the worms which inhabit the animal body.

Anther (Gr. ἄνθος, *anthos*, a flower). The top of the stamen, or male part of a flower, containing the pollen or fertilising dust.

Antherid'ium (*Anther*). A structure in some flowerless plants, supposed to be the analogue of an anther.

Anthocar'pous (Gr. ἄνθος, *anthos*, a flower ; καρπός, *karpos*, a fruit). In *botany*, a term applied to fruits which are formed of masses of inflorescence in a state of cohesion, as the fir-cone and pine-apple.

Anthocy'anine (Gr. ἄνθος, *anthos*, a

flower ; κυανος, *ku'anos*, blue).
Blue colouring matter of plants.

Anth'olites (Gr. *ανθος, anthos,* a
flower ; λιθος, *lithos,* a stone). The
fossil impressions of flowers.

Anthol'ogy (Gr. *ανθος, anthos,* a
flower ; λογος, *logos,* discourse). A
description of flowers.

Anthoxan'thine (Gr. *ανθος, anthos,* a
flower ; ξανθος, *xanthos,* yellow).
Yellow colouring matter of plants.

Anthozo'a (Gr. *ανθος, anthos,* a flower ;
ζωον, *zōon,* an animal). Animal
flowers ; the class of polypes including the actinia and allied
species, which resemble flowers.

Anth'racite (Gr. *ανθραξ, anthrax,* a
coal). A peculiar shining kind of coal.

Anthracothe'rium (Gr. *ανθραξ, anthrax,* coal ; θηριον, *thērion,* a
beast). A fossil pachydermatous
animal found in the coal-formation.

Anthro'poid (Gr. *ανθρωπος, anthrōpos,*
a man, *i.e.* human being ; ειδος,
eidos, form). Resembling man.

Anthropol'ogy (Gr. *ανθρωπος, anthrōpos,* a man ; λογος, *logos,* discourse). A description of the human
body or of the human species.

Anthropomor'phous (Gr. *ανθρωπος,
anthrōpos,* a man ; μορφη, *morphē,*
form). Resembling man.

Anthropoph'agous (Gr. *ανθρωπος,
anthrōpos,* a man ; φαγω, *phagō,*
I eat). Eating men ; cannibal.

Anthropos'ophy (Gr. *ανθρωπος, anthrōpos,* a man ; σοφια, *soph'ia,*
wisdom). The knowledge of the
nature of man.

Anti (Gr. *αντι, anti*). A Greek preposition used in composition, signifying against.

Antiarthrit'ic (Gr. *αντι, anti,* against;
αρθριτις, *arthri'tis,* gout). Curing
gout.

Antiasthmat'ic (Gr. *αντι, anti,*
against ; *asthma*). Curing or preventing asthma.

Antibra'chial (Lat. *antibra'chium,* the
forearm). Belonging to the fore-
arm.

Antibra'chium (Lat.). The forearm,
from the elbow to the wrist.

Anticli'nal (Gr. *αντι, anti,* against ;
κλινω, *klinō,* I bend). Inclining in

opposite directions, like the ridge of
a house.

An'tidote (Gr. *αντι, anti,* against ;
διδωμι, *didōmi,* I give). A remedy
to counteract poisons or anything
noxious.

Antife'brile (Gr. *αντι, anti,* against ;
Lat. *febris,* fever). Removing
fever.

Antilith'ic (Gr. *αντι, anti,* against ;
λιθος, *lithos,* a stone). Preventing the formation of calculi.

Antip'athy (Gr. *αντι, anti,* against ;
παθος, *pathos,* suffering or passion).
A strong dislike or repugnance.

Antiperiod'ic (Gr. *αντι, anti,* against;
periodic). Preventing or curing
diseases which recur at regular
periods, as ague.

Antiperistal'tic (Gr. *αντι, anti,*
against; περι, *peri,* around; στελλω,
stellō, I send). A term applied to
an unnatural or reversed action of
the alimentary canal.

Antiphlogis'tic (Gr. *αντι, anti,*
against ; φλοξ, *phlox,* flame).
Diminishing inflammation.

Antip'odes (Gr. *αντι, anti,* against ;
πους, *pous,* a foot). The inhabitants of the opposite side of the
globe, whose feet are, as it were,
applied *against* ours.

Antis'cians (Gr. *αντι, anti,* against ;
σκια, *skia,* a shadow). The inhabitants of the earth on different
sides of the equator, whose shadows
at noon are cast in contrary directions.

Antiscorbu'tic (Gr. *αντι, anti,* against;
Lat. *scorbūtus,* scurvy). Preventing or curing scurvy.

Antisep'tic (Gr. *αντι, anti,* against ;
σηπω, *sēpō,* I make putrid). Preventing putrefaction.

Antispasmod'ic (Gr. *αντι, anti,* against;
σπαω, *spaō,* I draw). Preventing
spasms or convulsions.

Antith'esis (Gr. *αντι, anti,* against ;
τιθημι, *tithēmi,* I place). Opposition or contrast, especially of words
or ideas.

Antit'ragus (Gr. *αντι, anti,* opposite ;
tragus). A projecting part of the
outer ear opposite the tragus.

Antit'ropous (Gr. *αντι, anti,* opposite;

τρεπω, *trepō*, I turn). In *botany*, applied to the position of the embryo in a seed in which the nucleus is erect, the embryo being consequently inverted.

Ant'lia (Gr. ἀντλια, *ant'lia*, a baling-out). The spiral apparatus by which butterflies and other insects pump up the juices of plants.

Ant'orbital (Lat. *ante*, before; *orbit*). In front of the orbits.

A'orist (Gr. ἀ, *a*, not; ὁριζω, *hori'zō*, I limit or define). In *grammar*, a tense which expresses past action without reference to duration or time.

Antrum (Lat. a cave). In *anatomy*, a term used to designate certain cavities of the body.

Aor'ta (Gr. ἀειρω, *aei'rō*, I take up or carry). The great vessel which, arising from the left ventricle of the heart, carries the blood to all parts of the body.

Aor'tic (Gr. ἀορτη, *aortē*, the aorta). Belonging to the aorta.

Aorti'tis (Lat. *aorta*; *itis*, denoting inflammation). Inflammation of the aorta.

Ape'rient (Lat. *aperio*, I open). Opening; laxative.

Ap'erture (Lat. *aperio*, I open). An opening; in *geometry*, the space between two straight lines forming an angle; in *optics*, the hole next the object-glass of a telescope or microscope through which the light enters the instrument.

Apet'alous (Gr. ἀ, *a*, not; πεταλον, *pet'alon*, a flower-leaf or petal). Having no distinction of sepals and petals.

Apex (Lat.). The top or highest point of anything.

Aphæ'resis (Gr. ἀπο, *apo*, from; αἱρεω, *haireō*, I take). In *grammar*, the taking a letter or syllable from the beginning of a word.

Aphanip'tera (Gr. ἀ, *a*, not; φαινω, *phainō*, I show; πτερον, *pteron*, a wing). An order of insects with rudimentary wings only, as the flea.

Aphe'lion (Gr. ἀπο, *apo*, from; ἡλιος, *hēlios*, the sun). The point in the orbit of a planet which is most distant from the sun.

Aphlogis'tic (Gr. ἀ, *a*, not; φλογιζω, *phlogizō*, I set on fire). Flameless; burning without flame.

Apho'nia (Gr. ἀ, *a*, not; φωνη, *phōnē*, voice). Loss of voice.

Aph'orism (Gr. ἀπο, *apo*, from; ὁριζω, *hori'zō*, I limit). A principle or precept expressed in a few words.

Aphthæ (Gr. ἀπτω, *haptō*, I fasten upon). Small white ulcers on the inside of the mouth.

Aphyl'lous (Gr. ἀ, *a*, not; φυλλον, *phullon*, a leaf). Leafless.

Ap'ical (Lat. *apex*, a top). Belonging to the top of a conical body.

Aplanat'ic (Gr. ἀ, *a*, not; πλαναομαι, *plana'omai*, I wander). Opposed to wandering; applied to lenses or combinations of lenses which correct the effects of spherical aberration of light.

Aplas'tic (Gr. ἀ, *a*, not; πλασσω, *plassō*, I form). Incapable of being moulded or organised.

Apnœ'a (Gr. ἀ, *a*, not; πνεω, *pneō*, I breathe). Loss of breath; suffocation.

Ap'o (Gr. ἀπο, *apo*). A Greek preposition in compound words, signifying from.

Apocar'pous (Gr. ἀπο, *apo*, from; καρπος, *karpos*, fruit). Applied to flowers and fruits in which the carpels are separate or only partially united.

Ap'odal (Gr. ἀ, *a*, not; πους, *pous*, a foot). Without feet. Apodal fishes have no ventral fins, which are the anologues of feet.

Ap'ogee (Gr. ἀπο, *apo*, from; γη, *gē*, the earth). The point in the orbit of a planet which is most distant from the earth or the moon.

Aponeuro'sis (Gr. ἀπο, *apo*, from; νευρον, *neuron*, a string or tendon). The membranous spreading out of a tendon.

Apoph'ysis (Gr. ἀπο, *apo*, from; φυω, *phuō*, I grow). A prominent elevation from the surface of a bone.

Apoplec'tic (Gr. ἀπο, *apo*, from; πλησσω, *plēssō*, I strike). Relating to apoplexy.

Ap'oplexy (Gr. ἀπο, *apo*, from; πλησσω, *plēssō*, I strike). A disease in which consciousness of the power of voluntary motion is abolished, from injury within the brain.

Appara'tus (Lat. *ad*, to ; *paro*, I make). An instrument or organ for the performance of any operation or function.

Ap'plicate (Lat. *ad*, to ; *plico*, I fold). In *geometry*, a straight line drawn across a curve so as to be bisected by the diameter.

Ap'sides (Gr. ἅπτω, *haptō*, I touch). The points in the path of the moon or a planet when it is respectively nearest to and most distant from the earth.

Ap'terous (Gr. ἀ, *a*, not ; πτερον, *pteron*, a wing). Without wings.

Ap'tote (Gr. ἀ, *a*, not; πτωσις, *ptōsis*, case). In *grammar*, applied to nouns which have no distinction of cases.

Apyret'ic (Gr. ἀ, *a*, not ; πυρεσσω, *puressō*, I have a fever). Without fever.

Apyrex'ia (Gr. ἀ, *a*, not ; πυρεσσω, *puressō*, I have a fever). Freedom from fever.

Aqua fortis (Lat. strong water). A name for nitric acid.

Aqua regia (Lat. royal water). A mixture of nitric and hydrochloric acids, used to dissolve gold.

Aqua vitæ (Lat. water of life). A name for strong spirits.

Aquat'ic (Lat. *aqua*, water). Belonging to, or living or growing in water.

A'queous (Lat. *aqua*, water). Watery ; consisting of or having the properties of water ; made with water.

Ar'able (Lat. *aro*, I plough). Capable of being cultivated by the plough.

Arach'nida (Gr. ἀραχνη, *arachnē*, a spider). A class of invertebrate animals, including spiders, scorpions, and mites.

Arachni'tis (*Arachnoid; itis*, denoting inflammation). Inflammation of the arachnoid membrane of the brain.

Arach'noid (Gr. ἀραχνη, *arachnē*, a spider or spider's web ; ειδος, *eidos*, form). A thin membrane covering the brain.

Ara'neiform (Lat. *ara'neus*, a spider ; *forma*, shape). Resembling a spider.

Arbor (Lat. a tree). In *mechanics*, the part of a machine which sustains the rest ; an axis or spindle.

Arbor vitæ (Lat. tree of life). In *anatomy*, a tree-like appearance of the brain-substance, seen when the cerebellum is cut transversely.

Arbores'cent (Lat. *arbor*, a tree). Resembling a tree ; becoming woody.

Arc (Lat. *arcus*, a bow). A part of the circumference of a circle or of a curved line.

Arca'num (Lat. *arca*, a chest). A secret.

Arch (Gr. ἀρχη, *archē*, the beginning or head). A prefix denoting eminence.

Archæol'ogy (Gr. ἀρχαιος, *archaios*, ancient ; λογος, *logos*, discourse). The science which describes antiquities.

Ar'chaism (Gr. ἀρχαιος, *archaios*, ancient). An ancient or disused word or expression.

Archenceph'ala (Gr. ἀρχος, *archos*, chief ; ἐγκεφαλος, *enkeph'alos*, the brain). Chief-brained : a term proposed by Professor Owen to denote the highest sub-class of the mammalia, comprising only man, from the superior development of his brain.

Ar'chetype (Gr. ἀρχη, *archē*, a beginning ; τυπος, *tupos*, a type). An original pattern or model.

Archime'des' screw. An instrument formed of a tube wound round a cylinder in the form of a screw, and used either for raising fluids or for propelling through water.

Arch'itecture (Gr. ἀρχος, *archos*, chief ; τεκτων, *toktōn*, a builder). The science of constructing houses, bridges, and other buildings, according to rule.

Arch'itrave (Gr. ἀρχος, *archos*, chief ; Lat. *trabs*, a beam). The lowest part of an entablature, being the chief beam resting immediately on the column.

Ar'ciform (Lat. *arcus*, a bow ; *forma*, shape). Like an arch.

Arctic (Gr. ἀρκτος, *arktos*, a bear, or the north pole). Relating to the north pole.

Ar'cuate (Lat. *arcus*, a bow). Shaped like a bow.

A'rea (Lat. an open space). A plain surface ; in *geometry*, the superficial contents of any figure.

Arena'ceous (Lat. *arēna*, sand). Sandy.

Are'nicole (Lat. *arēna*, sand ; *colo*, I inhabit). An animal which inhabits sand.

Are'ola (Lat. *area*, an open space). A small surface or space.

Are'olar (*Areola*). Containing little spaces ; applied to the connecting tissue of the body, which forms a number of little spaces or interstices.

Areom'eter (Gr. ἀραιος, *araios*, thin ; μετρον, *metron*, a measure). An instrument for measuring the specific gravity of liquids.

Argentif'erous (Lat. *argen'tum*, silver; *fero*, I produce). Producing or containing silver.

Argil (Gr. ἀργος, *argos*, white). Generally clay ; technically, pure clay or alumina.

Argilla'ceous (Lat. *argil'la*, white clay). Consisting of argil or clay, especially pure clay.

Aril. In *botany*, the expansion of the funiculus or placenta round the seed, as the mace of a nutmeg.

Aris'ta (Lat.). In *botany*, the beard of corn and other grasses.

Arithmetical mean. The middle term of three numbers in arithmetical progression.

Arithmetical progression. A series of quantities increasing or decreasing by the addition or subtraction of the same number.

Arithmetical ratio. The difference between any two terms in arithmetical progression.

Ar'mature (Lat. *arma*, arms). A supply of weapons ; applied, in *physics*, to two pieces of soft iron fastened to the poles of a magnet, and connected at their ends by a third piece, so as to increase its power.

Armil'lary (Lat. *armilla*, a bracelet). Like a bracelet ; generally applied to an artificial sphere composed of a number of circles of the mun-

dane sphere, placed in natural order.

Arrag'onite. A mineral consisting of carbonate of lime, with some carbonate of strontia.

Arrhi'zous (Gr. ἀ, *a*, not ; ῥιζα, *rhiza*, a root). Without roots.

Arse'niate (*Arsenic*). A salt of arsenic acid with a base.

Arsen'ic. In *chemistry*, applied to an acid containing an equivalent of metallic arsenic and five of oxygen.

Arse'nious (*Ar'senic*). In *chemistry*, applied to an acid containing an equivalent of metallic arsenic and three of oxygen ; the common arsenic of the shops.

Ar'senite (*Arsenic*). A salt formed of arsenious acid with a base.

Arte'rial (*Artery*). Belonging to an artery or to arteries.

Arteri'tis (Lat. *artēria*, an artery ; *itis*, denoting inflammation). Inflammation of arteries.

Ar'tery (Gr. ἀηρ, *aēr*, air ; τηρεω, *tēreō*, I keep ; because originally supposed to contain air). A vessel or tube which conveys blood in a direction from the heart to all parts of the body.

Arte'sian (Lat. *Artois*, a province of France). Artesian wells, supposed to have been first made in Artois, are perpendicular borings to a considerable depth in the earth for procuring water.

Arthrit'ic (*Arthritis*). Relating to inflammation of the joints, or gout.

Arthri'tis (Gr. ἀρθρον, *arthron*, a joint ; term. *itis*, inflammation). Any inflammation of the joints ; but specially applied to gout.

Arthro'dia (Gr. ἀρθροω, *arthroō*, I fit by joints). A joint in which the head of one bone is received into the socket of another ; a ball-and-socket joint.

Arthrodyn'ia (Gr. ἀρθρον, *arthron*, a joint ; ὀδυνη, *odu'nē*, pain). Pain in the joints.

Arthropod'aria (Gr. ἀρθρον, *arthron*, a joint ; πους, *pous*, a foot). A term applied to those invertebrate animals which have jointed limbs, including insects, myriapods, arachnides, and crustacea.

Artic'ular (Lat. *artic'ulus*, a joint). Belonging to joints.

Articula'ta (Lat. *artic'ulus*, a joint). A division of the animal kingdom, including the invertebrates with jointed bodies.

Artic'ulate (Lat. *artic'ulus*, a joint). To join together ; jointed or having joints.

Articula'tion (Lat. *artic'ulus*, a joint). A connection by joint ; also speech, because composed of sounds joined together.

Artiodac'tyle (Gr. *ἀρτιος, ar'tios*, even, *δακτυλος, dak'tulos*, a finger). Having an even number of toes.

Aryte'noid (Gr. *ἀρυταινα, arutai'na*, a pitcher ; *εἶδος, eidos*, shape). Shaped like a pitcher ; applied to two small cartilages at the top of the larynx.

Asbes'tos (Gr. *ἀ, a*, not ; *σβεννυμι, sbennūmi*, I extinguish). A fibrous variety of hornblende, capable of resisting heat.

As'caris (Gr. *ἀσκαριζω, askari'zō*, I leap). A small intestinal worm.

Ascen'sion (Lat. *ascen'do*, I rise). A rising ; in *astronomy*, right ascension denotes the distance of a heavenly body from the point of the spring equinox, measured on the celestial equator.

A'scian (Gr. *ἀ, a*, not ; *σκια, skia*, a shadow). Having no shadow at noon : applied to the inhabitants of the torrid zone, who, at certain times, have no shadow at noon.

Ascid'ian (Gr. *ἀσκος, askos*, a leather bottle; *εἶδος, eidos*, form). Acephalous or headless mollusca, shaped like a leather bottle.

Ascid'ium (Gr. *ἀσκος, askos*, a leather bottle). In *botany*, a form of leaf in which the stalk is hollowed out and closed by the blade as a lid.

Asci'tes (Gr. *ἀσκος, askos*, a leather bag). A collection of fluid in the abdomen.

Asex'ual (Gr. *ἀ, a*, not ; Lat. *sexus*, sex). Without distinct sexes.

Ashlar. In *architecture*, the facing of square stones on the front of a building; freestones roughly squared in the quarry.

Asper'ity (Lat. *asper*, rough). Roughness.

Asper'mous (Gr. *ἀ, a*, not ; *σπερμα, sperma*, seed). Without seed.

Asphyx'ia (Gr. *ἀ, a*, not ; *σφυζω, sphuzō*, I beat, as the pulse). Originally, failure of the pulse ; but now applied to the symptoms of suffocation produced by an accumulation of carbonic acid in the blood.

Assay (Fr. *essayer*, to try). To try the quality of metals.

Assimila'tion (Lat. *ad*, to ; *sim'ilis*, like). The process by which a substance or thing is rendered similar in form and property to that with which it comes into contact.

As'sonance (Lat. *ad*, to ; *sonus*, sound). Resemblance in sound or termination without making rhyme.

Astat'ic (Gr. *ἀ, a*, not ; *ἱστημι, histe'mi*, I fix or make to stand). Not moving ; applied to a magnetic needle which is not affected by the magnetism of the earth.

Asteracan'thus (Gr. *ἀστηρ, astēr*, a star ; *ἀκανθα, akan'tha*, a thorn). A genus of fossil fin-spines of fishes, having star-like tubercles on their surface.

As'teroid (Gr. *ἀστηρ, astēr*, a star ; *εἶδος, eidos*, form or likeness). A name applied to the small planets of the group which revolves between Mars and Jupiter ; also to star-like echinoderms.

Asterophyl'lites (Gr. *ἀστηρ, astēr*, a star ; *φυλλον, phullon*, a leaf). In *geology*, the fossil remains of some plants found in the coal-measure, lias, and oolite, having leaves arranged in star-like whorls.

Asthen'ia (Gr. *ἀ, a*, not ; *σθενος, sthen'os*, strength). Want of strength.

Asthen'ic (Gr. *ἀ, a*, not ; *σθενος, sthen'os*, strength). Characterised by want of strength.

Astheno'pia (Gr. *ἀ, a*, not ; *σθενος, sthen'os*, strength ; *ὠψ, ōps*, the eye). Weakness of vision.

Asthma (Gr. *ἀω, aō*, I blow). A difficulty of breathing, occurring in paroxysms, with intervals of freedom.

Asthmat'ic (Gr. ἀσθμα, *asthma*). Belonging to, or having asthma.

As'tomous (Gr. ἀ, *a*, not ; στομα, *stoma*, a mouth). Without a mouth.

Astrag'alus (Gr. ἀστραγαλος, *astra'-galos*, an ankle-bone). The bone of the foot which forms part of the ankle-joint.

As'tral (Gr. ἀστρον, *astron*, a star). Belonging to stars.

Astric'tion (Lat. *ad*, to ; *stringo*, I bind). The act of binding.

Astrin'gent (Lat. *ad*, to ; *stringo*, I tie fast). Binding or contracting.

As'trolabe (Gr. ἀστρον, *astron*, a star ; λαβειν, *labein*, to take). An instrument formerly used for taking the altitude of the sun or stars.

Astrol'ogy (Gr. ἀστρον, *astron*, a star; λογος, *logos*, a word or description). The science which pretends to teach the effects and influence of the stars.

Astrom'oter (Gr. ἀστρον, *astron*, a star ; μετρον, *metron*, a measure). An instrument for ascertaining the relative brightness of stars.

Astronom'ical (Gr. ἀστρον, *astron*, a star ; νομος, *nomos*, a law). Belonging to astronomy.

Astron'omy (Gr. ἀστρον, *astron*, a star ; νομος, *nomos*, a law). The science which describes the magnitude, position, motion, &c., of the heavenly bodies, as taught by observation and mathematical calculation.

Asymmet'rical (Gr. ἀ, *a*, not ; συν, *sun*, with ; μετρον, *metron*, a measure). Not consisting of similar parts on each side.

Asym'ptote (Gr. ἀ, *a*, not ; συν, *sun*, with ; πτοω, *ptoō*, I fall). A line approaching a curve, but never meeting it.

Atax'ic (Gr. ἀ, *a*, not ; τασσω, *tassō*, I put in order). Wanting order ; irregular.

Ate (Lat. term. *atus*). In *chemistry*, a termination applied to compounds of which the acid contains the largest quantity of oxygen.

Atelec'tasis (Gr. ἀ, *a*, not ; τελος,

telos, an end ; ἐκτεινω, *ektei'nō*, I stretch out). Imperfect expansion.

Atheric'era (Gr. ἀθηρ, *athēr*, a spike of corn ; κερας, *keras*, a horn.) A section of dipterous insects, having only two or three joints to the antennæ.

Ather'mancy (Gr. ἀ, *a*, not ; θερμαινω, *thermai'nō*, I make warm). The property of transmitting the light but not the heat of the sun.

Ather'manous (Gr. ἀ, *a*, not ; θερμαινω, *thermai'nō*, I make warm). Incapable of transmitting heat.

Athero'ma (Gr. ἀθαρα, *athara*, or ἀθηρη, *athērē*, a porridge of meal). A diseased state of blood-vessels and other structures of the body, characterised by a soft pulpy deposit.

Atlas (Gr. Ἀτλας, *Atlas*, a mythological personage, who was said to carry the world on his shoulders). The first vertebra of the neck ; so called because the head rests on it.

Atmom'eter (Gr. ἀτμος, *atmos*, vapour ; μετρον, *metron*, a measure). An instrument for measuring the amount of evaporation from a moist surface in a given time.

At'mosphere (Gr. ἀτμος, *atmos*, vapour ; σφαιρα, *sphaira*, a ball or globe). The mass of air surrounding the earth ; also applied to any gas surrounding an animal or other body.

Atmospheric Pressure. The weight of the atmosphere on a surface ; the mean being 14·7 pounds to the square inch.

At'oll. A coral island, consisting of a circular belt or ring of coral, with a lagoon or lake in the centre.

Atom (Gr. ἀ, *a*, not ; τεμνω, *temnō*, I cut). A particle of matter which can no longer be diminished in size.

Atom'ic (Gr. ἀτομος, *at'omos*, an atom). Relating to atoms.

Atomic Theory. An hypothesis in chemistry, which teaches that the atoms of elementary substances become combined in certain definite proportions.

Aton'ic (Gr. ἀ, *a*, not ; τεινω, *teino*,

I stretch or tighten). Weakened; characterised by want of energy.

At'ony (Gr. ἀ, a, not; τεινω, teinō, I stretch or tighten). Want of power.

At'rophy (Gr. ἀ, a, not; τρεφω, trephō, I nourish). Want of nourishment; a wasting.

At'ropous (Gr. ἀ, a, not; τρεπω, trepō, I turn). Not turned; in *botany*, applied to that form of the ovule or seed, in which its parts have undergone no change of position during growth.

Atten'uant (Lat. ad, to; ten'uis, thin). Making thin; diluting.

Atten'uate (Lat. ad, to; ten'uis, thin). To make thin.

Attol'lent (Lat. ad, to; tollo, I raise). Lifting up.

Attrac'tion (Lat. ad, to; traho, I draw). A drawing towards; the tendency of bodies to unite or cohere.

At'trahent (Lat. ad, to; traho, I draw). Drawing towards.

Attrit'ion (Lat. ad, to; tero, I rub). The act of wearing by. rubbing together.

Aud'itory (Lat. au'dio, I hear). Belonging to the sense or organ of hearing.

Aug'ite (Gr. ἀυγη, augē, bright light). A mineral, closely allied to hornblende, entering into the composition of many trap and volcanic rocks.

Au'ricle (Lat. auric'ula, a little ear). The external part of the ear; also a part on each side of the heart, from resembling the ears of animals.

Auric'ular (Lat. auric'ula, a little ear). Belonging to an auricle.

Auric'ulate (Lat. auric'ula). Shaped like a little ear; in *botany*, applied to leaves which have the lobes at the base forming distinct segments like little ears.

Auric'ulo-ventric'ular. Belonging to, or lying between the auricles and ventricles of the heart.

Aurif'erous (Lat. aurum, gold; fero, I produce). Yielding or producing gold.

Au'riform (Lat. auris, an ear; forma, form). Shaped like an ear.

Ausculta'tion (Lat. ausculto, I listen). The act of listening: applied, in

medicine, to a means of distinguishing the condition of internal parts by listening to the sounds which are produced in them.

Austral (Lat. auster, the south wind). Belonging to the south: applied to that pole of the magnet which points to the south.

Autoch'thon (Gr. αὐτος, autos. self; χθων, chthōn, the earth). Originating from the earth of the country; indigenous.

Autog'enous (Gr. αὐτος, autos, self; γενναω, gennaō, I produce). Self-produced: applied to those parts of a vertebra which are developed from independent centres of ossification.

Au'tograph (Gr. αὐτος, autos, himself; γραφω, graphō, I write). The actual signature of an individual.

Autograph'ic Telegraph. An electric telegraph for transmitting messages in the handwriting of the person sending them.

Automat'ic (Gr. αὐτος, autos, self; μαω, maō, I move). Having mechanical movement, as an automaton: applied, in *physiology*, to muscular movements produced independently of the will; self-moving.

Autom'aton (Gr. αὐτος, autos, self; μαω, maō, I move). A machine which, by means of mechanical contrivances, imitates the motion of living animals.

Au'topsy (Gr. αὐτος, autos, self: ὀψις, opsis, sight). Direct or personal observation; applied especially to an examination of the body after death.

Auxil'iary (Lat. auxil'ium, help). Aiding; taking a share of labour.

Av'alanche (Fr.) An accumulation of snow, or of snow and ice, descending from mountains.

Aves (Lat. birds). A class of oviparous vertebrate animals with double circulation, mostly organised for flight.

Avic'ula (Lat. a little bird). An unequal valved shell, fixing itself by a byssus.

Avic'uloid (Avic'ula; Gr. ἐιδος, eidos, form). Like anavicula.

Ax'ial (*Axis*). In the direction of the axis.

Ax'il (Lat. *axilla*, the armpit). In *botany*, the angle formed by a leaf with the stem.

Axil'la (Lat.) The armpit.

Axil'lary (Lat. *axilla*, the armpit). Belonging to the armpit; in *botany*, growing in the angle formed by a leaf with the stem.

Ax'iom (Gr. ἀξιοω, *axioō*, I think worthy). A self-evident truth, incapable of being made plainer by reasoning.

Axis (Lat. *axis*, an axletree). A straight line passing through the centre of a body ; a pivot on which anything turns ; the second vertebra of the neck, because the head turns on it.

Az'imuth (Arab. *samatha*, to go towards). The direction of an object in reference to the cardinal points, or to the plane of the meridian.

Az'imuth Compass. An instrument consisting of a magnetic bar or needle balanced on a vertical pivot, so as to turn freely in an horizontal plane.

Azo'ic (Gr. ἀ, *a*, not ; ζωον, *zōon*, an animal). Without animals ; applied to the lowest or primary geological strata, in which no remains of animals are found.

Az'ote (Gr. ἀ, *a*, not ; ζωη, *zōē*, life). A name for nitrogen gas, because it will not support animal life.

Az'otised (*Azote*). Containing azote of nitrogen.

Az'ygos (Gr. ἀ, *a*, not ; ζυγον, *zugon*, a yoke). Without a fellow ; having no corresponding symmetrical part.

B.

Baccate (Lat. *bacca*, a berry). Resembling a berry.

Baily's Beads. In *astronomy*, an appearance as of a string of beads round the sun in an eclipse.

Bal'anoid (Gr. βαλανος, *bal'anos*, an acorn). A family of cirripeds or barnacles, having shells arranged conically, like an acorn.

Balsam (Gr. βαλσαμον, *bal'samon*). A natural mixture of resin with a volatile oil.

Barb'ule (Lat. *barba*, a beard). A little beard.

Barilla (Spanish). An impure carbonate of soda.

Barom'eter (Gr. βαρος, *baros*, weight; μετρον, *metron*, a measure). An instrument for measuring the weight or pressure of the air.

Basalt'. A close-grained rock of the trappean group, dark-coloured, often arranged in more or less regular columns.

Base (Gr. βασις, *basis*, a foundation). The lower part of anything, or that on which it rests ; in *chemistry*, a substance which, when combined with an acid, forms a salt.

Basement Membrane. A fine, transparent layer, lying underneath the epithelium of mucous and serous membranes, and beneath the epidermis of the skin.

Ba'sic (*Base*). In *chemistry*, having a large proportion of base ; basic water is water which appears to act as a base in the formation of certain salts.

Bas'ilar (Lat. *basis*, a base). Basic ; belonging to the base of the skull; applied especially to an artery of the brain.

Basin (Fr. *bassin*). A hollow vessel ; in *geology*, a hollow or trough formed of rocks older than the deposit contained in it.

Basioccip'ital (Lat. *basis*, a base ; *oc'ciput*, the back of the head). A bone of the head of lower vertebrate animals, answering to a part of the occipital bone in man.

Bathymet'rical (Gr. βαθυς, *bathus*, deep ; μετρον, *metron*, a measure). Relating to the distribution of plants and animals along the bottom of the sea, according to the depth which they inhabit.

Batra'chia (Gr. βατραχος, bat'rachos, a frog). The order of reptiles of which the frog is the type.

Batra'chian (Gr. βατραχος, bat'rachos, a frog). Belonging to the order of animals of which the frog is the type.

Bat'tery. In *chemistry*, an apparatus of coated jars for electrical action, or of portions of zinc and copper, used for producing electro-chemical or voltaic action.

Belem'nite (Gr. βελεμνον, belemnon, a dart). Arrow-head ; also called thunderbolt ; a fossil shell of the cephalopod order, found in chalk and limestone.

Bell-metal. An alloy of copper and tin used in making bells.

Ben'zoate (*Benzoin*). A salt formed of benzoic acid with a base.

Bergmehl (Swedish, mountain-meal). A whitish, mealy earth, containing infusorial animalcules, said to be eaten by the Finns and Laplanders in scarcity.

Bi (Lat. *bis*, twice). A prefix signifying twice or twofold.

Biba'sic (Lat. *bis*, twice ; *base*). In *chemistry*, applied to acids which unite with two equivalents of base to form salts.

Bib'ulous (Lat. *bibo*, I drink). Spongy ; having the property of imbibing moisture.

Bicar'bonate (Lat. *bis*, twice ; *carbonate*). A carbonate containing two equivalents of carbonic acid, to one of base.

Bicen'tral (Lat. *bis*, twice ; *centrum*, a centre). Having two centres.

Bi'ceps (Lat. *bis*, twice ; *cap'ut*, a head). Having two heads ; in *anatomy*, applied to certain muscles.

Bichlo'ride (Lat. *bis*, twice ; *chlorine*). A compound consisting of two equivalents of chlorine with one of another element.

Bicip'ital (Lat. *bis*, twice ; *cap'ut*, a head). Belonging to that which has two heads.

Bicusp'id (Lat. *bis*, twice ; *cuspis*, the point of a spear). Having two points or fangs.

Bidens (Lat. *bis*, twice ; *dens*, a tooth). Having two teeth or prongs.

Bien'nial (Lat. *bis*, twice ; *annus*, a year). Continuing two years ; or occurring every second year.

Bifid (Lat. *bis*, twice ; *findo*, I cleave). Cleft in two parts.

Bi'furcated (Lat. *bis*, twice ; *furca*, a fork). Divided into two prongs or forks.

Bifurca'tion (Lat. *bis*, double ; *furca*, a fork). A division into two branches.

Bigem'inal (Lat. *bis*, twice ; *gem'ini*, twins). Arranged in two pairs.

Bi'hamate (Lat. *bis*, twice ; *hamus*, a hook). Having two hooks.

Bi'jugate (Lat. *bis*, twice ; *jugum*, a yoke). In *botany*, having two pairs of leaflets.

Bila'biate (Lat. *bis*, twice ; *la'bium*, a lip). Having two lips.

Bilat'eral (Lat. *bis*, twice ; *latus*, a side). Having two sides.

Bil'iary (Lat. *bilis*, bile). Belonging to or containing bile.

Bilit'eral (Lat. *bis*, twice ; *lit'era*, a letter). Containing two letters.

Bilo'bed (Lat. *bis*, twice ; Gr. λοβος, lobos, a lobe). Having two lobes.

Biloc'ular (Lat. *bis*, twice ; *loc'ulus*, a little place). Containing two cells.

Bi'manous (Lat. *bis*, twice ; *manus*, a hand). Having two hands : applied in *zoology* to man.

Bi'nary (Lat. *bini*, two and two). Arranged in couples.

Bi'nary Theory of Salts. In *chemistry*, a theory which supposes that oxygen salts are constituted on the same plan as haloid salts (as chloride of sodium), of a metal in union with a salt-radical.

Bi'nate (Lat. *bini*, two and two). In *botany*, applied to compound leaves, the leaflets of which come off in two from a single point.

Binax'ial (Lat. *bini*, two and two ; *axis*). Having two axes.

Binoc'ular (Lat. *bini*, two and two ; *oc'ulus*, an eye). Having two eyes ; also applied to optical instruments that have two apertures, so that both eyes may be used at once.

Bino'mial (Lat. *bis*, twice ; *nomen*, a

name). In *algebra*, applied to a term consisting of two quantities joined by the sign + *plus* or − *minus*.

Binox'alate (Lat. *bis*, twice; *oxalic* acid). An oxalate containing two equivalents of acid to one of base.

Binox'ide (Lat. *bis*, twice; *oxygen*). A term applied in *chemistry* to the second degree of oxidation of a metal or other substance.

Bipar'tite (Lat. *bis*, twice; *pars*, a part). Having two corresponding parts.

Biped (Lat. *bis*, twice; *pes*, a foot). Having two feet.

Bipen'nate (Lat. *bis*, twice; *penna*, a wing). Having two wings; or wing-like leaves on each side of a stem.

Bipin'nate (Lat. *bis*, twice; *pinnate*). Doubly pinnate; applied to compound leaves, of which the leaflets are pinnate.

Biquad'rate (Lat. *bis*, twice; *quadra*, a square). In *mathematics*, the fourth power of a number, or the square multiplied by the square.

Bira'mous (Lat. *bis*, twice; *ramus*, a branch). Having two branches.

Bisect' (Lat. *bis*, twice; *seco*, I cut). To divide into two equal parts.

Bise'rial (Lat. *bis*, twice; *series*, an order or row). Arranged into two series or courses.

Biser'rate (Lat. *bis*, twice; *serra*, a saw). Doubly serrated; applied to the edges of leaves which are doubly marked like the teeth of a saw.

Bisul'cate (Lat. *bis*, twice; *sulcus*, a furrow). Cleft in two; having cloven feet.

Bisul'phate (Lat. *bis*, twice; *sulphuric* acid). A sulphate containing two equivalents of sulphuric acid to one of base.

Biter'nate (Lat. *bis*, twice; *terni*, three and three). In *botany*, applied to compound leaves, which form three leaflets on each secondary vein.

Bituber'culate (Lat. *bis*, twice; *tuber'culum*, a tubercle). Having two tubercles.

Bituminif'erous (Lat. *bitu'men*, min-

eral pitch or tar; *fero*, I produce). Yielding bitumen.

Bitu'minous. Having the property of or containing bitumen.

Bivalve (Lat. *bis*, twice; *valva*, folding-doors). Having a shell of two valves, closing with a hinge.

Black flux. A mixture of carbonate of potash and charcoal, used in chemical operations.

Blaste'ma (Gr. βλαστανω, *blas'tanō*, I bud forth). Material exuded from the blood through the minute vessels or capillaries, and capable of organisation.

Blas'toderm (Gr. βλαστος, *blastos*, a bud; δερμα, *derma*, a skin). The germinal disc which forms on the ovum or egg in the early stage of incubation.

Blende (German *blenden*, to dazzle). A term applied to minerals having a peculiar lustre or glimmer.

Blow-pipe. An instrument by which a current of air is driven on the flame of a lamp or candle, thereby producing an increased heat.

Boiling-point. The temperature at which a substance boils; it varies greatly for different substances, but is constant for the same, under the same circumstances.

Bole (Gr. βωλος, *bōlos*, a clod). A friable clayey slate or earth, usually coloured with oxide of iron.

Borate (*Borax*). A salt formed of boracic acid with a base.

Bo'real (Gr. βορεας, *boreas*, the north wind). Belonging to the north or north wind; applied to the pole of a magnet which points to the north.

Borboryg'mus (Gr. βορβορυγμος, *borborug'mos*). The sound caused by wind within the intestines.

Bot'any (Gr. βοτανη, *bot'anē*, a plant). The science which describes vegetables. *Descriptive botany* teaches the description and naming of plants; *geographical botany*, the manner in which plants are distributed on the earth; *palæontological botany* comprehends the study of fossil plants; *physiological botany* describes the functions of plants and their organs; *structural* *botany* teaches the structure of the various parts of plants ; *systematic* or *taxological botany*, the arrangement and classification of plants.

Bothren'chyma (Gr. βοθρος, *bothros*, a pit ; ἐγχυμα, *en'chuma*, any thing poured in, a tissue). A vegetable tissue, consisting of cylindrical cells marked by pits resembling dots.

Botryoid'al (Gr. βοτρυς, *botrus*, a bunch of grapes ; ἐιδος, *cidos*, shape). Resembling a cluster of grapes.

Boulder. A rounded or water-worn block of stone.

Boustrophe'don (Gr. βους, *bous*, an ox ; στρεφω, *strephō*, I turn). A form of writing alternately from left to right, and from right to left, like ploughing, used by the ancient Greeks.

Bo'viform (Lat. *bos*, an ox ; *forma*, shape). Resembling the ox.

Bovine (Lat. *bos*, an ox). Belonging to oxen and cows.

Brachely'tra (Gr. βραχυς, *brachus*, short ; ἐλυτρον, *elu'tron*, a case). A family of beetles characterised by the shortness of their elytra or outer wings.

Bra'chial (Lat. *bra'chium*, the arm). Belonging to the arm.

Bra'chio-cephal'ic (Lat. *bra'chium*, the arm ; Gr. κεφαλη, *keph'alē*, the head). Belonging to the arm and the head : applied to an artery of the body.

Bra'chiopods (Gr. βραχιων, *bra'chiōn*, an arm ; πους, *pous*, a foot). A genus of molluscous invertebrate animals, so called because their feet, or organs of progressive motion, resemble arms.

Brachyu'ra (Gr. βραχυς, *brachus*, short ; οὐρα, *oura*, a tail). A class of crustacea with short tails, as the crab.

Bract (Lat. *brac'tea*, a thin leaf of metal). In *botany*, a leaf from the axil or angle of which a flower-bud arises.

Bractlet (*Bract*). A little bract ; any rudimentary leaf on a flower-stem between the bract and the calyx.

Bran'chiæ (Gr. βραγχια, *bran'chia*, gills). The gills or breathing organs of animals which live entirely in water ; they are analogous to lungs in air-breathing animals.

Bran'chial (Gr. βραγχια, *bran'chia*, gills). Belonging to the branchiæ or gills.

Bran'chiopods (Gr. βραγχια, *bran'chia*, gills ; πους, *pous*, a foot). Crustaceous animals which have gills attached to the feet.

Branchios'tegal (Gr. βραγχια, *bran'chia*, gills ; στεγω, *stegō*, I cover). Covering gills : applied to certain rays or bent bones which support a membrane covering in the gills of fishes.

Branchios'tegous (Gr. βραγχια, *bran'chia*, gills ; στεγος, *stegos*, a covering). Having covered gills.

Brassica'ceous (Lat. *bras'sica*, a cabbage). Belonging to the order of plants of which the cabbage is a type.

Brec'cia (Italian, a crumb). A term applied to rocks composed of agglutinated angular fragments.

Brevipen'nes (Lat. *brevis*, short ; *penna*, a feather). A family of grallæ or stilt-birds, characterised by the shortness of their wings, as the ostrich and emeu.

Bro'mate (*Bromic* acid). A salt formed by the combination of bromic acid with a base.

Bron'chia (Gr. βρογχος, *bronchos*, the windpipe). The smaller tubes into which the windpipe divides in entering the lung.

Bron'chial (Gr. βρογχος, *bronchos*, the windpipe). Belonging to the divisions of the windpipe.

Bronchi'tis (Gr. βρογχος, *bronchos*, the windpipe ; term. ιτις, *itis*, denoting inflammation). Inflammation of the tubes into which the windpipe divides.

Bron'chocele (Gr. βρογχος, *bronchos*, the windpipe ; κηλη, *kēlē*, a tumour). A kind of tumour on the front part of the neck.

Bronchoph'ony (Gr. βρογχος, *bronchos*, the windpipe ; φωνη, *phōnē*, sound). The sound produced by the passage of air through the bronchi.

Bronchot'omy (Gr. βρογχος, *bronchos*, the windpipe ; τεμνω, *temnō*, I cut).

An operation in which the windpipe is cut open.

Bronch′us (Gr. βρογχος, *bronchos*, the throat or windpipe). One of the large or primary divisions of the trachea or windpipe.

Bryozo′a or **Bryozoa′ria** (Gr. βρυος, *bruos*, moss ; ζωον, *zōon*, an animal). A term denoting the minute mollusca which live united in masses in a branched and moss-like manner.

Buccal (Lat. *bucca*, the cheek). Belonging to the cheek, or to the cavity of the mouth.

Buc′cina′tor (Lat. *buc′cina*, a kind of trumpet). A muscle forming a large part of the cheek, so called from its use in blowing wind-instruments.

Buffy Coat. The viscid layer formed on the surface of blood in inflammatory diseases.

Bulb (Lat. *bulbus*). In *botany*, a part of a plant, generally beneath the ground, formed of layers of scales in the manner of a bud, as the onion ; in *anatomy*, applied to various parts from their shape.

Bulbif′erous (Lat. *bulbus*, a bulb ; *fero*, I bear). Producing bulbs.

Bulblet (*Bulb*). A little bulb.

Bulbous (Lat. *bulbus*, a bulb). Containing bulbs.

Bulim′ia (Gr. βου, *bou*, a prefix signifying large or enormous ; λιμος, *limos*, hunger). Excessive appetite for food.

Bulwark-plains. In *astronomy*, circular areas in the moon enclosed by a ring of mountain-ridges.

Bunter (Germ.) A term in *geology* for new red-sandstone, from its variegated appearance.

Bursa (Lat. a purse). In *anatomy*, a closed sac containing synovial fluid.

Butyra′ceous (Lat. *butyrum*, butter). Having the properties of, or containing butter.

Butyr′ic (Lat. *butyrum*, butter). Belonging to butter ; applied to an acid formed in butter.

Byssus (Gr. βυσσος, *bussos*, fine flax). The thread or fibres by which the Mollusca attach themselves to rocks. Silky tufts of fungus from damp and decaying substances.

C.

Cachec′tic (Gr. κακος, *kakos*, bad ; ἑξις, *hexis*, habit). Belonging to, or having, a vitiated state of the body.

Cachex′ia (Gr. κακος, *kakos*, bad ; ἑξις, *hexis*, habit). A deranged or vitiated state of the constitution.

Cacoe′thes (Gr. κακος, *kakos*, bad ; ἠθος, *ēthos*, custom). A bad habit or disposition.

Cacoph′ony (Gr. κακος, *kakos*, bad ; φωνη, *phōnē*, voice). A disagreeable sound, produced by the meeting of harsh letters.

Cacoplas′tic (Gr. κακος, *kakos*, bad ; πλασσω, *plassō*, I form). Having a defective power of being organised or taking a definite form.

Cadaver′ic (Lat. *cadāver*, a carcase). Belonging to a dead body.

Cadu′cous (Lat. *cado*, I fall). Having a tendency to fall off.

Cæcal (*Cæcum*). Having a closed end ; belonging to the cæcum.

Cæcum (Lat. *cæcus*, blind). A tube with a closed end ; applied to a part of the intestinal canal.

Cænozo′ic, or **Cainozo′ic** (Gr. καινος, *kainos*, new ; ζωον, *zōon*, an animal). Applied in *geology* to the tertiary strata, which include the most recent remains of animals.

Caf′fein. A vegetable alkali found in tea and coffee.

Cal′amites (Lat. *cal′amus*, a reed). A genus of fossil stems, resembling gigantic reeds, occurring in the coal formations.

Calca′neal (Lat. *calx*, the heel). Belonging to the heel.

Cal′carate (Lat. *calcar*, a spur). Like or having a spur.

Calca'reo-arena'ceous. Consisting of lime, or chalk, and sand.

Calca'reous (Lat. *calx*, lime). Having the properties of or containing lime.

Cal'ceolate (Lat. *cal'ceus*, a shoe). Like a shoe or slipper.

Calcifica'tion (Lat. *calx*, lime ; *facio*, I make). A hardening by the deposition of salts of lime.

Cal'cify (Lat. *calx*, lime ; *facio*, I make). To change into lime or chalk ; to harden by the deposition of salts of lime.

Calcina'tion (Lat. *calx*, lime). The expelling by heat some volatile matter from a substance, as carbonate of lime (limestone) is reduced to lime by driving off the carbonic acid by heat.

Calci'ne (Lat. *calx*, lime). To drive off volatile matter by heat so as to render a substance friable, as in the operation of lime-burning.

Cal'culus (Lat. a pebble). In *mathematics*, a term applied to certain of the more abstruse branches of calculation ; in *medicine* a concretion formed within the body.

Calefa'cient (Lat. *calor*, heat ; *facio*, I make). Making warm ; heating.

Cal'endar (Lat. *calen'dæ*, the first day of the Roman months). A table of the days of each month, with the events connected with each.

Cal'enture (Span. *calentar'*, to heat). A violent ardent fever, principally affecting sailors in hot climates.

Cal'ibre (Fr.). The diameter of a round body ; the bore of a cylindrical tube, as of a gun.

Calic'iform (Lat. *calix*, a cup ; *forma*, shape). Shaped like a cup.

Calisthen'ics (Gr. καλος, *kalos*, beautiful ; σθενος, *sthen'os*, strength). Exercise of the body and limbs to promote strength and graceful movements.

Callos'ity (Lat. *callus*, hardness). A hardness.

Callus (Lat.). A hard deposit ; also applied to the excess of bony matter which is often formed in the process of union of broken bones.

Calor'ic (Lat. *calor*, heat). The principle of heat ; the cause of the effects or phenomena popularly recognised as heat.

Calorifa'cient (Lat. *calor*, heat ; *facio*, I make). Producing heat ; furnishing material for the production of heat.

Calorif'ic (Lat. *calor*, heat ; *facio*, I make). Producing heat.

Calorim'eter (Lat. *calor*, heat ; Gr. μετρον, *metron*, a measure). An instrument for measuring the relative quantities of heat contained in bodies.

Cal'otype (Gr. καλος, *kalos*, beautiful ; τυπος, *tupos*, a type or impression). A process of photography, in which the picture is produced by the rapid action of light on paper prepared with iodide of silver and gallonitrate of silver.

Calyc'ifloral (Lat. *calyx*, a cup or calyx ; *flos*, a flower). A subdivision of exogenous plants, including those which are provided with both calyx and corolla, the stamens being perigynous or epigynous.

Calyp'tra (Gr. καλυπτω, *kaluptō*, I cover). An appendage of the theca in mosses, covering it at first.

Calyp'trate (Gr. καλυπτρα, *kaluptra*, a covering). Having a calyptra or covering ; in *botany*, applied to the calyx of plants when it comes off like an extinguisher.

Calyx (Gr. καλυξ, *calux*, a shell, or unopened flower). The row of leaf-like organs, generally green, which immediately surrounds a flower.

Cam'bium. In *botany*, the mucilaginous fluid which lies between the young wood and the bark of a tree.

Cam'era Luc'ida (Lat. a bright chamber). An apparatus for facilitating the delineation of objects, by producing a reflected picture of them on paper by means of a prism.

Cam'era Obscu'ra (Lat. a dark chamber). An apparatus in which, the images of objects are received through a double convex glass, and exhibited in the interior of the machine on a plane or curved surface.

Campan'ulate (Lat. *compana*, a bell). Shaped like a bell.

Campylit'ropous (Gr. καμπυλος, *kam'-pulos*, curved; τρεπω, *trepō*, I turn). In *botany*, applied to an ovule bent down on itself till the apex touches the base.

Canalic'ulus (Lat. *canālis*, a channel; *ulus*, denoting smallness). A little channel.

Can'cellated (Lat. *cancelli*, cross-bar, or lattice-work). Resembling lattice-work : applied to the least compact structure of bones.

Cancel'li (Lat. lattice-work). In *anatomy*, the network which forms the less compact part of bones.

Canic'ular (Lat. *canic'ulus*, a small dog). Belonging to the dog-star.

Cani'ne (Lat. *canis*, a dog). Belonging or having relation to a dog.

Cannel-coal. A compact brittle variety of coal, breaking with a conchoidal fracture, and not soiling the fingers.

Can'nula (Gr. καννα, *kanna*, a reed or cane ; *ula*, implying smallness). A small pipe.

Can'tharis (Gr. κανθαρος, *kan'tharos*, a kind of beetle). The Spanish fly, an insect of the beetle tribe : used for producing blisters.

Canthus. The angle or corner of the eye.

Caoutchouc, or India-rubber. The produce of several trees in tropical countries, which produce a juice that hardens on exposure to air.

Capac'ity (Lat. *capio*, I receive). The power of containing ; in *chemistry*, applied to the proportion in which bodies take in and contain caloric ; the space included within the cubic boundaries of a body.

Cap'illary (Lat. *capil'lus*, a hair). Resembling or having relation to fine hairs, or to the minute blood-vessels.

Cap'itate (Lat. *caput*, a head). Ending in a knob, like the head of a pin.

Capit'ulum (Lat. *caput*, a head). A little head ; in *botany*, a flower-head, composed of a number of florets arranged without stems on the summit of a single peduncle.

Ca'priform (Lat. *caper*, a goat; *forma*, shape). Resembling a goat.

Cap'sular (Lat. *cap'sula*, a capsule). Belonging or having relation to a capsule.

Capsule (Lat. *cap'sula*, a little chest). In *chemistry*, a clay saucer for roasting ; in *botany*, a form of dry fruit containing many seeds ; in *anatomy*, a membranous bag inclosing an organ.

Car'amel. Burnt sugar.

Car'apace (Gr. καραβος, *kar'abos*, a stag-beetle or crab). The bony shield-like structure which protects the upper part of the turtle and tortoise ; also the shell covering the crab, formed by the union of the head with the thorax.

Carb'ide (*Carbon*). A compound of carbon with hydrogen or a metal.

Carbona'ceous (Lat. *carbo*, a coal). Belonging to or containing carbon or charcoal.

Car'bonate (Lat. *carbo*, a coal). A salt formed by the union of carbonic acid with a base.

Carbon'ic (Lat. *carbo*, a coal). Belonging to, or containing carbon or charcoal.

Carbonif'erous (Lat. *carbo*, coal ; *fero*, I bear). Producing or yielding coal.

Carbonisa'tion (Lat. *carbo*, coal). The process of burning a substance until nothing but the carbon or charcoal is left.

Car'bonise (Lat. *carbo*, coal). To turn into coal.

Car'buncle (Lat. *carbo*, a coal). A painful form of excrescence or growth on the skin.

Car'buret (*Carbon*). A compound of carbon with hydrogen or a metal.

Carcino'ma (Gr. καρκινος, *kar'kinos*, a crab). A form of cancer.

Carcinomatous (Gr. καρκινωμα, *karkinōma*, a cancer). Consisting of or belonging to the form of cancer called carcinoma.

Car'dia (Gr. καρδια, *kar'dia*, the heart). The opening in the stomach which admits the food : a term originating in the former confusion of ideas between the heart and the stomach.

Car'diac (Gr. καρδια, kar'dia, the heart). Belonging to the heart ; or to the upper orifice of the stomach.

Car'diaci (Gr. καρδια, kar'dia, the heart). A term proposed to be applied to the diseases of the heart.

Cardial'gia (Gr. καρδια, kar'dia, the heart ; dλγος, algos, pain). Pain in the stomach.

Car'dinal (Lat. cardo, a hinge). In astronomy, applied to the four principal intersections of the horizon with the meridian, or North, South, East, and West ; in zoology, belonging to or connected with the hinge in bivalve molluscs.

Cardi'tis (Gr. καρδια, kar'dia, the heart ; itis, denoting inflammation). Inflammation of the heart.

Ca'ries (Lat., the state of worm-eaten wood). Ulceration of the substance of bones.

Ca'rious (Lat. caries). Affected with caries.

Carmin'ative (Lat. carmen, a poem or song). A medicine used to relieve pain in the stomach and flatulence ; so called because it acts as incantations (carmina) or charms were supposed to act.

Carna'ria (Lat. caro, flesh). An order of mammalian animals which live on flesh, as the lion, tiger, &c.

Carnifica'tion (Lat. caro, flesh ; facio, I make). Conversion into a substance resembling flesh.

Carniv'ora (Lat. caro, flesh ; voro, I devour). See Carnaria.

Carniv'orous (Lat. caro, flesh ; voro, I devour). Living on animal food.

Carot'id (Gr. καρα, kara, the head ; ους, ous, the ear). A name given to the arteries which proceed to the head.

Carpal (Carpus). Belonging to the wrist.

Carpel (Gr. καρπος, karpos, fruit). A name given to the separate pistils of which a compound fruit is formed.

Carpel'lary (Carpel). Belonging to a carpel.

Carp'ology (Gr. καρπος, karpos, a fruit ; λογος, logos, discourse). The description and classification of fruits.

Carp'ophore (Gr. καρπος, karpos, fruit ; φερω, pherō, I carry). The axis or stalk which supports the achænia of which a cremocarp is formed.

Carpus (Gr. καρπος, karpos, the wrist). The wrist.

Car'polithes (Gr. καρπος, karpos, fruit ; λιθος, lithos, a stone). In geology, the general term for fossil fruits.

Car'tilage (Lat. cartila'go). Gristle.

Cartilag'inous (Lat. cartila'go, cartilage). Belonging to or consisting of gristle ; applied also to certain fishes, the skeleton of which is of a gristly consistence.

Car'uncle (Lat. caro, flesh). A small fleshy excrescence.

Caryat'ides (Gr. Καρυαι, Car'uai, a city of Laconia). In architecture, female figures used to support entablatures ; so called from the women of Caryæ (Καρυαι), when the city was taken by the Athenians, being represented in this posture to perpetuate the memory of the event.

Caryop'sis (Gr. καρυον, kar'uon, a walnut ; οψις, opsis, appearance). A form of dry fruit, consisting of one cell, not splitting, and containing a seed which is adherent to the pericarp.

Ca'sein (Lat. ca'scum, cheese). A peculiar compound substance, the characteristic component of milk, and the principal ingredient in cheese.

Cat'aclysm (Gr. κατακλυζω, kataclu'zō, I inundate). A deluge or inundation.

Catalepsy (Gr. κατα, kata, down ; ληψις, lēpsis, a seizing). A sudden suppression of consciousness, in which the body retains the position in which it was when the attack commenced.

Catal'ysis (Gr. κατα, kata, down ; λυω, luō, I loosen). A term applied to certain chemical phenomena, in which changes in the composition of substances are effected by the presence of another body, which itself remains unaltered.

Catalyt'ic (Gr. κατα, kata, down; λυω, luō, I loosen). Relating to catalysis.

Cat'aplasm (Gr. κατα, kata, down, or on; πλασσω, plassō, I mould). A poultice.

Cat'aract (Gr. καταρρηγνυμι, katarrhegnu'mi, I break down). A waterfall; in medicine, a disease of the eyes, consisting in opacity of the crystalline leus.

Catar'rh (Gr. κατα, kata, down; ρεω, rheō, I flow). A disorder attended with increased secretion from the nose and fauces; a cold.

Catar'rhal (Gr. κατα, kata, down; ρεω, rheō, I flow). Belonging to catarrh.

Catastal'tic (Gr. κατα, down; στελλω, stellō, I send). Acting from above downwards, or from the centre to the circumference: applied to nervous action.

Catas'trophē (Gr. κατα, down or over; στρεφω, strephō, I turn). In geology, a supposed change in the globe from some sudden violent physical action.

Catena'rian (Lat. caténa, a chain). Relating to or resembling a chain.

Cate'nopores (Lat. caténa, a chain; porus, a pore). Chainpore coral: a form of fossil coral.

Cathar'tic (Gr. καθαιρω, kathai'rō, I clean or purge). Purgative.

Cath'ode (Gr. κατα, kata, down; ὁδος, hodos, a way). The surface at which electricity passes out of a body.

Cat'ion (Gr. κατα, kata, down; ιων, iōn, going). A name given by Dr. Faraday to those substances which appear at the cathode.

Catop'trics (Gr. κατοπτρον, katoptron, a mirror). That part of optics which explains the phenomena of reflected light.

Cauca'sian (Cau'casus). A term properly denoting the peoples dwelling about the Caucasus, but applied also as the name of a class to most of the European and several Asiatic nations.

Cauda equi'na (Lat. a horse's tail).

The brush-like collection of nerves which terminates the spinal marrow.

Caudal (Lat. cauda, a tail). Belonging to the tail.

Caudate (Lat. cauda, a tail). Having a tail.

Caul'icle (Lat. caulis, a stalk; cle, denoting smallness). In botany, a term sometimes applied to the neck of the embryonic plant.

Caul'inary (Lat. caulis, a stem). In botany, applied to the leaves of mosses when produced on the stem.

Caul'ine (Lat. caulis, a stem). Belonging to a stem; applied to the leaves growing from the main axis of a plant.

Caustic (Gr. καιω, kai'ō, I burn). Burning; in surgery, destroying animal textures by powerful chemical action.

Cau'terise (Gr. καιω, kai'ō, I burn). To destroy animal tissues by heat, as with a hot iron.

Cau'tery (Gr. καιω, kai'ō, I burn). The destroying animal tissues by the application of heat; an iron instrument for the purpose.

Cav'ernous (Lat. caver'na, a cavern). Full of caverns; or like a cavern.

Celes'tial (Lat. cœlum, heaven). Belonging to the sky or visible heaven.

Cell (Lat. cella, a store-house or chamber). In physiology, a minute bag or vesicle.

Cel'lular (Lat. cel'lula, a little cell). Consisting of or containing cells; applied to the connecting tissue of the different parts of the body, which form cells or interstices.

Cel'lulose (Lat. cel'lula, a cell). A compound of carbon, hydrogen, and oxygen, forming the fundamental material of the structure of plants.

Cent'igrade (Lat. centum, a hundred; gradus, a degree). Consisting of a hundred degrees; the scale on which thermometers are constructed in France.

Cent'igramme (Fr. cent, a hundred; gramme, a weight so called). A French weight, the hundredth part of a *gramme* : about $\frac{3}{10}$ths of a grain avoirdupois.

Cent'ilitre (Fr. *cent*, a hundred ; *litre*, a quart, or $1\frac{3}{4}$ English pints). The hundredth part of a *litre* : about $\frac{1}{7}$th of an English pint.

Cent'ime'tre (Fr. *cent*, a hundred ; *mètre*, a measure equal to $3\frac{7}{25}$ English feet). The hundredth part of a mètre : equal to a little more than $\frac{39}{100}$ths of an English inch.

Cent'ipede (Lat. *centum*, a hundred ; *pes*, a foot). Having a hundred feet : applied to certain insect-like animals which have a large number of feet.

Cen'trical (Lat. *centrum*, a centre). Having coinciding centres ; centrical interposition, in *astronomy*, is the appearance presented in eclipses when the centres of the discs coincide, the margin of the larger disc being left free.

Centrif'ugal (Lat. *centrum*, the centre; *fugio*, I flee). Having a tendency to fly off in a direction from the centre; in *botany*, applied to plants in which the expansion of flowers commences at the top and proceeds downwards.

Centrip'etal (Lat. *centrum*, a centre ; *peto*, I seek). Having a tendency towards the centre; in *botany*, applied to plants in which the flowers expand from below upwards.

Cephalal'gia (Gr. κεφαλη, *keph'alē*, the head ; ἀλγος, *algos*, pain). Headache.

Cephal'ic (Gr. κεφαλη, *keph'alē*, the head). Belonging to the head.

Cephal'ici (Gr. κεφαλη, *keph'alē*, the head). A term proposed to be given to diseases seated in the head.

Ceph'alopods (Gr. κεφαλη, *keph'alē*, the head ; πους, *pous*, a foot). A class of molluscous invertebrate animals, which have their organs of motion arranged round the head, as the cuttle-fish.

Cephalotho'rax (Gr. κεφαλη, *keph'alē*, the head ; θωραξ, *thōrax*, a breastplate). The anterior part of the external skeleton of arachnida, consisting of the head and chest united in one mass.

Cerate (Lat. *cera*, wax). An ointment consisting of wax and oil.

Cer'atites (Gr. κερας, *keras*, a horn). A genus of fossil cephalopods in the triassic strata.

Cer'ato- (Gr. κερας, *keras*, a horn). In *anatomy*, a prefix in compound words signifying connection with the cornua or horns of the hyoid bone.

Cer'atose (Gr. κερας, *keras*, a horn). Horny ; applied to sponges, of which the hard part is of a horny consistence.

Cercæ (Gr. κερκος, *kerkos*, a tail). The feelers projecting from the hind part of the body in some insects.

Cer'eal (Lat. *Ceres*, the goddess of corn). Belonging to, or producing eatable grain.

Cerebel'lar (*Cerebellum*). Belonging to the cerebellum or little brain.

Cerebel'lum (Lat. *cer'ebrum*, the brain; *ellum*, signifying smallness). The little brain ; a portion of the mass within the skull, situated at the lower and back part.

Cer'ebral (Lat. *cer'ebrum*, the brain). Belonging to the brain.

Cer'ebric (Lat. *cer'ebrum*, the brain). Belonging to or produced from the brain.

Cereb'riform (Lat. *cer'ebrum*, the brain ; *forma*, shape). Shaped like the brain.

Cerebri'tis (Lat. *cer'ebrum*, the brain ; *itis*, denoting inflammation). Inflammation of the brain.

Cer'ebroid (Lat. *cer'ebrum*, the brain ; Gr. εἰδος, *eidos*, shape). Like or analogous to a brain.

Cer'ebro-spi'nal (Lat. *cer'ebrum*, the brain ; *spina*, the spine). Belonging to or consisting of the brain and spinal cord.

Cer'ebrum (Lat). The brain proper.

Ceru'minous (Lat. *cerumen*, the wax of the ear). Belonging to the wax contained in the ear.

Ceru'lean (Lat. *cœlum*, the sky). Sky-coloured ; blue.

Cervi'cal (Lat. *cervix*, the neck). Belonging to the neck.

Ces'toid (Gr. κεστος, *kestos*, a girdle ; εἰδος, *eidos*, form). Like a girdle ;

applied to intestinal worms with long flat bodies, as the tape-worm.

Cestra′cionts (Gr. κεστρα, *kestra*, a kind of fish). A family of fishes, mostly fossil, of which the Port Jackson shark is a type.

Ceta′ceous (Gr. κητος, *kētos*, a whale). Belonging to the order of mammalian animals of which the whale is a type.

Chala′za (Gr. χαλαζα, *chala′za*, a small tubercle). The twisted membranous cord attached at each end of the yolk of an egg; in *botany*, an expansion at the base of an ovule, uniting the coverings with the nucleus.

Chalyb′eate (Gr. χαλυψ, *chalubs*, steel). Containing iron.

Chame′leon (Gr. χαμαι, *chamai*, on the ground; λεων, *leōn*, a lion). A kind of lizard; in *chemistry*, a manganate of potassa, from the changes in colour which its solution undergoes.

Cha′os (Gr. χαος, *chaos*, void space, or unformed mass). A mass of matter without arrangement.

Cheirop′tera (Gr. χειρ, *cheir*, a hand; πτερον, *pteron*, a wing). Wing-handed animals; applied to an order of mammalian animals, of which the bat is an example, in which the toes of the fore-limbs are connected by a membrane, so as to serve as wings.

Che′late (*chēlē*). Having chelæ or two-cleft claws.

Chele (Gr. χηλη, *chēlē*, a hoof or claw). The two-cleft claws of the crustacea, scorpions, &c.

Chelic′era (Gr. χηλη, *chēlē*, a claw; κερας, *keras*, a horn). The prehensile claws of the scorpion.

Chelo′nia (Gr. χελωνη, *chelōnē*, a tortoise). The order of reptiles including tortoises and turtles.

Chem′ical (Gr. χεω, *cheō*, I pour). Belonging to chemistry.

Chem′istry (Gr. χεω, *cheō*, I pour). The science which has for its object the study of the nature and properties of all the materials which enter into the composition of the earth, sea, and air, and of the beings inhabiting them.

Chert. A term applied to flinty portions occurring in limestone and other rocks.

Chia′ro-oscu′ro (Italian, *chia′ro*, clear; *oscu′ro*, dark). A drawing in black and white; the art of advantageously distributing the lights and shadows in a picture.

Chilogna′tha (Gr. χειλος, *cheilos*, a lip; γναθος, *gnathos*, a jaw). A family of myriapodous invertebrate animals, having a pair of stout horny mandibles with sharp toothed edges.

Chilop′oda (Gr. χειλος, *cheilos*, a lip; πους, *pous*, a foot). A family of myriapodous invertebrate animals, having an additional lip formed by the second pair of legs, containing each a canal for the discharge of a poisonous liquid, as the centipede.

Chirur′gical (Gr. χειρ, *cheir*, a hand; εργον, *eryon*, work). Relating to surgery, or that branch of medicine which treats diseases and injuries by manual operations and instruments.

Chi′tine (Gr. χιτων, *chitōn*, a coat). The hardening substance of the covering of insects.

Chi′tinous (*Chitine*). Consisting of, or of the nature of, chitine.

Chlo′rate (*Chlorine*; term. *ate*). A compound of chloric acid with a base.

Chlo′ride (*Chlorine*; term. *ide*). A compound of chlorine with a metal or other elementary substance.

Chlo′rine (Gr. χλωρος, *chlōros*, yellowish green). An elementary gas, so called from its yellow colour.

Chlo′rite (Gr. χλωρος, *chlōros*, yellowish-green). A mineral occurring in the granite and metamorphic rocks, often disseminated through or coating the laminæ.

Chlorom′etry (*Chlorine*; Gr. μετρον, *metron*, a measure). The process of testing the quantity of chlorine contained in chloride of lime or any other bleaching material.

Chlo′rophyll (Gr. χλωρος, *chlōros*, yellowish-green; φυλλον, *phullon*, a leaf). The green colouring matter of the leaves of plants.

Chloro'sis (Gr. χλωρος, *chlōros*, yellowish-green). A diseased state, characterised by poverty of blood, and in which a greenish colour of the skin is a prominent symptom.

Chlorot'ic (Gr. χλωρος, *chlōros*, yellowish-green). Relating to or having chlorosis.

Choke-damp. Carbonic acid gas disengaged in mines.

Cho'lagogue (Gr. χολη, *cholē*, bile; ἀγω, *agō*, I lead). Having the property of causing an evacuation of bile.

Choled'ochus (Gr. χολη, *cholē*, bile; δεχομαι, *dech'omai*, I receive). Receiving bile; applied to the tube formed by the junction of the cystic and hepatic ducts.

Chol'era (Gr. χολη, *cholē*, bile: ρεω, *rheō*, I flow). An epidemic disease, characterised by diarrhœa and vomiting, and symptoms of depression of the powers of life.

Choles'terin (Gr. χολη, *cholē*, bile; στερεος, *ster'eos*, solid). A substance having the properties of fat, found principally in bile.

Chondrin (Gr. χονδρος, *chondros*, a cartilage or gristle). A substance somewhat resembling gelatine or animal jelly, produced by the action of hot water on cartilage.

Chon'drites (Lat. *chondrus*, a kind of sea-weed). Fossil marine plants in the chalk and other formations.

Chondropteryg'ii (Gr. χονδρος, *chondros*, cartilage or gristle; πτερυγιον, *pteru'gion*, a little wing). An order of fishes, the fin-bones of which are composed of gristle only.

Chord (Gr. χορδη, *chordē*, a string). In *geometry*, a line extending from one end of the arc of a circle to the other; in *music*, the union of two or more sounds uttered at once, forming a harmony.

Chor'ea (Gr. χορος, *choros*, a dance). The disease commonly called St. Vitus's Dance, consisting of involuntary movements of the muscles, consciousness being retained.

Cho'rion (Gr. χωρεω, *chōreū*, I contain). The external membrane which covers the fœtus.

Cho'risis (Gr. χωριζω, *chōri'zō*, I separate). A separation: in *botany*, applied to the increase in number of the parts of a flower produced by the splitting of organs during their development.

Chorog'raphy (Gr. χωρος, *chōros*, a place or region; γραφω, *graphō*, I write or describe). The description of a region or country.

Chor'oid (Gr. χωριον, *chōrion*, the chorion; εἰδος, *eidos*, shape). Resembling the chorion: applied to a coat of the eye, also to a network of blood-vessels in the brain.

Chro'mate (Gr. χρωμα, *chrōma*, colour). A compound of chromic acid with a base.

Chromat'ic (Gr. χρωμα, *chroma*, colour). Relating to colour; in *music*, the chromatic scale is that which proceeds by semitonic intervals.

Chro'matrope (Gr. χρωμα, *chroma*, colour; τρεπω, *trepō*, I turn). An optical apparatus for exhibiting the appearance of a stream of colours, by the revolution of a double set of coloured circular arcs.

Chro'mogen (Gr. χρωμα, *chrōma*, colour; γεννaω, *gennaō*, I produce). The colouring matter of plants.

Chronol'ogy (Gr. χρονος, *chronos*, time; λογος, *logos*, a word or description). The arrangement of events in order of time.

Chronom'eter (Gr. χρονος, *chronos*, time; μετρον, *metron*, a measure). An instrument for measuring time.

Chronomet'ric (Gr. χρονος, *chronos*, time; μετρον, *metron*, a measure). Relating to or employed in the measure of time.

Chro'tici (Gr. χρως, *chrōs*, the skin). A term proposed to be applied to diseases of the skin.

Chrys'alis (Gr. χρυσος, *chrusos*, gold). The form which certain insects assume between the caterpillar and the winged states; so called because yellow in some.

Chyle (Gr. χυλος, *chulos*, juice). The milky liquid prepared from the food, to be absorbed by the lacteal vessels, and supplied to the blood for nutriment.

Chyliferous (Lat. *chylus*, chyle; *fero*, I carry). Carrying chyle.

Chylific (Lat. *chylus*, chyle; *facio*, I make). Making chyle; especially applied to a part of the digestive apparatus of insects.

Chylifica'tion (Lat. *chylus*, chyle; *facio*, I make). The process of making chyle.

Chylopoiet'ic (Gr. χυλος, *chulos*, juice or chyle; ποιεω, *poi'eō*, I make). Making chyle: commonly applied to the stomach and intestines.

Chyme (Gr. χυμος, *chumos*, juice). The pulpy mass formed by digestion of the food in the stomach.

Cicatri'cula (Lat. *cicátrix*, a scar; *ula*, denoting smallness). A spot resembling a small scar.

Cicatrisa'tion (Lat *cicátrix*, a scar). The process of healing a wound.

Cic'atrise (Lat. *cicátrix*, a scar). To heal a wound, or induce the formation of a scar.

Cica'trix (Lat.) The scar left after the healing of a wound.

Cil'ia (Lat. *cil'ium*, an eyelash). In *anatomy*, the eyelashes; also certain minute bodies projecting from various parts of animals, and having waving motion; in *botany*, hairs on the margin of a body.

Cil'iary (Lat. *cil'ium*, an eyelash). Belonging to the eyelashes or eyelids, or to the minute vibratory bodies called cilia.

Cil'iated (*Cil'ia*). Provided with vibratile cilia: applied to a form of epithelium.

Ciliobra'chiate (Lat. *cil'ium; bra'chium*, an arm). Having the arms provided with cilia; applied to a class of polypes.

Cil'iograde (Lat. *cil'ium; gra'dior*, I step). Swimming by the action of cilia.

Cinen'chyma (Gr. κινεω, *ki'neō*, I move; εγχυμα, *en'chuma*, a tissue). A name given to the laticiferous vessels of plants.

Cineri'tious (Lat. *cinis*, ashes). Resembling ashes; grey.

Cin'nabar. A crystalline sulphide of mercury.

Cir'cinate (Lat. *cir'cino*, I turn round). Curled round like a shepherd's crook or a crosier.

Cir'culate (Lat. *cir'culus*, a circle). To move in such a manner as to return to the starting point, as the blood does.

Circula'tion (Lat. *cir'culus*, a circle). A motion in a circle; the process by which a moving body returns to the point from which it started.

Circum. A Latin preposition, used as a prefix in compound words, signifying around.

Circumduc'tion (Lat. *circum*, around; *duco*, I lead). A leading round; in *physiology*, a motion in which a bone is made to describe a cone, the apex of which is at the joint; as with the arm.

Cir'cumflex (Lat. *circum*, around; *flecto*, I bend). Bent round; in *anatomy*, applied to certain vessels and nerves, from their course.

Circumgyra'tion (Lat. *circum*, about; *gyrus*, a circle). Motion in a circle.

Circumnav'igate (Lat. *circum*, around; *navis*, a ship). To sail round.

Circumpo'lar (Lat. *circum*, around; *polus*, the pole). Round the pole: a term applied to the stars near the North Pole.

Circumscis'sile (Lat. *circum*, around; *scindo*, I cut). In *botany*, applied to a form of dehiscence or opening of fruits, in which the upper part separates like a lid, as if cut off.

Cirrho'se (Lat. *cirrhus*, a curl or tendril). Having or giving off tendrils.

Cirrho'sis (Gr. κιρρος, *kirrhos*, tawny). A term applied to a diseased state of the liver.

Cirri (Lat. *cirrus*, a lock of hair or curl). The curled filaments acting as feet to barnacles; in *botany*, tendrils.

Cirrig'erous (Lat. *cirrus*, a curl; *gero*, I bear). Supporting cirri or curled filaments.

Cir'rigrade (Lat. *cirrus; gra'dior*, I step). Moving by means of cirri.

Cir'ripeds (Lat. *cirrus; pes*, a foot). See Cirropods.

Cir'ropods (Lat. *cirrus*, a fringe ; Gr. πους, *pous*, a foot). A class of invertebrate animals with curled jointed feet.

Ci'tigrade (Lat. *citus*, quick ; *gradus*, a step). Moving quickly.

Ci'trate (Lat. *citrus*, a citron or lemon). A compound of citric acid with a base.

Cit'ric (Lat. *citrus*, a lemon). Belonging to or existing in lemons ; applied to an acid found in lemons and some other fruits.

Cladoc'era (Gr. κλαδος, *klados*, a branch ; κερας, *keras*, a horn). Having branched horns : applied to a family of crustaceous animals with branched antennæ.

Clairvoy'ance (Fr. *clair*, clear; *voir*, to see). A state in which persons pretend to see that which, under ordinary circumstances, is not apparent to the eye.

Clarifica'tion (Lat. *clarus*, clear ; *facio*, I make). A making clear.

Class (Lat. *classis*). A group of things or beings, having some conspicuous mark of similarity, but capable, on further examination, of being subdivided into other groups or orders.

Classifica'tion (Lat. *classis*, a class ; *facio*, I make). An arrangement into classes.

Cla'vate (Lat. *clavus*, a club). Club-shaped.

Clavicor'nes (Lat. *clavus*, a club ; *cornu*, a horn). A family of insects whose antennæ end in a club-shaped enlargement, as the necrophorus or burying beetle.

Clay. In *geology*, a fine impalpable sediment from water, nearly entirely consisting of aluminous and flinty particles, forming a tough plastic mass.

Cleav'age. A tendency to split in certain fixed directions.

Clep'sydra (Gr. κλεπτω, *klepto*, I steal or hide; ὑδωρ, *hudor*, water). An instrument in which time was attempted to be measured by the flow of water ; a water-clock.

Climac'teric (Gr. κλιμαξ, *klimax*, a ladder). A period of human life in which a marked change is supposed to take place in the constitution.

Climatolog'ical (*Climate ;* λογος, *logos*, discourse). Relating to climate, or to a description of climates.

Climatol'ogy (Gr. κλιμα, *klima*, a region ; λογος, *logos*, discourse). The description of the general phenomena of the climate or state of weather of different countries.

Clin'ical (Gr. κλινη, *kline*, a bed). Belonging to a bed; in *medicine*, applied to instruction derived from the actual observation of patients.

Cli'noid (Gr. κλινη, *kline*, a bed or couch ; ειδος, *eidos*, form). Like a couch ; in *anatomy*, applied to certain processes of bone, from an imagined resemblance to a couch.

Clinom'eter (Gr. κλινω, *klino*, I bend or slope; μετρον, *metron*, a measure). An instrument for ascertaining the angle at which geological strata are inclined.

Cloa'ca (Lat. a sink). The common excretory outlet of birds and some other animals.

Clon'ic (Gr. κλονεω, *klon'eo*, I agitate). Applied to spasm or convulsion which rapidly alternates with relaxation.

Clove (Lat. *clavus*, a nail). A bulblet formed in the axil of a leaf which is still part of a bulb, as in garlic.

Clyp'eiform (Lat. *clyp'eus*, a shield ; *forma*, form). Like a shield.

Clyster (Gr. κλυζω, *kluzo*, I wash). A liquid substance thrown into the lower intestine.

Coag'ulable (Lat. *con*, together ; *ago*, I drive). Capable of being congealed, or changed from a liquid to a thick state.

Coag'ulate (Lat. *con*, together ; *ago*, I drive). To turn from a fluid to a thick state.

Coagula'tion (Lat. *con*, together ; *ago*, I drive). A turning from a fluid to a thick or solid state.

Coal-formation. The strata or layers of the crust of the earth in which coal is found.

Coales'cent (Lat. *coales'co*, I grow

D

together). Growing together or uniting.

Coalit'ion (Lat. *coales'co*, I grow together). A union of separate bodies or parts in one mass.

Coapta'tion (Lat. *con*, together ; *apto*, I fit). A fitting together.

Coarcta'tion (Lat. *con*, together; *arctus*, narrow). A narrowing or compression.

Coch'lea (Gr. κοχλος, *kochlos*, a shell-fish with a spiral shell). In *anatomy*, a part of the internal ear, of a conical form, marked by a spiral groove.

Coeffic'ient (Lat. *con*, together ; *effic'io*, I effect or make up). That which unites with something else to produce a result.

Cœlelmin'tha (Gr. κοιλος, *koilos*, hollow ; ἑλμινς, *helmins*, a worm). The intestinal worms which have an alimentary tube.

Cœ'liac (Gr. κοιλια, *koilia*, the belly). Belonging to the abdomen.

Coer'cive (Lat. *con*, together ; *arceo*, I drive). Driving together; applied to the force which brings about the recomposition of bodies after separation into their elements.

Cohe'sion (Lat. *con*, together ; *hæ'reo*, I stick). The property by which bodies stick together.

Coleop'tera (Gr. κολεος, *kol'eos*, a sheath ; πτερον, *pteron*, a wing). Having sheathed wings : applied to an order of insects of which beetles are the type, in which the outer or upper wings form sheaths for the inner or lower.

Coleorhi'za (Gr. κολεος, *kol'eos*, a sheath ; ριζα, *rhiza*, a root). The sheath which covers the bundle of young roots in endogens.

Col'ic (Gr. κωλον, *kōlon*, a part of the large intestine). In *anatomy*, belonging to the colon ; in *medicine*, a painful disorder of the intestines.

Collapse (Lat. *con*, together ; *labor*, I glide or fall). To fall together ; a falling together.

Collat'eral (Lat. *con*, together ; *latus*, a side). Placed side by side ; descending from the children of a common ancestor.

Collen'chyma (Gr. κολλα, *kolla*, glue ; ἐγχυμα, *en'chuma*, a tissue). In *botany*, the substance lying between and uniting cells.

Collima'tion (Lat. *con*, with ; *limes*, a limit). The art of aiming at a mark ; in *astronomy*, the line of collimation is the line of sight that passes through the point of intersection of the wires fixed in the focus of the object-glass and the centre of that glass.

Colliq'uative (Lat. *con*, with ; *liq'ueo*, I melt). Melting ; applied to diseases attended with profuse loss of the animal fluids.

Collis'ion (Lat. *con*, together ; *lædo*, I strike). A striking together.

Collo'dion (Gr. κολλα, *kolla*, glue). A solution of gun-cotton in a mixture of ether and alcohol.

Collum (Lat. a neck). In *botany*, the portion between the plumule and the radicle.

Collyr'ium (Gr. κολλυρα, *collu'ra*, eye-salve). A wash for the eyes.

Co'lolites (Gr. κωλον, *kōlon*, one of the intestines ; λιθος, *lithos*, a stone). In *geology*, a name given to tortuous masses and impressions, resembling the intestines of fishes.

Columel'la (Lat. a little column). In *conchology*, the central pillar round which a spiral shell is wound ; in *anatomy*, applied to the central part or axis of the cochlea of the ear.

Colum'næ Car'neæ (Lat. fleshy columns). Small rounded muscular bands covering the inner surface of the ventricles of the heart.

Colum'nar (Lat. *colum'na*, a column). Arranged in columns.

Coma (Gr. κωμα, *kōma*, a sound sleep). A state of complete insensibility, with loss of power of speech and motion.

Coma (Gr. κομη, *komē*, hair). The nebulous or hazy appearance which surrounds a comet.

Combina'tion (Lat. *con*, with ; *bini*, two and two). Union of different substances into a new compound.

Combus'tible (Lat. *combūro*, I burn up). Capable of being burned.

Combus'tion (Lat. *combūro*, I burn up).

A burning ; the process in which, by the aid of heat, a substance unites with oxygen, or sometimes with chlorine.

Com'et (Gr. κομη, *komē*, hair). A body revolving round the sun in an elliptical orbit, and having generally a tail or train of light, whence its name.

Com'ma (Gr. κοπτω, *koptō*, I cut). In *music*, an interval between two sounds, distinguishable by the ear.

Commen'surable, or **Commen'surate** (Lat. *con*, together ; *mensu'ra*, a measure). Having a common measure ; applied to two or more numbers capable of being divided by the same quantity without leaving a remainder.

Com'minute (Lat. *con*, together ; *minuo*, I lessen). To break into small pieces ; to reduce to powder.

Com'missure (Lat. *con*, together ; *mitto*, I send). A joining together ; a joint or seam.

Com'mutator (Lat. *con*, with ; *muto*, I change). That which changes one with another : an apparatus to control and modify the course of an electric current.

Co'mose (Lat. *coma*, hair). Hairy.

Compat'ible (Lat. *con*, with ; *pa'tior*, I suffer or endure). In *logic*, expressing two views of one object at the same time ; in *chemistry* and *pharmacy*, not decomposing each other.

Compensa'tion Balance. In a watch or chronometer, a contrivance for correcting errors caused by variations of temperature, by means of bars of two or more metals of different powers of expansion.

Com'plement (Lat. *com'pleo*, I fill up). That which is required to fill up or complete some quantity or thing.

Com'plex (Lat. *con*, with ; *plecto*, I weave). Made up of two or more parts.

Complica'tion (Lat. *con*, together ; *plico*, I fold or weave). An interweaving or involving together ; in *medicine*, applied to a disease which appears during the presence of another.

Compo'nent (Lat. *con*, together ; *pono*, I put). Making up a compound body.

Com'posite (Lat. *con*, together ; *pono*, I put). Formed of things placed together ; in *architecture*, applied to an order the characteristics of which are made up from other orders ; in *arithmetic*, applied to numbers which can be divided exactly by a whole number greater than unity.

Compres'sible (Lat. *con*, together ; *premo*, I press). Capable of being pressed together into a smaller space.

Compres'sor (Lat. *con*, together ; *premo*, I press). That which presses together : an apparatus for exercising pressure on bodies viewed through a microscope.

Con'cave (Lat. *con*, with ; *cavus*, hollow). Sinking into a depression in which a rounded body would lie.

Con'cavo-con'vex. Concave on one surface and convex on the other.

Concen'trate (Lat. *con*, together ; *centrum*, a centre). To bring to a common centre ; to increase the strength of a compound fluid by evaporating the water contained in it.

Concen'tric (Lat. *con*, together ; *centrum*, a centre). Having a common centre.

Conchif'erous (Lat. *concha*, a shell ; *fero*, I bear). Shell-fish ; especially those with bivalve shells.

Conchoi'dal (Gr. κογχη, *konchē*, a shell ; ειδος, *eidos*, form). Like a shell.

Conchol'ogy (Gr. κογχη, *konche*, a shell ; λογος, *logos*, a word or description). The science which describes shells.

Conchyliom'etry (Gr. κογχυλιον, *konchu'lion*, a shell ; μετρον, *metron*, a measure). The art of measuring shells or their curves.

Concoc'tion (Lat. *con*, implying perfection ; *coquo*, I cook). A digestion, or ripening.

Concom'itant (Lat. *con*, with ; *comes*, a companion). Accompanying.

Concord (Lat. *con*, with ; *cor*, the

heart). Agreement; in *music*, the union of two or more sounds so as to produce an agreeable impression on the ear.

Con'crete (Lat. *con*, together; *cresco*, I grow). Grown together, or united; in *logic*, applied to a term which includes both the subject and its quality; in *architecture*, a mass of lime, sand, and gravel, or broken stones, commonly used for the foundation of buildings.

Concre'tion (Lat. *con*, together; *cresco*, I grow). The act of growing together, or becoming consistent or hard; a mass formed by the union of particles.

Concre'tionary Deposits. In *geology*, the recent alluvial strata, including calcareous and other deposits from springs.

Condensa'tion (Lat. *con*, together; *densus*, thick). The act of making dense, or of causing the particles of a body to approach each other more closely; the state of being made dense.

Condens'e (Lat. *con*, with; *densus*, thick). To make dense or thick, by forcing the particles of a body into a smaller compass.

Condens'er (Lat. *con*, with; *densus*, thick). An instrument or apparatus by which gases or vapours may be condensed.

Conduc'tion (Lat *con*, with; *duco*, I lead). A leading; the property by which heat, electricity, &c., is transmitted without a change in the particles of the conducting body.

Conduc'tor (Lat. *con*, together; *duco*, I lead). A leader; in *natural philosophy*, a body that receives and communicates electricity or heat.

Condu'plicate (Lat. *con*, together; *duplex*, double). Double, or folded over together; applied in *botany* to leaves, when folded together from the midrib.

Con'dyle (Gr. κονδυλος, *kon'dulos*, a knuckle). A rounded projection at the end of a bone; a knuckle.

Con'dyloid (Gr. κονδυλος, *kon'dulos*, a knuckle; ειδος, *eidos*, form). Resembling a condyle: applied especially to the projection by which the lower jaw is articulated with the head.

Con'dylopods (Gr. κονδυλος, *kon'dulos*, a knuckle; πους, *pous*, a foot). Articulated animals with jointed legs, as insects and crustacea.

Cone (Gr. κωνος, *kōnos*). A body with a circular base, ending in a point at the top; in *botany*, a mass of hard scales or bracts covering naked seeds.

Confer'væ (Lat.). Plants consisting merely of round or cylindrical cells united into a filament.

Confer'void (Lat. *conferva*, a kind of water plant; Gr. ειδος, *eidos*, form). Resembling conferva; a kind of fresh-water plant consisting of jointed stems.

Configura'tion (Lat. *con*, together; *figu'ra*, a figure). The shape or outline of a body.

Con'fluent (Lat. *con*, together; *fluo*, I flow). Flowing or running together: applied to the union of parts originally separate.

Conform'able (Lat. *con*, together; *forma*, form). In *geology*, applied to strata or groups of strata lying in parallel order one above another.

Conforma'tion (Lat. *con*, together; *forma*, form). The manner in which a body is formed; structure.

Congela'tion (Lat. *con*, together; *gelo*, I freeze). The process of passing from a fluid to a solid state, as water becomes converted into ice.

Congen'erate (Lat. *con*, together; *genus*, a kind). Of the same kind or nature, or having the same action.

Congen'ital (Lat. *con*, with; *gignor*, I am born). Born with; belonging to an individual from birth.

Conge'ries (Lat. *con*, together; *gero*, I bear). A mass of things heaped up together.

Conges'tion (Lat. *con*, together; *gero*, I bear). An accumulation of blood or other fluid in the vessels.

Conges'tive (Lat. *con*, together; *gero*,

I bear). Belonging to or attended by congestion.

Con'globate (Lat. *con*, together ; *globus*, a ball). Gathered into a round mass or ball.

Conglom'erate (Lat. *con*, together ; *glomus*, a ball). Gathered into a ball or mass. Applied to works composed of rounded fragments.

Con'ic (Gr. κωνος, *kōnos*, a cone). Having the form of or belonging to a cone.

Con'ic Sections. The figures formed by the division of a cone by a plane : they are five in number—the triangle, circle, ellipse or oval, parabola, and hyperbola.

Coniferous (Lat. *conus*, a cone ; *fero*, I bear). Bearing cones : an order of plants, of which the fir, pine, and juniper are examples ; so called because their fruit is in the form of a cone.

Coniros'tres (Lat. *conus*, a cone ; *rostrum*, a beak). A tribe of insessorial or perching birds having strong conical beaks, of which the finches, crows, and hornbills are examples.

Con'jugate Foci. In *optics*, when part of the rays falling on a lens are refracted so as to meet in another focus than the principal focus, then the two foci are called conjugate foci.

Conjunc'tion (Lat. *con*, together ; *jungo*, I join). A joining ; in *astronomy*, the meeting of two or more stars or planets in the same degree of the zodiac ; a planet is in conjunction with the sun, when it appears in the same straight line from the earth.

Conjuncti'va (Lat. *con*, together ; *jungo*, I join). The fine membrane covering the front of the eye, which is a continuation of the mucous membrane lining the eyelids.

Con'nate (Lat. *con*, together ; *nascor*, I am born). Growing together.

Connec'tive (Lat. *con*, together ; *necto*, I knit). Connecting or joining together ; in *botany*, the mass of cellular tissue and spiral vessels generally connecting the lobes of the anther.

Co'noid (Gr. κωνος, *kōnos*, a cone ; ειδος, *eidos*, shape). Like a cone ; in *geometry*, the solid figure formed by the revolution of a conic section round its axis.

Conserva'trix (Lat. *conser'vo*, I preserve). Preserving : applied, in the expression *vis conservatrix naturæ*, to the power which the body has of resisting hurtful influences.

Consol'idate (Lat. *con*, together ; *sol'idus*, solid or firm). To make or become firm and hard.

Con'sonance (Lat. *con*, together ; *sonus*, a sound). A sounding together ; in *music*, an accord of sounds which produces an agreeable sensation in the ear.

Constella'tion (Lat. *con*, together ; *stella*, a star). A cluster or assemblage of stars.

Constit'uent (Lat. *con*, together ; *stat'uo*, I place). Forming an essential or necessary part of anything.

Constitutional Diseases. Diseases which become developed under the influence of agents acting within the body.

Constric'tor (Lat. *con*, together ; *stringo*, I bind). A binder or drawer together : applied in *anatomy* to muscles which close any orifice.

Consump'tion (Lat. *consu'mo*, I consume). A consuming or destruction ; in *medicine*, a gradual decay of the body, especially attended with a disease of the lungs.

Contact Theory. In *electrical science*, the hypothesis of Volta, by which any two different conductors of electricity placed in contact with each other produce a decomposition and mutual transference of their electric fluids.

Conta'gion (Lat. *con*, together; *tango*, I touch). A touching; in *medicine*, the communication of disease by touching the sick or his clothes, &c.

Conta'gious (Lat. *con*, together; *tango*, I touch). Capable of being communicated by touch, or containing communicable matter.

Con'tinent (Lat. *con*, together; *teneo*, I hold). In *geography*, a large connected tract of land.

Contort'ed (Lat. *con*, together; *tor'queo*, I twist). Twisted.

Contor'tion (Lat *con*, together; *tor'queo*, I twist). A twisting out of the natural situation.

Contor'tive (Lat. *contor'queo*, I twist together). In *botany*, applied to the arrangement of a flower-bud in which the edges of the parts alternately overlap, while each part is twisted on its axis.

Contra. A Latin preposition signifying against, used in composition.

Contrac'tile (Lat. *con*, together; *traho*, I draw). Having the property of contracting or drawing together.

Contractil'ity (Lat. *con*, together; *traho*, I draw). The property by which bodies shrink or contract.

Contu'se (Lat. *con*, together; *tundo*, I beat). To beat or bruise.

Contu'sion (Lat. *con*, together; *tundo*, I beat). The act of beating or bruising; a bruise.

Convales'cence (Lat. *con*, together; *valeo*, I am in health). The recovery of health after illness.

Convec'tion (Lat. *con*, with; *veho*, I carry). The power which fluids have of transmitting heat or electricity by currents.

Conver'ge (Lat. *con*, together; *vergo*, I incline). To tend to one point.

Con'verse (Lat. *con*, with; *verto*, I turn). In *mathematics* or *logic*, a proposition formed by inverting or interchanging the terms of another.

Con'vex (Lat. *convex'us*). Rising into a spherical or rounded form.

Con'volute (Lat. *con*, together; *volvo*, I roll). Rolled together; applied to leaves rolled together in the bud in a single coil.

Convolu'tion (Lat. *con*, together; *volvo*, I roll). A rolling together: in *anatomy*, applied to the windings of the brain and the intestines.

Convul'sion (Lat. *con*, together; *vello*, I pull). General involuntary contraction of the muscles.

Co-or'dinates (Lat. *con*, together; *or'-dino*, I put in order.) In *geometry*, a system of lines to which points under consideration are referred, and by means of which their position is determined.

Coper'nican (*Copernicus*, an astronomer). In *astronomy*, applied to the system proposed by Copernicus, who taught that the earth revolves round the sun.

Cop'rolites (Gr. κοπρος, *kopros*, dung; λιθος, *lithos*, a stone). Fossilised excrements of animals.

Cor'acoid (Gr. κοραξ, *korax*, a crow; ειδος, *eidos*, shape). Resembling a crow's beak : applied to a process of the shoulder-blade, which attains a large size in birds and reptiles.

Coral (Gr. κοραλλιον, *koral'lion*). A general term for all calcareous structures formed by the action of marine polypes or zoophytes.

Cor'alloid (*Coral*; Gr. ειδος, *eidos*, shape). Resembling coral.

Cord'ate (Lat. *cor*, the heart). Shaped like a heart.

Cord'iform (Lat. *cor*, the heart; *forma*, form). Shaped like a heart.

Coria'ceous (Lat. *co'rium*, leather). Resembling leather; tough.

Co'rium (Lat. skin or leather). The true skin, lying beneath the cuticle.

Corm (Gr. κορμος, *kormos*, a stem or log). In *botany*, a thickened underground stem.

Corm'ogen (Gr. κορμος, *kormos*, a corm; γενναω, *gennaō*, I produce). Producing corms; applied to plants which produce stems composed of both vessels and cells.

Cornbrash. A coarse shelly limestone in the upper oolite.

Cor'nea (Lat. *cornu*, a horn). The horny membrane : a part of the eye, so called from its resembling transparent horn.

Cor'neous (Lat. *cornu*, a horn). Horny.

Cor'neule (*Cornea*; *ule*, denoting smallness). A little cornea; such as covers each segment of the compound eyes of insects.

Cor'nice (Gr. κορωνις, *koro'nis*, a crown). The highest part of the entablature of a column; any series of ornamental work that crowns a wall externally or internally.

Cor'nua (Plural of Lat. *cornu*, a horn). Horns : applied in *anatomy* to certain parts from their position.

Corol'la (Lat. *coro'na*, a crown). The inner whorl or row, generally coloured, of the leaves which form a flower.

Cor'ollary (Lat. *corol'la*, a crown). A conclusion drawn from something already demonstrated.

Corollifio'ral (*Corolla* ; *flos*, a flower). A sub-class of exogenous plants which have both calyx and corolla, the petals being united, and the stamens hypogynous.

Coro'na (Lat. a crown). In *anatomy*, the upper surface of the molar teeth ; in *botany*, the circumference or margin of a radiated compound flower ; in *optics*, a halo or luminous circle round the sun, moon, or stars.

Coro'nal (Lat. *coro'na*, a crown). Belonging to the top of the head.

Cor'onary (Lat. *coro'na*, a crown). Belonging to a crown ; applied in *anatomy*, to the vessels which supply the heart with blood for its nutrition, also to vessels of the lips and stomach.

Coro'niform (Lat. *coro'na*, a crown ; *forma*, shape). Like a crown.

Coro'noid (Gr. κορωνη, *korūnē*, a crow ; ειδος, *eidos*, form). Resembling a crow's beak ; in *anatomy*, applied to certain processes of bones from their shape.

Cor'pus (Lat.) A body : applied in *anatomy* to several parts of the body.

Corpus'cle (Lat. *corpus'culum*, a little body, from *corpus*, a body). A small particle.

Corpus'cular (Lat. *corpus'culum*, a little body). Relating to small particles ; applied to a theory of light, which supposes it to consist of minute particles emitted from luminous bodies.

Correla'tion (Lat. *con*, together ; *relātus*, brought). A mutual or reciprocal relation.

Corro'de (Lat. *con*, together ; *rodo*, I gnaw). To eat or wear away by degrees.

Corro'sion (Lat. *con ; rodo*, I gnaw). A wearing away, as of metals, by the action of acids.

Corro'sive (Lat. *con ; rodo*, I gnaw). Having the property of gradually eating or wearing away.

Cor'rugate (Lat. *con ; ruga*, a wrinkle). To draw into folds or wrinkles.

Cort'ical (Lat. *cortex*, bark). Belonging to or forming the external covering.

Corusca'tion (Lat. *corusco*, I flash). A flash of light.

Cor'ymb (Gr. κορυμβος, *kor'umbos*, a cluster). A form of inflorescence consisting of a raceme or panicle in which the lower flowers have short pedicels, and the upper short ones, so that all form a nearly level surface.

Cose'cant (Lat. *con ; seco*, I cut). The secant of the complement of an arc of a circle.

Co'sine (Lat. *con*, with ; *sine*). The sine of the complement of the arc of a circle.

Cos'mical (Gr. κοσμος, *kosmos*, the universe). Relating to the universe.

Cosmog'ony (Gr. κοσμος, *kosmos*, the world or universe ; γενναω, *gennaō*, I produce). The science which treats of the orgin or formation of the universe.

Cosmog'raphy (Gr. κοσμος, *kosmos*, the universe ; γραφω, *graphō*, I write). A description of the universe.

Cosmol'ogy (Gr. κοσμος, *kosmos*, the universe ; λεγω, *legū*, I describe). The science of the universe, or of the formation and arrangement of its compouent parts.

Cosmora'ma (Gr. κοσμος, *kosmos*, the universe ; ὁραω, *horaō*, I see). A view, or series of views, of the world.

Cosmos (Gr. κοσμος, *kosmos*, order or arrangement ; also the world.) The universe ; the whole created things constituting the perceptible world.

Cos'mosphere (Gr. κοσμος, *kosmos*, the world ; σφαιρα, *sphaira*, a sphere).

An instrument for showing the position of the earth with respect to the fixed stars.

Costal (Lat. *costa*, a rib). Belonging to the ribs.

Cotan'gent (Lat. *con*, with ; *tango*, I touch). The tangent of the complement of an arc of a circle.

Coti'dal (Lat. *con*, with ; *tide*). Having tides at the same time.

Cotyle'don (Gr. κοτυληδων, *kotulēdŏn*, a cup-like hollow). In *botany*, the seed lobe which first appears above ground ; in *anatomy*, applied to the portions of which the placentæ of some animals are formed.

Cot'yloid (Gr. κοτυλη, *kot'ulē*, a cup or socket ; ειδος, *eidos*, shape). Resembling the socket of a joint.

Coup (Fr.). A blow or stroke.

Coup d'œil (Fr., stroke of the eye). A general view.

Coup de soleil (Fr., stroke of the sun). A disease produced by exposure of the head to the rays of the sun.

Coxal'gia (Lat. *coxa*, the hip ; Gr. αλγος, *algos*, pain). Pain in the hip.

Cra'nial (Lat. *cra'nium*, the skull). Of or belonging to the skull.

Craniol'ogy (Gr. κρανιον, *kra'nion*, the skull ; λογος, *logos*, a description). A description of the skull.

Crasis (Gr. κεραννυμι, *kerannu'mi*, I mix). A mixture : applied to the just mixture of the fluids of the body : in *grammar*, the union of two short vowels into a long one or a diphthong.

Crassament'um (Lat. *crassus*, thick). The thick part or clot of blood.

Crater (Gr. κρατηρ, *kratēr*, a large cup). The mouth of a volcano.

Crayon (Fr. *craie*, chalk). A coloured stone or earth used in drawing; a kind of pencil made of the same.

Cre'asote (Gr. κρεας, *kreas*, flesh ; σωζω, *sōzō*, I preserve). An oily liquid consisting of carbon, oxygen, and hydrogen, obtained from tar, and named from its property of preserving animal substances.

Cre'atin (Gr. κρεας, *kreas*, flesh). A substance obtained from flesh, believed to be its essential element.

Creat'inin (Gr. κρεας, *kreas*, flesh). A modified form of creatin.

Crem'ocarp (Gr. κρεμαω, *kremaō*, I suspend ; καρπος, *karpos*, fruit). A fruit consisting of two achænia united by their faces, and covered by the tube of the calyx.

Cre'nate (Lat. *crena*, a notch). Notched ; in *botany*, applied to leaves having superficial rounded divisions at their edges.

Crep'itant (Lat. *crep'ito*, I crackle). Crackling or snapping.

Crep'itate (Lat. *crep'ito*, I crackle). To crackle.

Crep'itus (Lat.). A crackling sound.

Crepus'cular (Lat. *crepus'culum*, twilight). Of or relating to twilight.

Crepuscula'ria (Lat. *crepus'culum*, twilight). A family of lepidopterous or scaly-winged insects, which mostly fly by twilight, as the sphinxes or hawk-moths.

Creta'ceous (Lat. *creta*, chalk). Of or relating to chalk.

Cret'inism. The state of a Cretin : a diseased state characterised by imbecility of mind and body, common in Switzerland and some other mountainous countries.

Crib'riform (Lat. *cribrum*, a sieve ; *forma*, shape). Like a sieve.

Cri'coid (Gr. κρικος, *krikos*, a ring ; ειδος, *eidos*, shape). Like a ring.

Cri'noid (Gr. κρινος, *krinos*, a lily ; ειδος, *eidos*, shape). Like a lily : applied to certain fossil echinodermatous invertebrates supported on jointed stalks.

Cri'sis (Gr. κρινω, *krinō*, I judge or determine). That state of a disease or other affair, in which it has arrived at its height, and must soon change; in *medicine*, generally applied to the change itself.

Cris'ta (Lat. a crest). In *anatomy*, a term applied to several processes of bones.

Crit'ical (Gr. κρινω, *krinō*, I judge or determine). Relating to judging ; in *medicine*, marking or producing a change in a disease.

Crocodil'ia (*Crocodile*). The class of reptiles of which the crocodile is the type.

Crop. In *geology*, the edge of an inclined stratum when it comes to the surface.

Cru'cial (Lat. *crux*, a cross). Transverse ; like a cross ; in *experimental science*, searching, decisive.

Cru'cible (Lat. *cru'cio*, I torment). A vessel of clay, sand, and ground ware, or other material capable of enduring heat : used in chemistry and manufactures.

Crucif'erous (Lat. *crux*, a cross ; *fero*, I bear). Bearing a cross : applied to an order of plants, the four petals of the flowers of which are arranged in the form of a cross.

Cru'ciform (Lat. *crux*, a cross ; *forma*, shape). Shaped or arranged like a cross.

Crudity (Lat. *crudus*, raw). Rawness ; undigested substance.

Crura (Lat. *crus*, a leg). Legs ; in *anatomy*, applied fancifully to projections of some parts of the body.

Crural (Lat. *crus*, a leg). Of or belonging to the legs.

Crusta petrosa (Lat. a strong crust). A bony layer which covers the fangs of the teeth.

Crusta'ceous (Lat. *crusta*, a crust or shell). Having a crust : applied to a class of invertebrate animals, of which the lobster is an example, which have hard jointed shells.

Cryoph'orus (Gr. κρυος, *kruos*, ice ; φερω, *pherō*, I bear). An instrument for freezing water by its own evaporation.

Crypt (Gr. κρυπτω, *kruptō*, I hide). A hidden recess ; in *anatomy*, applied to some of the minute cavities or simple glands of mucous membranes.

Cryptobranch'iate (Gr. κρυπτω, *kruptō*, I hide ; βραγχια, *branchia*, gills). Not having conspicuous gills ; applied to certain articulated and molluscous animals.

Cryptogam'ia (Lat. κρυπτω, *kruptō*, I hide ; γαμος, *gamos*, marriage). An order of plants in which the distinction of sexes is not obvious.

Crystal (Gr. κρυσταλλος, *krustal'los*, ice). A geometrical figure, assumed by most substances under favourable circumstances ; also a general name for some transparent mineral substances.

Crys'talline (Gr. κρυσταλλος, *krustal'los*, ice or crystal). Consisting of or resembling crystal : applied to a lens of the eye.

Crystallisa'tion (Gr. κρυσταλλος, *krustal'los*, ice or crystal). The assuming of crystalline or geometrical forms by substances.

Crystallog'raphy (Gr. κρυσταλλος, *krustal'los*, ice or crystal ; γραφω, *graphō*, I write). The science which describes crystals.

Cten'oid (Gr. κτεις, *kteis*, a comb ; ειδος, *eidos*, form). An order of fishes having scales jagged like the teeth of a comb.

Ctenoptych'ius (Gr. κτεις, *kteis*, a comb ; πτυχη, *ptuchē*, a wrinkle). A genus of fossil teeth distinguished by the serrated margin of their cutting edges.

Cube (Gr. κυβος, *kubos*, a solid square). In *geometry*, a solid body having six equal sides with equal angles ; in *arithmetic*, the product of a number multiplied twice into itself.

Cubic (Gr. κυβος, *kubos*, a cube). Having the property of, or capable of being contained in, a cube.

Cu'bital (Lat. *cubitus*, the elbow). Of or belonging to the elbow.

Cu'boid (Gr. κυβος, *kubos*, a cube ; ειδος, *eidos*, shape). Like a cube or die.

Cucul'late (Lat. *cucul'lus*, a hood). Like a hood.

Cul-de-sac (French). A passage closed at one end.

Cul'minate (Lat. *culmen*, a top). To become vertical, or gain the extreme point of height.

Cultriros'tres (Lat. *culter*, a ploughshare ; *rostrum*, a beak). A family of grallæ or stilt-birds, having a long, thick, stout beak, including cranes, herons, and storks.

Cum'brian (*Cumbria*, Wales). A name given to the strata which lie beneath the true Silurian system, from their occurring largely in Wales and Cumberland.

Cu'neate (Lat. *cu'neus*, a wedge). Like a wedge.

Cu'neiform (Lat. *cu'neus*, a wedge; *forma*, shape). Like a wedge.

Cupel (Lat. *cupel'la*, a little cup). A kind of cup used in chemistry, which, when heated, absorbs the refuse matter of the metals placed in it for purification.

Cupella'tion (Lat. *cupel'la*, a little cup). The process of refining, especially gold and silver, by means of a cupel.

Cu'pola. A spherical or spheroidal covering to a building.

Cuprif'erous (Lat. *cuprum*, copper; *fero*, I bear). Yielding copper.

Curso'res (Lat. *curro*, I run). An order of birds constituted for running only, as the ostrich : also a division of spiders which have the legs adapted for running.

Curvicau'date (Lat. *curvus*, curved; *cauda*, a tail). Having a bent tail.

Curvifo'liate (Lat. *curvus*, curved; *fo'lium*, a leaf). Having bent leaves.

Curvilin'ear (Lat. *curvus*, crooked; *lin'ea*, a line). Having or moving in a curved line or curved lines.

Curviner'vate (Lat. *curvus*, curved; *nervus*, a nerve). Having the veins or nervures curved.

Curviros'tral (Lat. *curvus*, crooked; *rostrum*, a beak). Having a bent beak.

Cusp'idate (Lat. *cuspis*, the point of a weapon). Pointed : applied in *anatomy* to the canine or eye-teeth.

Cuta'neous (Lat. *cutis*, the skin). Of or belonging to the skin.

Cu'ticle (Lat. *cutis*, the skin). The external or scarf skin, a membrane covering the true skin.

Cutis (Lat.) The skin.

Cy'anate. A compound of cyanic acid with a base.

Cyan'ic (Gr. κυανος, *ku'anos*, blue). Relating to blue; applied to a series of colours having blue as the type.

Cy'anide (*Cyan'ogen*; terminal *ide*). A compound of cyanogen with an elementary substance.

Cyan'ogen (Gr. κυανος, *ku'anos*, blue; γενναω, *gennaō*, I produce). A gas consisting of carbon and nitrogen : it enters into the composition of hydrocyanic acid, and has its name from the blue colour produced by its compounds with certain salts of iron.

Cyano'sis (Gr. κυανος, *ku'anos*, blue). A diseased condition, arising from a defect in the formation of the heart, and characterised by blueness of the skin.

Cyan'otype (*Cyanogen*; Gr. τυπος, *tupos*, an impression). A photograph prepared by washing paper with cyanide of potassium.

Cyca'deous. Belonging to the order of plants which has the palm-tree as a type.

Cyc'adites (*Cycas*). Fossil plants allied to the cycas and zamia.

Cycle (Gr. κυκλος, *kuklos*, a circle). A series of numbers, as of years, in which, after a certain round has passed, a similar course commences.

Cyc'lical (Gr. κυκλος, *kuklos*, a circle). Belonging to a cycle.

Cyclobran'chiate (Gr. κυκλος, *kuklos*, a circle; βραγχια, *bran'chia*, gills). Having the gills disposed in a circle : applied to an order of gasteropods.

Cy'cloid (Gr. κυκλος, *kuklos*, a circle; ειδος, *eidos*, form). Resembling a circle; applied to an order of fishes having smooth round scales, simple at the margin.

Cycloneu'rous (Gr. κυκλος, *kuklos*, a circle; νευρον, *neuron*, a nerve). Having the nervous system in the form of a circle; as in some of the radiated invertebrate animals.

Cyclopæ'dia (Gr. κυκλος, *kuklos*, a circle; παιδεια, *paidei'a*, instruction). A work which contains an account of all the arts and sciences, or of all that relates to any particular department.

Cyclop'teris (Gr. κυκλος, *kuklos*, a circle; πτερις, *pteris*, a fern). A genus of fossil fern-like plants, with circular leaflets.

Cyclo'sis (Gr. κυκλος, *kuklos*, a circle). Motion in a circle : applied to a movement of fluid observed in some parts of plants.

Cyclos'tomous (Gr. κυκλος, *kuklos*, a circle; στομα, *stoma*, a mouth.

Having a circular mouth, as certain fishes.

Cyl'inder (Gr. κυλινδω, *kulin'dō*, I roll). A roller; a body produced by the revolution of a right-angled parallelogram round one of its sides.

Cyme (Gr. κυμα, *kuma*, a wave ?). In *botany*, a form of inflorescence resembling a corymb, but branched, so as to have in part the character of an umbel.

Cynan'che (Gr. κυων, *kuōn*, a dog; αγχω, *anchō*, I strangle). Quinsy.

Cyn'osure (Gr. κυων, *kuōn*, a dog; ουρα, *oura*, a tail). The dog's tail: a constellation of seven stars near the north pole; generally called Ursa Minor, or Charles's wain.

Cyst (Gr. κυστις, *kustis*, a bladder). A small bladder; generally applied to small sacs or bags containing matter of various kinds in disease.

Cyst'ic (Gr. κυστις, *kustis*, a bladder). Belonging to, or resembling a cyst or bladder : applied to a class of parasitic animals ; also to a duct or tube proceeding from the gall-bladder.

Cystid'eæ (Gr. κυστις, *kustis*, a bladder). A family of fossil echinoderms, of a bladder-like shape.

Cy'toblast (Gr. κυτος, *kutos*, a cell ; βλαστανω, *blas'tanō*, I bud forth). The nucleus of animal and vegetable cells.

Cytoblaste'ma (Gr. κυτος, *kutos*, a cell ; βλαστανω, *blas'tanō*, I bud forth). The viscid fluid in which animal and vegetable cells are produced, and by which they are held together.

Cytogen'esis (Gr. κυτας, *kutos*, a cell ; γενεσις, *gen'esis*, origin). The development of cells in animal and vegetable structures.

D.

Dac'tyl (Gr. δακτυλος, *dak'tulos*, a finger). A foot in verse, consisting of a long syllable followed by two short ones, like the joints of a finger.

Daguer'reotype. A picture produced according to the process invented by M. Daguerre, by the action of light on iodide of silver.

Da'ta (Lat. *do*, I give). Things given ; facts or quantities, the existence of which is admitted as a foundation for the discovery of other results.

Da'tive (Lat. *do*, I give). Giving ; that case or part of nouns which conveys with it the idea of giving or acquisition.

Debac'le (Fr.). In *geology*, a sudden flood or rush of water which breaks down opposing barriers.

Debil'ity (Lat. *debilis*, weak). Weakness.

De'bris (Fr. waste). Fragments ; broken pieces ; in *geology*, generally applied to the larger fragments.

Deca (Gr. δεκα, *deka*, ten). A prefix in compound words, signifying ten.

Decade (Gr. δεκα, *deka*, ten). A collection of ten.

Dec'agon (Gr. δεκα, *deka*, ten ; γωνια, *gōnia*, an angle). A figure having ten sides and ten angles.

Dec'agramme (Gr. δεκα, *deka*, ten ; Fr. *gramme*, a weight so called). A French weight, consisting of ten *grammes*, or nearly 154½ grains.

Decagyn'ia (Gr. δεκα, *deka*, ten ; γυνη, *gunē*, a female). An order of plants in the Linnæan system, having ten pistils.

Decahed'ron (Gr. δεκα, *deka*, ten ; εδρα, *hedra*, a base). A solid having ten sides.

Dec'alitre (Gr. δεκα, *deka*, ten ; Fr. *litre*, a quart, or 1¾ English pints). A measure of ten *litres*.

Dec'alogue (Gr. δεκα, *deka*, ten ; λογος, *logos*, a word). The ten commandments.

Dec'ametre (Gr. δεκα, *deka*, ten ; Fr. *mètre*, a measure equal to 3⅓ English feet). A measure of ten *mètres*.

Decan'dria (Gr. δεκα, *deka*, ten ; ανηρ, *anēr*, a man). A class of plants in the Linnæan system, having ten stamens.

Decap'oda (Gr. δεκα, *deka*, ten ; πους, *pous*, a foot). Animals having ten feet.

Decar'bonize (Lat. *de*, from ; *carbon*). To remove carbon from a body.

Dec'astyle (Gr. δεκα, *deka*, ten ; στυλος, *stulos*, a column). Having ten pillars or columns.

Decay (Lat. *de*, down ; *cado*, I fall). A slow destruction ; a decomposition of moist organic matter exposed to air, by means of oxygen, without sensible increase of heat.

Decem (Lat. ten). A prefix in compound words, signifying ten.

Decen'nial (Lat. *decem*, ten ; *annus*, a year). Occurring every ten years ; lasting ten years

Decid'uous (Lat. *de*, down ; *cado*, I fall). Apt to fall off.

Dec'igramme (Lat. *decem*, ten ; Fr. *gramme*). A tenth of a *gramme ;* about $1\frac{27}{100}$ English grains.

Dec'ilitre (Lat. *decem*, ten ; Fr. *litre*, a quart, or $\frac{3}{4}$ English pint). A tenth of a *litre.*

Dec'imal (Lat. *decem*, ten). Relating to the number ten ; increasing or diminishing tenfold.

Dec'imetre (Lat. *decem*, ten ; Fr. *mètre*, a measure equal to $3\frac{7}{25}$ English feet). A tenth part of a *mètre ;* nearly 4 English inches.

Declen'sion (Lat. *decli'no*, I bend down). A descent or slope ; the variation in a noun produced by a change of the ending of the word.

Decli'nal (Lat. *decli'no*, I bend down). Bending down or sloping ; in *geology*, applied to the slope of strata from an axis.

Declina'tion (Lat. *decli'no*, I bend down). A variation from a fixed line or point : as of a heavenly body from the equator, or of a magnetic needle from the true meridian.

Decoction (Lat. *de*, down ; *co'quo*, I cook). The art of boiling a substance in water ; fluid impregnated with any substance by boiling.

Decol'lated (Lat. *de*, off ; *collum*, a neck). Having the apex or head worn off.

Decolorisa'tion (Lat. *de*, from ; *color*, colour). Removal of colour.

Decol'orise (Lat. *de*, from ; *color*, colour). To remove colour.

Decompo'se (Lat. *de*, from ; *compo'no*, I put together). To separate the constituent parts of a body from each other.

Decom'position (Lat. *de*, from ; *compo'no*, I put together). The separation of a body into its constituent parts or elements.

Decomposition of Forces. The term applied to the division of any force into several others, the result of which is equal to the force decomposed.

Decomposition of Light. The separation of a beam into the several rays producing prismatic colours.

Decompound' (Lat. *de*, from ; *compo'no*, I put together). In *botany*, applied to leaves, of which the petiole is so divided that each part forms a compound leaf.

Decor'ticate (Lat. *de*, from ; *cortex*, bark). To strip off the bark or outer covering.

De'crement (Lat. *decres'co*, I grow less). The quantity by which anything is lessened.

Decrep'itation (Lat. *de*, from ; *crep'itus*, a crackling). A roasting with a crackling noise, produced by a series of small explosions from sudden expansion by heat.

Decu'bitus (Lat. *de*, down ; *cumbo*, I lie). A lying down ; position in bed.

Decum'bent (Lat. *decumbo*, I lie down). Lying down ; in *botany*, applied to stems which lie on the ground, but rise towards their end.

Decuss'ate (Lat. *decus'so*, I cut across). To intersect or cross, like the strokes of the letter X.

Decuss'ation (Lat. *decus'so*, I cut across). An intersection or crossing.

Defeca'tion (Lat. *de*, from ; *fæx*, dregs or refuse matter). Purification from dregs ; expulsion of adventitious matter.

Def'erent (Lat. *de*, from ; *fero*, I carry). Carrying away.

Def'inite (Lat. *de*, down ; *finio*, I limit). In *logic*, marking out a particular class ; in *botany*, applied to inflorescence when it ends in a single flower, which is the first on the stem to expand.

De'flagrate (Lat. *de*, down ; *flagro*, I burn). To burn rapidly.

Deflec'ted (Lat. *de*, down ; *flecto*, I bend). Bent down.

Deflec'tion (Lat. *de*, from ; *flecto*, I bend). A bending or turning aside from the direct course.

Deflec'tive (Lat. *de*, from *flecto*, I bend). Bending or turning aside.

Deflux'ion (Lat. *de*, down ; *fluo*, I flow). A flowing down.

Degen'eration (Lat. *de*, down ; *genus*, a kind). A growing worse or inferior ; a falling from the normal or healthy state to one which is inferior.

Deglutit'ion (Lat. *de*, down ; *glutio*, I swallow). The act of swallowing.

Degrada'tion (Lat. *de*, down ; *gradus*, a step). In *geology*, a removing or casting down step by step.

Degree (Lat. *de*, from ; *gradus*, a step). A step ; in *geometry*, the three hundred and sixtieth part of the circumference of a circle.

Dehis'cence (Lat. *dehis'co*, I gape). A gaping or opening ; the splitting open of a bag containing eggs, or of a fruit containing seeds.

Dehis'cent (Lat. *dehis'co*, I gape). Opening like the pod of a plant.

Delete'rious (Gr. δηλεομαι, *dēleomai*, I destroy). Destructive ; injurious ; poisonous.

Deliques'cence (Lat. *de*, down ; *liques'co*, I melt). A melting ; the process by which saline matters attract water from the air, and thus become melted.

Deliq'uium (Lat. want or defect). A failure of power ; fainting.

Delir'ium (Lat. *deli'ro*, I dote or rave). A wandering of the ideas of the mind.

Delta (the Greek letter Δ). A piece of land enclosed within two mouths of a river which branches before reaching the sea : originally applied to the land enclosed between the mouths of the Nile.

Del'toid (Gr. Δελτα, the letter delta or Δ ; ειδος, *eidos*, shape). Resembling the letter Δ or delta ; triangular.

Demen'tia (Lat. *de*, from ; *mens*, the mind). Want of intellect ; a form of insanity characterised by a rapid succession of imperfect and unconnected ideas, with loss of reflection and attention.

Demi (Lat. *dimid'ium*, half). A prefix in compound words, signifying half.

Demot'ic (Gr. δημος, *dēmos*, people). Belonging to the people : applied to the alphabet used by the people, as distinguished from that used by a certain class ; as among the Egyptians.

Demul'cent (Lat. *de*, from ; *mul'ceo*, I soothe or soften). Softening or soothing.

De'nary (Lat. *deni*, a series of tens). Containing tens; having the number tens as the characteristic.

Dendriform (Gr. δενδρον, *dendron*, a tree ; Lat. *forma*, shape). Resembling a tree.

Dendrit'ic (Gr. δενδρον, *dendron*, a tree). Resembling a tree or shrub ; branch-like.

Den'droid (Gr. δενδρον, *dendron*, a tree ; ειδος, *eidos*, shape). Resembling a tree.

Den'drolite (Gr. δενδρον, *dendron*, a tree ; λιθος, *lithos*, a stone). A fossil plant or part of a plant.

Dendrom'eter (Gr. δενδρον, *dendron*, a tree ; μετρον, *metron*, a measure). An instrument for measuring trees.

Dens'ity (Lat. *densus*, thick). Thickness ; the quantity of matter in a substance, compared with that in an equal volume of another substance.

Dental (Lat. *dens*, a tooth). Belonging to the teeth ; formed by the teeth.

Dental Formula. A formula used to denote the number of the different kinds of teeth in an animal.

Dent'ary (Lat. *dens*, a tooth). A bone in the head of fishes and reptiles, which supports the teeth.

Dentate (Lat. *dens*, a tooth). Having tooth-like projections.

Den'ticle (Lat. *dens*, a tooth ; *cle*, denoting smallness). A little tooth, or projection like a tooth.

Dentic'ulate (Lat. *dens*, a tooth).

Having small teeth, or projections like teeth.

Den'tifrice (Lat. *dens*, a tooth ; *frico*, I rub). A substance used in cleaning teeth ; tooth-powder.

Dentig'erous (Lat. *dens*, a tooth ; *gero*, I bear). Bearing teeth.

Den'tine (Lat. *dens*, a tooth). The part of a tooth commonly known as ivory.

Dentiros'tres (Lat. *dens*, a tooth ; *rostrum*, a beak). A family of birds of the passerine order, having the upper bill notched towards the point.

Dentit'ion (Lat. *dens*, a tooth). The process of breeding or cutting teeth.

Denuda'tion (Lat. *de*, from ; *nudus*, bare). A stripping bare.

Deo'dorise (Lat. *de*, from ; *odor*, smell). To deprive of smell.

Deodorisa'tion (Lat. *de*, from ; *odor*, smell). A depriving of smell.

Deox'idate, or **Deox'idise**, or **Deoxyg'- enate** (Lat. *de*, from ; *oxidate*, to charge with oxygen). To deprive of oxygen.

Dephlogis'ticated. Deprived of phlogiston, the supposed principle of inflammability : a term formerly applied to oxygen gas.

Depi'latory (Lat. *de*, from ; *pilus*, hair). Having the property of removing hair.

Deple'tion (Lat. *de*, from ; *pleo*, I fill). Emptying ; diminishing the quantity contained.

Depos'it (Lat. *de*, down ; *pono*, I put). Any thing or substance thrown down, as from fluid in which it has been suspended.

Deprava'tion (Lat. *de*, down ; *pravus*, bad). A making bad or worse.

Depres'sion (Lat. *de*, down ; *prem'o*, I press). A pressing down ; a sinking in or down.

Depres'sor (Lat. *de*, down ; *prem'o*, I press). That which depresses or draws down : applied to certain muscles.

De'purate (Lat. *de*, from ; *purus*, pure). To render free from impurities.

Depura'tion (Lat. *de*, from ; *purus*, pure). Purification ; rendering free from impurities.

Derby-spar. Fluoride of calcium, or fluorspar.

Deriva'tion (Lat. *de*, from ; *rivus*, a stream). In *grammar*, the tracing a word to the source from which it has been obtained.

Deriv'ative (Lat. *de*, from ; *rivus*, a stream). Turning aside, or drawing away from another part, as applied to medicines ; in *grammar*, a word which has its origin in another word.

Derma (Gr. δερμα, *derma*, skin). The true skin.

Der'mal (Gr. δερμα, *derma*, skin). Belonging to or formed of skin.

Dermatol'ogy (Gr. δερμα, *derma*, the skin ; λογος, *logos*, discourse). A description of the skin.

Dermone'ural (Gr. δερμα, *derma*, the skin ; νευρον, *neuron*, a nerve). A name given to the outer or upper row of spines on the back of a fish, from their connection with the skin, and their position in respect to the part of the skeleton which protects the nervous system.

Dermoskel'eton (Gr. δερμα, *derma*, skin ; σκελετον, *skel'eton*). A skin skeleton ; the external covering, more or less hard, of many invertebrate animals ; also the skeleton formed of bones connected with the skin in fishes and some other vertebrates.

Desic'cate (Lat. *de*, from ; *siccus*, dry). To make dry.

Desicca'tion (Lat. *de*, from ; *siccus*, dry). The act of making dry.

Desic'cative (Lat. *de*, from ; *siccus*, dry). Drying.

Desmog'raphy (Gr. δεσμος, *desmos*, a ligament ; γραφω, *graphō*, I write). A description of the ligaments of the body.

Desquama'tion (Lat. *de*, from ; *squama*, a scale). A throwing off in scales.

Deter'gent (Lat. *de*, from ; *tergo*, I wipe). Cleansing.

Deter'minate (Lat. *de*, from ; *ter'minus*, an end). Limited ; in *mathematics*, applied to problems that are capable of only one solution.

De'tonate (Lat. *de*, from ; *tono*, I

thunder). To explode, or cause to explode.

Detona'tion (Lat. *de*, from ; *tono*, I thunder). An explosion or sudden report.

De'trahent (Lat. *de*, down ; *traho*, I draw). Drawing down.

Detri'tus (Lat. *de*, down ; *tero*, I rub). That which is worn off from solid bodies, as rocks, by friction : generally applied to the more finely divided portions.

Detru'sion (Lat. *de*, from ; *trudo*, I thrust). A thrusting from or down.

Deu'tero- or **Deuto-** (Gr. δευτερος, *deu'teros*, second). A prefix, denoting the second degree of the word joined with it.

Deutox'ide (Gr. δευτερος, *deu'teros*, second ; *oxide*). The compound of a body with oxygen, containing the next greatest quantity of oxygen to the protoxide, or basic oxide.

Devel'opment (Fr. *developper*, to unfold). An unfolding ; the change which takes place in living bodies in their progress towards maturity.

Devo'nian (Devon). In *geology*, a term applied to the old red sandstone system, of which portions are particularly developed in Devonshire.

Dew-point. The temperature at which the watery vapour in the atmosphere begins to be deposited on the surface of the earth.

Dextrin (Lat. *dexter*, right). A substance resembling gum, and used in art as a substitute for it : so called from turning the plane in polarised light to the right hand.

Diabe'tes (Gr. δια, *dia*, through ; βαινω, *bainō*, I go). An immoderate flow of urine.

Diacous'tics (Gr. δια, *dia*, through ; ἀκουω, *akouō*, I hear). The science of refracted sounds.

Diadel'phia (Gr. δις, *dis*, double ; αδελφος, *adel'phos*, a brother). A class of plants in the Linnæan system, having the filaments of the stamens united into two parcels.

Diæ'resis (Gr. δια, *dia*, apart ; αἱρεω, *haireō*, I take). A separation ; in

grammar, the separation of a syllable into two ; or the mark ¨, which denotes that the vowel on which it is placed is separated from that which precedes it.

Diagno'sis (Gr. δια, *dia*, through or between ; γινωσκω, *ginōs'kō*, I know). A distinction or difference ; in *medicine*, the distinction of one disease from another.

Diag'onal (Gr. δια, *dia*, through ; γωνια, *gōnia*, an angle). A line drawn from one angle of a foursided figure to the opposite angle.

Di'agram (Gr. δια, *dia*, through ; γραφω, *graphō*, I write). A figure drawn for the purpose of giving a general idea of an object, without accuracy in minute details.

Di'alect (Gr. δια, *dia*, separate ; λεγω, *legō*, I speak). The form in which the parent language of a state is spoken in a province.

Dial'lage' (Gr. διαλλαγη, interchange). In *mineralogy*, a mineral consisting of silica and magnesia of a changeable colour ; in *rhetoric*, a figure by which arguments are placed in different points of view, and then brought to bear upon one point.

Diamagnet'ic (Gr. δια, *dia*, through ; μαγνης, *magnēs*, a magnet). A term applied to substances which, under the influence of magnetism, take a position at right angles to the magnetic meridian.

Diamag'netism (Gr. δια, *dia*, through ; μαγνης, *magnēs*, a magnet). A peculiar property of many bodies, which, not being themselves magnetic, are repelled by sufficiently powerful electro-magnets, and take a position at right angles to the magnetic equator.

Diam'eter (Gr. δια, *dia*, through ; μετρον, *metron*, a measure). A straight line passing through the centre of a body from one side to the other.

Dian'dria (Gr. δις, *dis*, double ; ἀνηρ, *anēr*, a man). A class of plants in the Linnæan system, having two stamens.

Diaph'anous (Gr. δια, *dia*, through ;

φαινω, *phainō*, I show). Allowing light to pass through, but not so as to form distinct images of objects.

Diaphore'sis (Gr. δια, *dia*, through ; φορεω, *phor'eō*, I carry). An increase of perspiration.

Diaphoret'ic (Gr. δια, *dia*, through ; φορεω, *phor'eō*, I carry). Producing an increase of perspiration.

Diaphragm (Gr. δια, *dia*, apart ; φρασσω, *phrassō*, I fence in). The midriff, or membranous and muscular partition which divides the chest from the abdomen ; a black perforated plate, used in optical instruments, for allowing only the central rays to reach the eye.

Diaphragmat'ic (Gr. διαφραγμα, *diaphragma*, the midriff). Belonging to the diaphragm.

Diaph'ysis (Gr. δια, *dia*, apart ; φυω, *phuō*, I grow). A term applied to the shaft of a long bone, of which the ends are completed by the addition of portions ossified separately.

Diapoph'ysis (Gr. δια, *dia*, apart ; απο, *apo*, from ; φυω, *phuō*, I grow). A name given to the transverse process of a vertebra in the archetype skeleton.

Diarrhœ'a (Gr. δια, *dia*, through ; ρεω, *rheō*, I flow). An excessive discharge from the bowels.

Diarthro'sis (Gr. δια, *dia*, through ; αρθρον, *arthron*, a joint). A moveable joint, such as those of the limbs or lower jaw.

Di'astase (Gr. διιστημι, *diüstēmi*, I separate). A peculiar azotised substance found in germinating seeds or buds in a state of development, and having the property of transforming starch into sugar.

Dias'tole' (Gr. δια, *dia*, apart ; στελλω, *stellō*, I send). In *physiology*, the dilatation or opening of the heart after contraction ; in *grammar*, a lengthening of a syllable.

Diather'mancy (Gr. δια, *dia*, through; θερμαινω, *thermai'nō*, I heat). The property which some substances possess of allowing rays of heat to pass through them, as light passes through glass.

Diather'manous (Gr. δια, *dia*, through ; θερμαινω, *thermai'nō*, I heat). Having the property of transmitting heat, as glass transmits light.

Diath'esis (Gr. δια, *dia*, apart ; τιθημι, *tithēmi*, I place). A particular state or disposition.

Diaton'ic (Gr. δια, *dia*, through ; τονος, *tonos*, sound). Ascending or descending from sound to sound.

Dibran'chiate (Gr. δις, *dis*, double ; βραγχια, *bran'chia*, gills). Having two gills : applied to an order of cephalopods.

Diceph'alous (Gr. δις, *dis*, twice ; κεφαλη, *keph'alē*, a head) Having two heads on one body.

Dichlamyd'eous (Gr. δις, *dis*, twice ; χλαμυς, *chlamus*, a garment). Having two coverings ; in *botany*, having calyx and corolla.

Dichobu'ne (Gr. διχα, *dicha*, doubly ; βουνος, *bounos*, a ridge). A genus of fossil quadrupeds, having deeply cleft ridges in the upper molar teeth.

Dichot'omous (Gr. διχα, *dicha*, doubly ; τεμνω, *temnō*, I cut). Dividing by pairs.

Dicœ'lous (Gr. δις, *dis*, double ; κοιλος, *koilos*, hollow). Having two cavities.

Dicotyle'donous (Gr. δις, *dis*, double ; κοτυληδων, *kotulēdōn*, a seed lobe or leaf). Having two cotyledons or seed-leaves.

Dic'tyogens (Gr. δικτυον, *dik'tuon*, a net ; γενναω, *genn'aō*, I produce). A sub-class of endogenous plants, having the veins of the leaves arranged in a net-work, like exogens, instead of parallel.

Dictyophyl'lum (Gr. δικτυον, *dik'tuon*, a net ; φυλλον, *phullon*, a leaf). Net-leaf : a genus provisionally including all unknown fossil dicotyledonous leaves of net-like structure.

Dicyn'odon (Gr. δις, *dis*, double ; κυων, *kuōn*, a dog ; οδους, *odous*, a tooth). Double canine-toothed : a provisional genus of reptiles with no teeth in the upper jaw, except two long tusks in sockets, curved downwards.

Didac'tyle (Gr. δις, *dis*, double ; δακτυλος, *dak'tulos*, a finger). Having two fingers or toes.

Didel'phic (Gr. δις, *dis*, double ; δελφυς, *delphus*, the womb). A term applied to a division of mammals of which the young are born prematurely, including the marsupiate and monotrematous animals.

Didynam'ia (Gr. δις, *dis*, double ; δυναμις, *du'namis*, power). A Linnæan class of plants, having four stamens, two long and two short.

Dielec'tric (Gr. δια, *dia*, between ; *electric*). A bad conductor of electricity.

Dietet'ic (Gr. διαιτα, *diai'ta*, food or diet). Relating to food or diet.

Differen'tial (Lat. *dis*, apart ; *fero*, I bear). Pointing out a distinction or difference : applied to a thermometer which shows the difference in the temperature of two portions of air ; also to an infinitely small quantity in arithmetic or algebra.

Differen'tiate (Lat. *differen'tia*, a difference). To establish a distinction or difference.

Diffrac'tion (Lat. *dis*, apart ; *frango*, I break). The turning aside of rays of light from their straight course, when made to pass by the boundaries of an opaque body.

Diffu'sible (Lat. *dis*, apart ; *fundo*, I pour). Capable of being poured or spread in all directions.

Diffu'sion (Lat. *dis*, apart ; *fundo*, I pour). A pouring or spreading in all directions.

Diffusion of Gases. The process by which gases mix with each other.

Digas'tric (Gr. δις, *dis*, double ; γαστηρ, *gastēr*, a belly). Having a double belly.

Diges'tion (Lat. *di*, apart ; *gero*, I bear or carry). A division or separation ; the process by which the nutritive parts of food are separated and rendered available for nutrition.

Diges'tive (Lat. *di'gero*, I digest). Relating to or promoting digestion.

Dig'it (Lat. *dig'itus*, a finger). A finger's breadth ; the twelfth part of the diameter of the sun or moon, used in measuring the extent of eclipses ; in *arithmetic*, a single figure.

Dig'itate (Lat. *dig'itus*, a finger). Arranged like fingers.

Dig'itigrade (Lat. *dig'itus*, a finger or toe ; *gradior*, I step). Walking on the toes, as the lion, cat, &c.

Digyn'ia (Gr. δις, *dis*, twice ; γυνη, *gunē*, a female). A Linnæan order of plants having two pistils.

Dihed'ral (Gr. δις, *dis*, double ; έδρα, *hedra*, a seat or face). Having two sides.

Dilata'tion (Lat. *dis*, apart ; *latus*, wide). A widening in all directions.

Di'luent (Lat. *di'luo*, I wash away). Making thin, or more liquid ; weakening in intensity.

Dilu'te (Lat. *di'luo*, I wash away). Reduced in strength ; rendered more liquid.

Dilu'vial (Lat. *dilu'vium*, a deluge). Relating to or produced by a deluge ; in *geology*, applied to those deposits which give indications of having been carried from a distance by a violent current of water.

Dilu'vium (Lat. *di'luo*, I wash away). In *geology*, a term applied to the results of extraordinary or violent agency of water.

Di'merous (Gr. δις, *dis*, double ; μερος, *meros*, a part). Having parts arranged in twos.

Dimid'iate (Lat. *dimia'ium*, half). Divided into two halves.

Dimorph'ism (Gr. δις, *dis*, double ; μορφη, *morphē*, form). The property of assuming two forms under different circumstances.

Dimor'phous (Gr. δις, *dis*, double ; μορφη, *morphē*, form). Having two forms.

Dimy'ary (Gr. δις, *dis*, double ; μυς, *mus*, a muscle). Applied to bivalve shells which are closed by two muscles.

Dinor'nis (Gr. δεινος, *deinos*, terrible ; όρνις, *ornis*, a bird). A gigantic extinct bird of New Zealand.

E

Dinosau'ria (Gr. δεινος, *deinos*, terrible; σαυρος, *sauros*, a lizard). Gigantic fossil animals of the saurian or lizard tribe.

Dinothe'rium (Gr. δεινος, *deinos*, terrible; θηριον, *thērion*, a beast). A gigantic fossil pachydermatous animal.

Diœ'cia (Gr. δις, *dis*, double; οικος, *oi'kos*, a house). A Linnæan class of plants, having male flowers on one plant, and female on another.

Diop'tric (Gr. δια, *dia*, through; οπτομαι, *op'tomai*, I see). Affording a medium for the sight: relating to the science of refracted light.

Diop'trics (Gr. δια, *dia*, through; οπτομαι, *op'tomai*, I see). The part of optics which describes the phenomena of the refraction of light.

Diora'ma (Gr. δια, *dia*, through; ὁραω, *hora'ō*, I see). An apparatus in which a picture is exhibited through a large aperture, partly by reflected, and partly by transmitted light.

Dip. The angle which the magnetic needle, freely poised, makes with the plane of the horizon; the inclination of a geological stratum or bed to the horizon.

Dipet'alous (Gr. δις, *dis*, double; πεταλον, *pet'alon*, a petal). Having two petals.

Diphthe'ria (Gr. διφθερα, *diph'thera*, leather). A disease characterised by the formation of a leathery membrane in the throat and fauces.

Diphtherit'ic (Gr. διφθερα, *diph'thera*, leather). Tough, like leather; attended with the formation of a leathery membrane.

Diphyl'lous (Gr. δις, *dis*, double; φυλλον, *phul'lon*, a leaf). Having two leaves.

Diphy'odonts (Gr. δις, *dis*, double; φυω, *phuā*, I produce; οδους, *odous*, a tooth). Animals which produce two sets of teeth in succession.

Dip'loe' (Gr. διπλους, *dip'lous*, double). The network of bone-tissue which fills up the interval between the two compact plates in the bones of the skull; in *botany*, the cellular substance of a leaf.

Diplo'ma (Gr. διπλοω, *dip'loō*, I double). Originally, a folded letter or writing; now applied to a letter or writing conferring some power, privilege, or dignity.

Diplo'pia (Gr. διπλους, *dip'lous*, double; οπτομαι, *op'tomai*, I see). Double vision; a state in which objects are seen double, from a disturbance of the combined action of the eyes.

Diplop'tera (Gr. διπλους, *dip'lous*, double; πτερον, *pter'on*, a wing). A family of hymenopterous or membrane-winged insects, having the fore-wings folded longitudinally, as the wasp.

Dip'terous (Gr. δις, *dis*, twice; πτερον, *pter'on*, a wing. Having two wings, as certain insects: in *botany*, applied to seeds which have the margin prolonged in the form of wings.

Dipteryg'ian (Gr. δις, *dis*, twice; πτερυγιον, *pteru'gion*, a fin). Having two fins.

Dip'tote (Gr. δις, *dis*, double; πιπτω, *piptō*, I fall). A noun having two cases only.

Disc. *See* Disk.

Disc'oid (Gr. δισκος, *diskos*, a quoit; ειδος, *eidos*, form). Shaped like a disk or quoit.

Discord (Lat. *dis*, separate; *cor*, the heart). Disagreement; in *music*, the mixed sound of notes, the vibrations producing which are not in a simple ratio to each other.

Discord'ant (Lat. *dis*, apart; *cor*, the heart). Disagreeing; in *geology*, applied to strata deposited horizontally on other strata which have been thrown into an oblique direction by disturbing causes.

Disep'alous (Gr. δις, *dis*, double; *sepal*). Having two sepals.

Disinfect (Lat. *dis*, from; *infect*). To purify from infection.

Disin'tegrate (Lat. *dis*, from; *in'teger*, entire). To break up into integrant parts, not by chemical action.

Disjunc'tive (Lat. *dis*, separate; *jungo*, I join). Separating; in *grammar*, uniting words or sentences, but disjoining the sense.

Disk (Gr. δισκος, *diskos*, a quoit). In *astronomy*, the surface of the sun, moon, or planet, as it appears to an observer on the earth ; in *botany*, a body seated between the base of the stamens and the base of the ovary ; also the central parts of a radiate compound flower.

Dis'locate (Lat. *dis*, from ; *locus*, a place). To put out of place.

Disloca'tion (Lat. *dis*, from ; *locus*, a place). A putting out of place.

Disper'mous (Gr. δις, *dis*, double ; σπερμα, *sperma*, a seed). Having two seeds.

Disper'sion (Lat. *dis*, apart ; *spargo*, I scatter). A scattering ; in *optics*, the separation of the coloured rays of light in passing through a prism, varying according to the refracting power of the material of which the prism is composed.

Disrup'ted (Lat. *dis*, apart ; *rumpo*, I break). Violently torn apart.

Disrup'tion (Lat. *dis*, apart ; *rumpo*, I break). A rending asunder ; in *geology*, a displacement in the crust of the earth by earthquakes, or other disturbing causes.

Dissec'tion (Lat. *dis*, apart ; *seco*, I cut). A cutting in pieces ; the cutting up an animal or vegetable to ascertain its structure.

Dissep'iment (Lat. *dis*, from ; *sepes*, a hedge). A partition in an ovary or fruit.

Dissolu'tion (Lat. *dis*, from ; *solvo*, I loosen). Melting ; the separation of the particles of a body from each other.

Dissolve (Lat. *dis*, apart ; *solvo*, I loosen). To melt ; to separate the particles of a substance from each other.

Dissyl'lable (Gr. δις, *dis*, double ; συλλαβη, *sul'labē*, a syllable). A word of two syllables.

Dis'tal (Lat. *dis*, apart ; *sto*, I stand). At a distance from a given line or point.

Dis'tichous (Gr. δις, *dis*, double ; στιχος, *stichos*, a row). Arranged in two rows.

Distil' (Lat. *dis*, from ; *stilla*, a drop). To let fall in drops ; to separate a lighter fluid from another by heat or evaporation, the vapour being cooled and falling in drops into a vessel placed to receive it.

Dis'tillation (Lat. *dis*, apart ; *stilla*, a drop). The process by which substances are separated which rise in vapour at different degrees of heat, or by which a volatile liquid is parted from a substance incapable of volatilisation.

Distor'tion (Lat. *dis*, apart ; *torqueo*, I twist). A twisting out of regular shape ; in *optics*, the change in the form of an image depending on the form of the lens.

Diu'resis (Gr. δια, *dia*, through ; ουρον, *ouron*, urine). An increased flow of urine.

Diuret'ic (Gr. δια, *dia*, through ; ουρον, *ouron*, urine). Increasing the secretion of urine.

Diur'nal (Lat. *diurnus*, daily). Relating to, or performed in a day.

Divarica'tion (Lat. *di*, apart ; *va'rico*, I straddle). A branching at an obtuse angle.

Divel'lent (Lat. *di*, apart ; *vello*, I pull). Drawing asunder.

Divertic'ulum (Lat. *di*, apart ; *verto*, I turn). A turning aside ; a short blind tube branching out of a larger one.

Divisibil'ity (Lat. *di'vido*, I divide). The property of bodies by which their parts are capable of being separated.

Dodeca- (Gr. δωδεκα, *dōdeka*, twelve). A prefix in compound words, signifying twelve.

Dodec'agon (Gr. δωδεκα, *dō'leka*, twelve ; γωνια, *gōnia*, an angle). A figure consisting of twelve equal sides and angles.

Dodecagyn'ia (Gr. δωδεκα, *dōdeka*, twelve ; γωνη, *gunē*, a female). An order of plants in the Linnæan system having twelve pistils.

Dodecahed'ron (Gr. δωδεκα, *dōdeka*, twelve ; ἑδρα, *hedra*, a seat or face) A solid figure having twelve equal bases or sides.

Dodecan'dria (Gr. δωδεκα, *dōdeka*,

twelve ; ἀνηρ, *anēr*, a man). A class of plants in the Linnæan system, having twelve stamens.

Dol'omite. A variety of magnesian limestone.

Dome (Lat. *domus*, a house). A house ; the external part of a spherical roof.

Domin'ical (Lat. *(dies) domin'ica*, Sunday). Belonging to Sunday; applied to the letter prefixed in Almanacks to the Sundays, from which the days of the week falling on the successive days of past or present years may be computed.

Dor'sal (Lat. *dorsum*, the back). Placed on, or belonging to, the back.

Dorsibran'chiate (Lat. *dorsum*, the back; Gr. βραγχια, *bran'chia*, gills). Having the branchia or breathing organs distributed on the back ; applied to certain mollusca.

Dorso. (Lat. *dorsum*, the back). In *anatomy*, a prefix in compound words signifying connection with, or relation to, the back.

Double Salt. A salt in which the acid is combined with two different bases.

Double Stars. Two stars placed so close together that to the naked eye they appear single.

Doublet. A magnifying glass, consisting of a combination of two plano-convex lenses.

Drastic (Gr. δραω, *draū*, I do or act). Acting powerfully ; applied to certain medicines.

Dropsy (Gr. ὑδωρ, *hudōr*, water ; ὀψις, *opsis*, an appearance). An unnatural collection of watery fluid in any part of the body.

Drupa'ceous (*Drupe*). Of the nature of a drupe ; bearing fruit in the form of a drupe.

Drupe (Gr. δρυππα, *druppa*, an overripe olive). A pulpy fruit without valves, containing a stone with a kernel, as the peach.

Du'al (Lat. *duo*, two). Relating to two ; applied to a form of nouns and verbs in which two persons or things are denoted, as in the Greek and some other languages.

Dual'ity (Lat. *duo*, two). The state of being two in number.

Duct (Lat. *duco*, I lead). A tube or vessel for conveying a fluid, especially a secretion from a gland.

Ductile (Lat. *duco*, I lead). Capable of being drawn out.

Ductil'ity (Lat. *duco*, I lead). The property which substances possess of being drawn out.

Duode'cimal (Lat. *duod'ecim*, twelve). Proceeding in a scale of twelves.

Duode'nary (Lat. *duode'ni*, twelve). Increasing in a twelvefold proportion.

Duode'num (Lat. *duode'ni*, twelve). The first portion of the small intestine ; which, in man, is twelve finger-breadths in length.

Du'plicate (Lat. *duplex*, double). Double ; duplicate proportion or ratio is the proportion or ratio of squares.

Dura Mater (Lat. hard mother : because the other membranes were supposed to proceed from it). The strong fibrous membrane which envelopes the brain and spinal cord.

Dura'men (Lat. *durus*, hard). The central or heart wood of an exogenous tree.

Dyke. A wall or fence ; in *geology*, applied to wall-like intrusions of igneous rock which fill up veins and fissures in the stratified system.

Dynam'ic (Gr. δυναμις, *du'namis*, power). Relating to strength or force.

Dynam'ics (Gr. δυναμις, *du'namis*, power). That part of natural philosophy which investigates the properties of bodies in motion.

Dynamom'eter (Gr. δυναμις, *du'namis*, power ; μετρον, *metron*, a measure). An instrument for measuring strength.

Dysæsthe'sia (Gr. δυς, *dus*, badly ; αισθανομαι, *aisthan'omai*, I feel). Impaired power of feeling.

Dys'entery (Gr. δυς, *dus*, badly ; ἐντερον, *en'teron*, an intestine). A discharge from the intestines accompanied by blood, mucus, or other morbid matter.

Dyspep'sia (Gr. δυς, *dus*, badly ; πεπτω, *peptō*, I digest). Indigestion ; difficulty of digestion.

Dyspha'gia (Gr. δυς, *dus*, badly ;

φαγω, *phagō*, I eat). Difficulty of swallowing.

Dyspnœ'a (Gr. δυς, *dus*, badly ; πνεω, *pneō*, I breathe). Difficult breathing.

E.

Earth. In *chemistry*, an oxide of a metal : but applied especially to the oxides and salts of barium, calcium, magnesium, and aluminium.

Ebrac'teate (Lat. *e*, from ; *brac'tea*, a bract). Without bracts.

Ebullit'ion (Lat. *e*, out ; *bulla*, a bubble). Boiling ; the formation by heat of bubbles of vapour within a liquid, which rise to the surface.

Eburna'tion (Lat. *ebur*, ivory). A rendering dense like ivory ; the excessive deposition of compact osseous matter which sometimes takes place in diseased states of bones.

Eccen'tric (Gr. ἐκ, *ek*, from ; κεντρον, *kentron*, a centre). Deviating from a centre ; incapable of being brought to a common centre.

Eccentric'ity (Gr. ἐκ, *ek*, from ; κεντρον, *kentron*, a centre). The state of being eccentric ; the distance between the centre of an ellipse and either of its foci.

Ecchymo'sis (Gr. ἐκ, *ek*, out ; χυμος, *chumos*, juice). An effusion of blood under the skin ; a bruise.

Eccoprot'ic (Gr. ἐκ, *ek*, out ; κοπρος, *kopros*, dung). Promoting the discharge from the bowels.

Ec'dysia (Gr. ἐκ, *ek*, out ; δυω, *duō*, I put on). A casting off or moulting.

Echinococ'cus (Gr. ἐχινος, *echi'nos*, a hedgehog ; κοκκος, *kokkos*, a berry). A parasitic animal, consisting of a membranous sac or bag, and provided with a series of minute hooks.

Echinoder'mata or **Echi'noderms** (Gr. ἐχινος, *echi'nos*, a hedgehog ; δερμα, *derma*, a skin). A class of invertebrate animals, the bodies of which are covered by a thick covering or shell, often with spikes.

Echom'eter (Gr. ἠχω, *ēchō*, sound ; μετρον, *metron*, a measure). An instrument for measuring the duration of sounds, and their intervals.

Eclamp'sia (Gr. ἐκ, *ek*, from ; λαμπω, *lampō*, I shine). An appearance of flashing of light which attends epilepsy ; but now applied to epilepsy or convulsive disease itself.

Eclec'tic (Gr. ἐκ, *ek*, out ; λεγω, *legō*, I choose). Selecting or choosing ; selected.

Eclips'e (Gr. ἐκ, *ek*, from ; λειπω, *leipō*, I leave). A failure ; an interception of the light of the sun, moon, or other luminous body.

Eclip'tic (Gr. ἐκ, *ek*, from ; λειπο, *leipō*, I fail). The circle of the heavens which forms the apparent annual path of the sun : so called because eclipses can only take place when the moon is very near it.

Ecliptic Limits. In *astronomy*, the limits within which an eclipse of the sun or moon may occur.

Econ'omy (Gr. οἰκος, *oikos*, a house ; νομος, *nomos*, a rule). The regulation of a family or household ; the operations of nature in the formation and preservation of animals and plants.

Ec'stacy (Gr. ἐκ, *ek*, out ; ἱστημι, *histēmi*, I make to stand). A state in which the senses are suspended in the contemplation of some extraordinary object.

Ecthlip'sis (Gr. ἐκ, *ek*, from ; θλιβω, *thlibō*, I press or rub). In *Latin grammar*, the cutting off in pronunciation the final syllable of a word ending in *m*, when the next word begins with a vowel.

Ecto- (Gr. ἐκτος, *ektos*, outside). A prefix in some compound words, signifying outside.

Ecto'pia (Gr. ἐκ, *ek*, out ; τοπος, *top'os*, a place). A displacement.

Ectro'pium (Gr. ἐκ, *ek*, out ; τρεπω, *trepō*, I turn). A disease in which the eyelashes are turned outwards.

Ec'zema (Gr. ἐκ, *ek*, out ; ζεω, *zeō*, I boil). An eruption on the skin, of small pustules, without fever, and not contagious.

Ede'ma, Edem'atous. See Œde'ma and Œdem'atous.

Eden'tate (Lat *e*, out ; *dens*, a tooth). Without teeth ; applied to an order of mammalian animals which have no front teeth.

Eden'tulous (Lat. *e*, out ; *dens*, a tooth). Without teeth.

Ed'ible (Lat. *edo*, I eat). Fit to be eaten as food.

Edrioph'thalmia (Gr. ἑδρα, *hedra*, a seat ; ὀφθαλμος, *ophthal'mos*, an eye). A section of crustaceous animals, having the eyes sessile, or not mounted on a foot-stalk.

E'duct (Lat. *e*, out ; *duco*, I lead). Any thing separated from another with which it was previously combined.

Efferves'cence (Lat. *ex*, out ; *fer'veo*, I boil). The escape of bubbles of gas from a fluid, not produced by heat.

Efflores'cence (Lat. *ex*, out ; *flus*, a flower). In *botany*, the time of flowering ; in *medicine*, an eruptive redness of the skin ; in *chemistry*, the formation of a dry powder in some salts on exposure to the air, by losing water of crystallisation.

Efflu'vium (Lat. *ex*, out ; *fluo*, I flow). A flowing out ; the minute particles which exhale or pass off into the air, from substances.

Ef'flux (Lat. *ex*, out ; *fluo*, I flow). A flowing out.

Effodien'tia (Lat. *effo'dio*, I dig out). Digging : applied to a family of edentate animals from their digging habits, as the armadillo.

Effu'sion (Lat. *ex*, out ; *fundo*, I pour). A pouring out ; the escape of a fluid from the vessel or cavity containing it.

Ei'dograph (Gr. εἱδος, *eidos*, form ; γραφω, *graphō*, I write). An instrument for copying designs.

Ejec'tion (Lat. *e*, out ; *jacio*, I cast). A casting out.

Elab'orate (Lat. *e*, out ; *labo'ro*, I labour). To produce by labour, or by successive operations.

Elain (Gr. ἐλαιον, *elai'on*, oil). The liquid principle of oils and fats.

Elas'tic (Gr. ἐλαυνω, *elau'nō*, I drive). Having the property of springing back to its original form after this has been altered.

Elastic'ity (Gr. ἐλαυνω, *elau'nō*, I drive). The property by which a body, after having been compressed, or having had its form changed, recovers its original shape on being released from the force applied to it.

El'ater (Gr. ἐλαυνω, *elau'nō*, I drive). A spiral fibre in the thecæ or seed-cases of some cryptogamic plants, serving to disperse the sporules by uncurling.

Elec'tive Affin'ity (Lat. *e'ligo*, I choose out). The disposition which bodies have to unite chemically with certain substances in preference to others.

Elec'tric (Gr. ἠλεκτρον, *elek'tron*, amber). Containing, pertaining to, derived from, or communicating electricity.

Elec'tricity (Gr. ἠλεκτρον, *elek'tron*, amber ; became first observed in amber). A series of phenomena (also their cause) in various substances ; supposed to be due to the presence of a compound fluid, which is developed by friction or other mechanical means.

Elec'tro-chem'istry. The science which explains the phenomena of the decomposing power of electric currents.

Electro-mag'netism. The branch of electrical science which explains the phenomena of the action of a voltaic current on the magnetic needle.

Elec'trify (*Electricity* ; Lat. *facio*, I make). To charge with, or affect by, electricity.

Elec'trode (*Electricity* ; Gr. ὁδος, *hodos*, a way). The termination of a voltaic battery, by which the electricity passes into or from the fluid in which it is placed.

Electrology (*Electricity* ; Gr. λογος,

logos, discourse). The department of physical science which treats of electricity.

Elec'tro-dynam'ic (*Electricity*; Gr. δυναμις, *du'namis*, power). Relating to electricity in motion, and producing its effects.

Electroly'sis (*Electricity*; Gr. λυω, *luō*, I loosen). Decomposition by an electric current.

Elec'tro-magnet'ic (*Electricity; mag'-net*). Relating to magnetism as connected with electricity.

Elec'trolyte (*Electricity*; Gr. λυω, *luō*, I loosen). A body capable of being decomposed by an electric current.

Elec'tro-metallur'gy (*Electricity*; Gr. μεταλλον, *metal'lon*, a metal; ἐργον, *ergon*, a work). The art of depositing metals from solutions of their salts, by the voltaic current, on other bodies.

Electrom'eter (*Electricity*; Gr. μετρον, *metron*, a measure). An instrument for measuring the intensity of the electricity of a body.

Elec'tro-mo'tive. Moving by means of electricity: applied by Volta to the power of decomposition by the electric current.

Elec'tro-neg'ative. Having negative electricity, and appearing at the positive pole of a voltaic battery.

Elec'troph'orus (*Electricity*; Gr. φερω, *pherō*, I bear). An apparatus for collecting electricity, for the purpose of fixing gaseous mixtures in close vessels.

Elec'tro-pla'ting. The process of depositing a coating of metal on some other metal or substance by means of electric action.

Elec'tro-pos'itive. Having positive electricity, and appearing at the negative pole of the voltaic battery.

Elec'troscope (*Electricity*; Gr. σκοπεω, *skop'eō*, I look). An instrument for measuring the intensity of electricity.

Electrostat'ic (*Electricity*; Gr. στατικος, *stat'ikos*, stationary). Relating to electricity in a state of equilibrium.

Elec'tro-teleg'raphy (*Electricity*; Gr. τηλε, *tēle*, far off; γραφω, *graphō*, I write). The application of electricity to the conveying of messages.

Elec'tro-type (*Electricity*; Gr. τυπος, *tupos*, a type). The process of copying medals, plates, &c., by means of depositing metals from a solution by a galvanic current.

Elec'tuary (Gr. ἐκ, *ek*, out; λειχω, *leichō*, I lick). A medicine made in the form of a confection.

El'ement (Lat. *elemen'tum*). The first principle or constituent part of anything; in *chemistry*, especially, any substance which has resisted all efforts to decompose it; in *anatomy*, the autogenous or primary part of a vertebra.

Elemen'tary (Lat. *elemen'tum*). Primary; incapable of further analysis.

Elephanti'asis (Gr. ἐλεφας, *el'ephas*, an elephant). A disease of the skin, attended with much thickening and the formation of tubercles.

Eleva'tion (Lat. *c*, from; *levo*, I raise.) A raising; in *astronomy*, the distance of a heavenly body above the horizon; in *trigonometry*, angle of elevation is the angle formed by two lines drawn in the same vertical plane from the observer's eye, one to the top of the object and the other parallel to the horizon; in *architecture*, a drawing of the front or a face of a building.

Eleva'tor (Lat. *e*, from; *levo*, I raise). A lifter or raiser.

Elim'inate (Lat. *e*, out; *limen*, a threshold). To thrust out; to remove or expel.

Elis'ion (Lat. *eli'do*, I strike out). A cutting off or suppression of a vowel at the end of a word.

El'lipse (Gr. ἐκ, *ek*, out; λειπω, *leipō*, I leave). An oval figure, produced by the section of a cone by a plane cutting both sides obliquely; in *grammar*, an omission of words.

Ellips'oid (*Ellipse*; Gr. εἰδος, *eidos*, form). A figure formed by the revolution of an ellipse round its axis.

Ellip'tic (Gr. ἐκ, ek, out ; λειπω, leipō, I leave). Relating to, or having the form of, an ellipse.

Elonga'tion (Lat. e, from ; longus, long). A lengthening or stretching ; in astronomy, the apparent recession of a planet from the sun.

Elutria'tion (Lat. e, from ; Gr. λουτρον, loutron, a bath). The process of removing lighter matter from a powdered solid substance by washing it with water, and pouring off the latter.

Ely'trum (Gr. ἐλυω, eluō, I roll over or cover). The outer sheath which protects the body and membranous wings in beetles.

Emana'tion (Lat. e, out ; mano, I flow). That which issues from any substance or body.

Emar'ginate (Lat. e, from ; margo, a margin). Having a piece apparently notched or bitten out of the margin.

Embank'ment. The act of surrounding by a bank ; a structure raised to protect lands from the overflow of rivers or the sea.

Emboss' (Fr. en, in ; bosse, a stud or knob). To form bosses or protuberances ; to ornament by the formation of ornaments in relief or projecting from the surface.

Em'bouchure (Fr. bouche, a mouth). The mouth of a river, &c.

Embroca'tion (Gr. ἐν, en, in ; βρεχω, brechō, I moisten). A mixture of oil, spirit, &c., with which any part of the body is rubbed.

Em'bryo (Gr. ἐμβρυον, em'bruon). The first or rudimentary form of an animal or vegetable.

Embryog'eny (Gr. ἐμβρυον, em'bruon, an embryo ; γενναω, gennaū, I produce). The development of the embryo.

Embryol'ogy (Gr. ἐμβρυον, em'bruon, an embryo ; λογος, logos, a description). A description of the foetus or embryo.

Emer'sion (Lat. emer'go, I issue out). In astronomy, the passage of a satellite out of the shadow of a planet.

Emet'ic (Gr. ἐμεω, em'eū, I vomit). Producing the act of vomiting.

Em'inence (Lat. emin'eo, I stand above others). In anatomy, a general term for a projection on a bone.

Emol'lient (Lat. e, from ; mol'lis, soft). Softening or relaxing.

Emphyse'ma (Gr. ἐν, en, in ; φυσαω, phusa'ō, I blow). Distension with air.

Empir'ic (Gr. ἐν, en, in ; πειρα, peira, experience). Properly, one who makes experiments ; a physician whose knowledge consists in observation alone ; but commonly applied to a quack.

Empir'ical (Gr. ἐν, en, in ; πειρα, peira, experience). Relating to or derived from simple experience or observation, without the aid of science.

Empir'icism (Gr. ἐν, en, in ; πειρα, peira, experience). Practice on the ground of experience alone.

Emprosthot'onos (Gr. ἐμπροσθεν, empros'then, before ; τεινω, tei'nō, I stretch). A form of tetanus in which the body is bent forward.

Empye'ma (Gr. ἐν, en, in ; πυον, puon, pus). A collection of pus in the cavity of the chest.

Empyreu'ma (Gr. ἐν, en, in ; πυρευω, pureuō, I set on fire). A disagreeable smell arising from the burning of animal and vegetable matter.

Empyreumat'ic (Gr. ἐν, en, in ; πυρευω, pureuō, I set on fire). Having the taste or smell of slightly burnt animal or vegetable substances.

Emul'gent (Lat. e, out ; mul'geo, I milk). Milking or drawing out : applied to the blood-vessels of the kidneys, which were supposed to strain the serum.

Emul'sion (Lat. e, from ; mul'geo, I milk). A milk-like substance, produced by rubbing oil with sugar or gum, &c., and water.

Emunc'tory (Lat. emun'go, I wipe out). Removing excreted matter.

Enai'ma (Gr. ἐν, en, in ; αἱμα, haima, blood). Having blood ; applied by Aristotle as a distinctive character of certain animals.

Enaliosau'rians (Gr. ἐν, en, in ; ἁλς, hals, the sea ; σαυρος, sauros, a lizard). A name given to some extinct gigantic lizards, supposed to have lived in the sea.

Enal'lage' (Gr. ἐν, *en*, in ; ἀλλαττω, *allat'tō*, I change). A figure in *grammar*, by which one word or mode of expression is substituted for another.

Enam'el. A compound of the nature of glass, but more fusible and opaque ; the smooth hard substance covering the crown of a tooth. ·

Enarthro'sis (Gr. ἐν, *en*, in ; ἀρθρον, *arthron*, a joint). The ball-and-socket joint, such as is formed by the head of the thigh-bone and the hip.

Encaustic (Gr. ἐν, *en*, in ; καιω, *kaiō*, I burn). Applied to a kind of painting in which colours are made permanent by being burned in.

Enceph'ala (Gr. ἐν, *en*, in ; κεφαλη, *keph'alē*, the head). Molluscous animals having a distinct head.

Encephali'tis (Gr. ἐγκεφαλον, *en-keph'alon*, the brain ; *itis*, denoting inflammation). Inflammation of the substance of the brain, or of the structures in general within the skull.

Enceph'alon (Gr. ἐν, *en*, in ; κεφαλη, *keph'alē*, the head). That part of the nervous system which is contained in the skull.

Enclit'ic (Gr. ἐν, *en*, on ; κλινω, *klinō*, I lean). Leaning on ; applied to certain words which throw their accent on the word immediately preceding, and thus, as it were, lean on it.

En'crinite (Gr. ἐν, *en*, in ; κρινον, *krinon*, a lily). A fossil radiated animal, resembling a lily.

Encysted (Gr. ἐν, *en*, in ; κυστις, *kustis*, a bladder or sac). Enclosed in a sac or bag.

Endeca-. *See* Hendeca-.

Endem'ic (Gr. ἐν, *en*, in ; δημος, *dēmos*, people). Among the people ; applied to diseases which habitually prevail in any locality.

Endermat'ic, or **Ender'mic** (Gr. ἐν, *en*, in ; δερμα, *derma*, the skin). A term applied to the administration of medicines by means of the skin.

Endo- (Gr. ἐνδον, *en'don*, within). A prefix to words, signifying within.

Endocar'dial (Gr. ἐνδον, *en'don*, within ; καρδια, *kar'dia*, the heart). Relating to the lining membrane of the heart.

Endocardi'tis (Gr. ἐνδον, *en'don*, within ; καρδια, *kar'dia*, the heart ; *itis*, denoting inflammation). Inflammation of the lining membrane of the heart.

Endocar'dium (Gr. ἐνδον, *en'don*, within ; καρδια, *kar'dia*, the heart). The membrane lining the interior of the heart.

En'docarp (Gr. ἐνδον, *en'don*, within ; καρπος, *karpos*, fruit). The membrane in some fruit, as apples, which lines the cavity containing the seeds.

Endogen (Gr. ἐνδον, *en'don*, within ; γενναω, *genna'ō*, I produce). A plant which grows by deposition of woody matter in the interior, without distinction of pith, wood, and bark.

Endog'enites (*Endogen*). Fossil stems exhibiting the endogenous structure.

En'dolymph (Gr. ἐνδον, *en'don*, within ; Lat. *lympha*, water). A watery fluid in the interior of the membranous labyrinth of the ear.

Endophlœ'um (Gr. ἐνδον, *en'don*, within ; φλοιος, *ph'oios*, bark). The inner layer of the bark of trees.

Endopleu'ra (Gr. ἐνδον, *en'don*, within ; πλευρα, *pleura*, a rib or membrane). The coat of the nucleus in the seed.

Endorhi'za (Gr. ἐνδον, *en'don*, within ; ῥιζα, *rhiza*, a root). Having a root within ; applied to plants of which the root bursts first through the coverings of the seed before elongating downwards.

Endoskel'eton (Gr. ἐνδον, *en'don*, within ; σκελετον, *skel'eton*, a framework of bone). An internal skeleton ; such as exists in vertebrate animals.

Endosmom'eter (Gr. ἐνδον, *en'don*, within ; ὠσμος, *ōsmos*, an impulse ; μετρον, *metron*, a measure). An instrument for measuring the intensity of endosmose.

En'dosmose (Gr. ἐνδον, *en'don*, within ; ὠθέω, ōthcū, I push). The process by which one fluid, separated from another by a membrane, mixes with it in a direction inwards from without.

Endos'teum (Gr. ἐνδον, en'don, within; ὀστεον, os'teon, a bone). The fine membrane lining the medullary canal of bones.

En'dostome (Gr. ἐνδον, en'don, within; στομα, stoma, a mouth). The inner aperture of an ovule.

Ene'ma (Gr. ἐν, en, in ; ἱημι, hiēmi, I send). A medicine thrown into the lower bowel.

Engineering. The art of constructing and using engines or machines.

Engor'gement (Fr. en, in ; gorge, the throat). A swallowing greedily ; but applied in medicine to an over-filled state of the vessels of a part.

Enneagyn'ia (Gr. ἐννεα, cn'nea, nine: γυνη, gunē, a female). An order of plants having nine pistils.

Ennean'dria (Gr. ἐννεα, en'nea, nine ; ἀνηρ, anēr, a male). A class of plants in the Linnæan system having nine stamens.

Enode (Lat. e, from ; nodus, a knot). Without knots or joints.

En'siform (Lat. ensis, a sword ; forma, shape). Like a sword.

Entab'lature (Lat. in, in ; tab'ula, a board or table). The structure which lies horizontally on columns, divided into architrave, frieze, and cornice.

Enter'ic (Gr. ἐντερον, en'teron, an intestine). Belonging to the intestines.

Enteri'tis (Gr. ἐντερον, en'teron, an intestine ; itis, denoting inflammation). Inflammation of the intestines.

En'terocele (Gr. ἐντερον, en'teron, an intestine ; κηλη, kēlē, a tumour). A hernial tumour containing intestine.

En'terolith (Gr. ἐντερον, en'teron, an intestine ; λιθος, lithos, a stone). A concretion resembling a stone, formed in the intestines.

Enthet'ic (Gr. ἐν, en, in ; τιθημι, tithēmi, I place). A term applied to diseases which become developed in the body after the introduction of a poison.

En'thymeme (Gr. ἐνθυμεομαι, enthu'meomai, I think). In rhetoric, an argument consisting of two propositions only, an antecedent and a consequent.

Ento- (Gr. ἐντος, en'tos, within). A prefix in compound words, signifying to the inner side.

En'tomoid (Gr. ἐντομον, en'tomon, insect, from ἐν, en, into ; τεμνω, temnō, I cut ; εἰδος, eidos, form). Resembling an insect.

Entomol'ogy (Gr. ἐντομον, en'tomon, an insect ; λογος, logos, a description). A description of insects.

Entomoph'agous (Gr. ἐντομον, en'tomon, an insect ; φαγω, phagō, I eat). Feeding on insects.

Entomos'traca (Gr. ἐντομον, en'tomon, an insect ; ὀστρακον, os'trakon, a shell). A section of minute crustaceous animals.

Entomot'omy (Gr. ἐντομον, en'tomon, an insect ; τεμνω, temnō, I cut). The dissection of insects.

Entomozo'ria (Gr. ἐν, en, into ; τεμνω, temnō, I cut ; ζωον, zōon, an animal). Invertebrate animals, having their bodies arranged in ring-like segments.

Entozo'on (Gr. ἐντος, en'tos, within ; ζωον, zōon, an animal). An animal which lives on the bodies of other animals : properly applied to those infesting the interior.

En'trochite (Gr. ἐν, en, in ; τροχος, trochos, a wheel). A name given in geology to the wheel-like joints of the encrinite.

Entro'pium (Gr. ἐν, en, in ; τρεπω, trepō, I turn). A turning of the eyelashes inwards towards the eye.

Enu'cleate (Lat. e, out of ; nu'cleus, a kernel). To remove as a kernel, from a nut.

E'ocene (Gr. ἠως, ēōs, the dawn ; καινος, kainos, new). Early ; applied to the earliest deposits in the tertiary geological strata.

Eol'ipile (Lat. Æ'olus, the god of the winds ; pila, a ball). An instrument consisting of a hollow metal ball, with a tube, used for exhibiting the elastic power of steam by filling the ball with water and heating it.

Ep'act (Gr. ἐπι, *ep'i*, on ; ἀγω, *agō*, I drive.) The number which denotes the age of the ecclesiastical moon on the first day of any year in a cycle of nineteen years.

Epen'thesis (Gr. ἐπι, *ep'i*, on ; ἐν, *en*, in ; τιθημι, *tithēmi*, I place). The insertion of a letter or syllable in the middle of a word.

Ephe'lis (Gr. ἐπι, *ep'i*, on ; ἡλιος, *hēlios*, the sun). Freckles ; an eruption of greyish or yellowish spots.

Ephem'eris (Gr. ἐπι, *ep'i*, on ; ἡμερα, *hēmera*, a day). A diary; an account of the daily positions of the planets.

Ep'i, or **ep-** (Gr. ἐπι, *ep'i*, on). A prefix in compound words, signifying upon.

Ep'ic (Gr. ἐπω, *ep'ō*, I speak). Narrative ; applied to poems which relate real or supposed events.

Ep'icarp (Gr. ἐπι, *ep'i*, on ; καρπος, *karpos*, a fruit). The outer skin of a fruit.

Ep'icene (Gr. ἐπι, *ep'i*, on; κοινος, *koinos*, common). Common ; applied to nouns which denote both the male and the female species.

Epicon'dyle (Gr. ἐπι, *ep'i*, on ; κονδυλος, *kon'dulos*, a knuckle). In *anatomy*, an additional condyle, a joint placed on a condyle.

Epicy'cle (Gr. ἐπι, *ep'i*, on ; κυκλος, *kuklos*, a circle). A small circle, of which the centre is in the circumference of a larger one.

Epicy'cloid (Gr. ἐπι, *ep'i*, on ; κυκλος, *kuklos*, a circle ; ἐιδος, *eidos*, form). A curve produced by the revolution of the circumference of a circle along the convex or concave side of another circle.

Epidem'ic (Gr. ἐπι, *ep'i*, on ; δημος, *dēmos*, the people). Attacking numbers of people in any locality at the same time, but of temporary duration, and not essentially connected with the locality.

Epidemiol'ogy (Gr. ἐπι, *ep'i*, on ; δημος, *dēmos*, the people ; λογος, *logos*, a description). The descrip-

tion or investigation of epidemic disease.

Epider'mal (*Epidermis*). Belonging to, or formed from the epidermis.

Epider'mis (Gr. ἐπι, *e'pi*, on ; δερμα, *derma*, the skin). The cuticle, or scarf-skin ; the external layer of the skin, or of the bark in plants.

Epigas'tric (Gr. ἐπι, *ep'i*, on ; γαστηρ, *gastēr*, the stomach). Belonging to the upper and anterior part of the abdomen ; over the stomach.

Epiglot'tis (Gr. ἐπι, *ep'i*, on ; γλωττα, *glōtta*, a tongue). A tongue-shaped projection lying over the entrance of the windpipe, and preventing the entrance of food or drink.

Epig'ynous (Gr. ἐπι, *ep'i*, on ; γυνη, *gunē*, a female). Growing on the top of the ovary in plants ; applied to stamens which are united both to the calyx and to the ovary.

Ep'ilepsy (Gr. ἐπι, *ep'i*, on ; ληψις, *lēpsis*, a seizing). The falling sickness; a sudden loss of sensation and voluntary power attended by convulsions, recurring at irregular intervals.

Epilep'tic (Gr. ἐπι, *ep'i*, on ; ληψις, *lēpsis*, a seizing). Subject to epilepsy.

Epilep'tiform (*Epilepsy ;* Lat. *forma*, form). Resembling epilepsy.

Epime'ral (Gr. ἐπι, *ep'i*, on ; μηρος, *mēros*, a thigh or limb). The part of the segment of an insect or other articulated animal which is above the joint of the limb.

Epipet'alous (Gr. ἐπι, *ep'i*, on ; πεταλον, *pet'alon*, a petal). Placed or growing on the petals.

Epiphlœ'um (Gr. ἐπι, *ep'i*, on ; φλοιος, *phloios*, bark). The layer of bark immediately beneath the epidermis.

Epiphyl'lous (Gr. ἐπι, *ep'i*, upon ; φυλλον, *phullon*, a leaf). Inserted on a leaf.

Epiph'ora Gr. ἐπι, *ep'i*, on ; φερω, *pher'ō*, I bear). Watery eye; a disease in which the tears flow over the cheek, from an obstruction in the canal which should carry them off.

Epiph'ysis (Gr. ἐπι, *ep'i*, on ; φυω, *phuō*, I grow). The end of a long bone, which is formed at first separately from the shaft, and afterwards is united to it.

Ep'iphyte (Gr. ἐπι, ep'i, on ; φυω, phuō, I grow). A plant which grows on or adheres to another vegetable, or to an animal.

Epip'loon (Gr. ἐπι, ep'i, on ; πλεω, pleō, I float). The caul ; a portion of the peritoneum, or lining membrane of the abdomen, which covers in front, and as it were floats on, the intestines.

Epispas'tic (Gr. ἐπι, ep'i, on ; σπαω, spaō, I draw). Drawing; blistering.

Ep'isperm (Gr. ἐπι, ep'i, on ; σπερμα, sperma, a seed). The outer covering of a seed.

Epistax'is (Gr. ἐπι, ep'i, on ; σταζω, stazō, I drop). Bleeding from the nose.

Epister'nal (Gr. ἐπι, ep'i, on ; στερνον, sternon, the breast). Situated on or above the sternum or breast-bone.

Epithe'lial (*Epithelium*). Belonging to, or formed of, epithelium.

Epithe'lium. A covering membrane in animals and vegetables, formed of the same structure as epidermis, but finer and thinner.

Ep'ithem (Gr. ἐπι, ep'i, on ; τιθημι, tithēmi, I place). A liquid in which cloths are dipped to be laid on any part of the body.

Epit'ome' (Gr. ἐπι, ep'i, on ; τεμνω, temnō, I cut). An abridgment of a book or writing.

Epizo'on (Gr. ἐπι, ep'i, on ; ζωον, zōon, an animal). An animal which fastens itself to the exterior of other animals and lives on them.

Epizoot'ic (Gr. ἐπι, ep'i, on ; ζωον, zōon, an animal). A term applied to diseases prevailing among animals, as epidemic diseases among men.

E'poch (Gr. ἐπι, ep'i, on ; ἐχω, ech'ō, I hold). A fixed point of time from which dates are numbered ; any fixed time or period.

Equa'tion (Lat. æquo, I make equal). A making equal ; in *algebra*, a form expressing the equality of two quantities ; in *astronomy*, the dif-

ference between real and apparent time or space.

Equa'tor (Lat. æquo, I make equal). A great imaginary circle, surrounding the earth at an equal distance from each pole.

Equato'rial (*Equator*). An astronomical instrument, capable of revolving on a fixed axis, coinciding in direction with that of the celestial sphere.

Equicru'ral (Lat. æquus, equal ; crus, a leg). Having equal legs ; or two sides of equal length, as a triangle.

Equidif'ferent (Lat. æquus, equal ; different). Having an equal difference ; applied to numbers in arithmetical progression, which increase or decrease by the addition or subtraction of the same number.

Equidis'tant (Lat. æquus, equal ; dis, from ; sto, I stand). At equal distances from some point.

Equilat'eral (Lat. æquus, equal ; latus, a side). Having all the sides equal.

Equilib'rium (Lat. æquus, equal ; libra, a balance). Equality of weight or force ; balance.

Equimul'tiple (Lat. æquus, equal ; multip'lico, I multiply). The product of multiplying a number by the same quantity as that by which some other number is also multiplied.

Equinoc'tial (Lat. æquus, equal ; nox, night). A term applied to the points at which the ecliptic intersects the celestial equator : so called from the days and nights being equal when the sun arrives in them.

Equinox'es (Lat. æquus, equal ; nox, night). The times at which the sun's centre is found in the equinoctial points, the days and nights being equal.

Eq'uipoise (Lat. æquus, equal ; Fr. prids, weight). Equality of weight ; equilibrium ; even balance.

Equirat'ional (Lat. æquus, equal ; ratio, a reckoning). Having an equal ratio ; applied to numbers in geometrical progression, which increase or decrease regularly by being multiplied or divided by the same number.

Equiv'alent (Lat. *æquus*, equal; *val'eo*, I am worth). Equal in value or power; in *chemistry*, a term applied to the numbers in which elements uniformly replace each other in combination.

Erec'tile (Lat. *e'rigo*, I raise up). Having the property of raising itself.

Erec'tor (Lat. *e'rigo*, I raise up). That which raises up: applied to some muscles.

Eremacau'sis (Gr. ἠρεμα, *ērema*, gradually; καιω, *kaii*, I burn). Slow combustion: the process by which the matters formed in the fermentation and putrefaction of animal and vegetable bodies combine gradually with the oxygen of the air.

Er'ethism (Gr. ἐρεθιζω, *erethi'zō*, I excite). Excitement; unnatural energy of action.

Er'gotism (*Ergot*, spurred rye). A diseased state, characterised by a kind of mortification, produced by eating spurred rye.

Ero'dent (Lat. *e*, out; *rodo*, I gnaw). Eating into; gnawing.

Ero'sion (Lat. *e*, from; *rodo*, I gnaw). The state of eating or being eaten away.

Errat'ic (Lat. *erro*, I wander). Wandering; not fixed; occurring in a casual manner.

Er'rhine (Gr. ἐν, *en*, in; ῥιν, *rhin*, the nose). Affecting the nose; producing discharge from the nose.

Eruct'ation (Lat. *eruc'to*, I belch). A bursting forth of wind from the stomach; or of gases or other matter from the earth.

Erup'tion (Lat. *e*, out; *rumpo*, I break). A breaking forth; a rash on the skin.

Erysip'elas (Gr. ἐρυω, *eruō*, I draw; πελας, *pelas*, near). A spreading inflammation of the skin; St. Anthony's fire.

Erythe'ma (Gr. ἐρυθρος, *eru'thros*, red). A superficial redness of the skin.

Esca'pement. An apparatus in clocks and watches for regulating the action of the pendulum or balance wheel.

Escar'pment. (Fr. *escarper*, to cut a slope.) Ground cut away nearly vertically about a military position; also a natural cutting away of the ground, as in ravines.

Eschar (Gr. ἐσχαρα, *es'chara*, a hearth or gridiron). A crust or scab produced by heat or caustics.

Escharot'ic (Gr. ἐσχαρα, *es'chara*, a hearth or gridiron). Producing an eschar or crust on the flesh.

Esophagot'omy (Gr. οἰσοφαγος, *oisoph'agos*, the œsophagus; τεμνω, *temnō*, I cut). The operation of making an incision or opening into the œsophagus.

Esoph'agus (Gr. οἰω, *oiō*, I carry; φαγω, *phag'ō*, I eat). The gullet, or tube which carries food to the stomach.

Esoter'ic (Gr. ἐσω, *esō*, within). Private; applied to the private instructions of Pythagoras.

Es'sence (Lat. *esse*, to be). The particular and distinguishing nature of a being or substance.

Essen'tial (Lat. *esse*, to be). Necessary to the constitution of a thing; specially distinctive.

Esthot'ics. *See Æsthet'ics.*

Estiva'tion (Lat. *æstas*, summer). The manner in which the petals of a flower are arranged within the bud.

Es'tuary (Lat. *æstus*, tide). An arm of the sea, or mouth of a river, where the tide meets the current.

Ethe'real (Gr. αἰθηρ, *aithēr*, ether). Relating to or formed of ether.

Etherisa'tion (*Ether*). The production of insensibility by inhaling the vapour of ether.

Eth'ical (Gr. ἠθος, *ēthos*, habit of men, manners). Relating to public manners or morals.

Eth'ics (Gr. ἠθος, *ēthos*, manners). The science of moral philosophy, or of the duties of men.

E'thmoid (Gr. ἠθμος, *ēthmos*, a sieve; εἰδος, *eidos*, form). Perforated with holes like a sieve.

Ethnol'ogy (Gr. ἐθνος, *ethnos*, a nation; λογος, *logos*, discourse). The science which describes the relation of the different varieties of mankind to each other.

E'tiolate. To whiten by excluding the rays of the sun.

E'tiology (Gr. αἰτια, *aitia*, a cause ; λογος, *logos*, a discourse). A description of causes ; in *medicine*, the department of the science which studies the agents by which diseases are produced.

Etymol'ogy (Gr. ἐτυμος, *et'umos*, true ; λογος, *logos*, a word). A description of the origin, derivation, and changes of words.

Et'ymon (Gr. ἐτυμος, *et'umos*, true). The root of a word, from which it is derived.

Eudiom'eter (Gr. εὐ. *eu.* well ; διος, *dios*, air ; μετρον, *metron*, a measure). An instrument for measuring the amount of oxygen contained in air or in gaseous mixtures.

Eudiom'etry (Gr. εὐ, *eu*, well ; διος, *dios*, air ; μετρον, *metron*, a measure). The art of measuring the quantity of oxygen in the air or in gaseous mixtures.

Eu'phemism (Gr. εὐ, *eu*, well ; φημι, *phēmi*, I speak). The substitution of a delicate or agreeable word for one which is offensive.

Euphon'ic (Gr. εὐ, *eu*, well ; φωνη, *phōnē*, voice). Having an agreeable sound.

Eu'phony (Gr. εὐ, *eu*, well ; φωνη, *phōnē*, voice). A combination of letters and syllables which is agreeable to the ear.

Eusta'chian Tube (*Eusta'chius*, a celebrated anatomist). The tube which connects the internal ear with the back part of the mouth.

Eusta'chian Valve. A fold of membrane lying between the anterior margin of the lower vena cava and the right auricles of the heart.

Evac'uant (Lat. *e*, from ; *vac'uo*, I empty). Emptying.

Evac'uate (Lat. *e*, out ; *vac'uo*, I empty). To empty or free from.

Evacua'tion (Lat. *e*, out ; *vac'uo*, I empty). An emptying or clearing.

Evap'orate (Lat. *e*, from ; *vapor*,

vapour). To pass off in vapour ; to convert into vapour.

Evap'oration (Lat. *e*, from ; *vapor*, vapour). The conversion of a fluid into vapour or steam ; the removal of fluid from any substance by converting it into vapour.

Evec'tion (Lat. *e*, out ; *veho*, I carry). A carrying out ; in *astronomy*, an inequality in the moon's place, produced by the mean progression of the apsides, and the variation of the excentricity.

Evolu'tion (Lat. *e*, out ; *volvo*, I roll). An unfolding or unrolling ; in *algebra* and *arithmetic*, the extraction of a root, or the unfolding of a number multiplied into itself any number of times ; in *military affairs*, changes in the position and arrangement of troops.

Evul'sion (Lat. *e*, from ; *vello*, I pluck). A pulling out by force.

Exacerba'tion (Lat. *ex*, from ; *acer'bus*, sharp). Irritation ; an increase in violence.

Exalbu'minous (Lat. *ex*, from ; *albu'men*). Without albumen.

Exan'thema (Gr. ἐξ, *ex*, out ; ἀνθος, *anthos*, a flower). An eruption : now applied to contagious diseases, attended by fever and by an eruption on the skin.

Excen'tric. *See* Eccen'tric.

Excentric'ity. *See* Eccentric'ity.

Excis'ion (Lat. *ex*, from ; *cædo*, I cut). A cutting off.

Excitabil'ity (Lat. *ex*, from ; *cito*, I provoke). The power of being roused to action.

Exci'tant (Lat. *ex*, from ; *cito*, I provoke). Calling into action ; stimulating.

Exci'to-mo'tor (Lat. *excito*, I excite ; *moveo*, I move). A term applied to those actions which arise from an impression made on the extremity of a nerve, conveyed to the spinal cord, and thence reflected, without sensation, to the nerves supplying the muscles of the part moved.

Excor'iate (Lat. *ex*, from ; *co'rium*, the skin). To strip off the skin.

Ex'crement (Lat. *ex*, from ; *cerno*, I separate). Refuse matter.

Excres'cence (Lat. *ex*, from; *cresco*, I grow). An unnatural or superfluous growth.

Excre'tion (Lat. *ex*, from ; *cerno*, I separate). A separation of fluids from the body by means of glands ; the fluids separated.

Excre'tory (Lat. *ex*, from ; *cerno*, I separate). Having the property of excreting or throwing off ; removing.

Exege'sis (Gr. ἐξηγεομαι, *exēgeomai*, I explain). An explanation.

Exfo'liate (Lat. *ex*, from ; *fo'lium*, a leaf). To separate in scales, as diseased bone, or the lamina of a mineral.

Exha'lant (Lat. *ex*, from ; *halo*, I breathe). Breathing out or evaporating.

Exhala'tion (Lat. *ex*, from ; *halo*, I breathe). The act of exhaling or sending forth in vapour; that which is emitted as vapour.

Exha'le (Lat. *ex*, from ; *halo*, I breathe). To breathe or send out vapour.

Exhau'st (Lat. *ex*, from ; *hau'rio*, I draw). To draw off; to empty by drawing out the contents.

Exocar'dial (Gr. ἐξω, *exō*, outside ; καρδια, *kar'dia*, the heart). Outside the heart.

Ex'ogen (Gr. ἐξω, *exō*, outside ; γενναω, *gennaō*, I produce). A plant which grows by additions to the outside of the wood.

Exog'enites (*Ex'ogen*). Fossil stems exhibiting the exogenous structure.

Exog'enous (Gr. ἐξω, *exō*, outside ; γενναω, *gennaō*, I produce). In *botany*, growing by addition to the outside; in *anatomy*, growing out from a bone already formed.

Exor'dium (Lat. *ex*, from ; *or'dior*, I begin). The introductory part of a discourse.

Exorhi'zal (Gr. ἐξω, *exō*, outside ; ριζα, *rhiza*, a root). A term applied to plants of which the roots do not burst through the coverings of the seed before growing downwards.

Exoskel'eton (Gr. ἐξω, *exō*, outside; σκελετον, *skel'eton*). An external skeleton, such as is found in many invertebrate animals ; also in those vertebrate animals which have ossified or bony plates connected with the skin.

Ex'osmose (Gr. ἐξ, *ēx*, out ; ὠθεω, *ōtheō*, I drive). The passage of one fluid to another through a membrane from within outwards.

Ex'ostome (Gr. ἐξω, *exō*, outwards ; στομα, *stoma*, a mouth). The outer aperture in the ovule of a plant, towards which the apex of the nucleus points.

Exosto'sis (Gr. ἐξ, *ex*, out ; ὀστεον, *os'teon*, a bone). An unnatural projection or growth from a bone.

Exoter'ic (Gr. ἐξω, *exō*, outside). External ; public.

Exothe'cium (Gr. ἐξω, *exō*, outside ; θηκιον, *the'kion*, a box). In *botany*, the outside covering of the anther, the inner being the endothecium.

Exot'ic (Gr. ἐξω, *exō*, outside). Brought from a foreign country.

Expansibil'ity (Lat. *ex*, out ; *pando*, I open). Capability of being enlarged or extended in all directions.

Expec'torant (Lat. *ex*, from ; *pectus*, the breast). Promoting discharge from the air-passages and lungs.

Expec'torate (Lat. *ex*, from ; *pectus*, the breast). To discharge from the air-passages or lungs.

Expira'tion (Lat. *ex*, from ; *spiro*, I breathe). A breathing out of air or vapour.

Expo'nent (Lat. *expo'no*, I set forth). A number or figure which, placed above and to the right hand of a number, denotes what root is to be extracted, or to what power it is to be raised : in the former case, fractions are used ; in the latter, whole numbers ; also the number which denotes the ratio between two quantities.

Expres'sion (Lat. *ex*, out ; *prem'o*, I press). A pressing out ; in *algebra*, any quantity, simple or compound.

Exsan'guine (Lat. *ex*, from ; *sanguis*, blood). Without blood ; deprived of blood.

Exsert'ed (Lat. *ex'serso*, I thrust out). In *botany*, extending beyond an organ.

Exsicca'tion (Lat. *ex*, from ; *siccus*, dry). Drying.

Exstip'ulate (Lat. *ex*, from ; *stipule*). Without stipules.

Exten'sor (Lat. *ex*, out ; *tendo*, I stretch). A stretcher out ; applied to certain muscles.

External Contact. In *astronomy*, the apparent touching of two disks at their edges, without interposition.

Extine (Lat. *ex*, out). The outer covering of the pollen-grain.

Extracell'ular (Lat. *extra*, beyond ; *cell'ula*, a cell). Without cells : applied to the formation of nuclei or cells in animal and vegetable matter, without the influence of a previously existing cell.

Extravas'cular (Lat. *extra*, beyond ; *vas'cular*). Without vessels.

Extrac'tion (Lat. *ex*, from ; *traho*, I draw). A drawing out.

Extrac'tive (Lat. *ex*, from ; *traho*, I draw). That which is drawn out : a term used in *chemistry* to denote matter of a peculiar kind obtained from substances by chemical operations.

Extravasa'tion (Lat. *extra*, out of ; *vas*, a vessel). The pouring of a fluid, as blood, out of its vessels.

Extro'rse (Lat. *extror'sum*, outwards). Turned outwards.

Exuda'tion (Lat. *ex*, out ; *sudo*, I perspire). A discharge of moisture through pores.

Exu'de (Lat. *ex*, out ; *sudo*, I perspire). To discharge through pores.

Exu'viæ (Lat. from *exuo*, I put off). Cast-off shells or skins of animals ; remains of animals found in the earth.

Eye-piece. The lens or combination of lenses in a microscope to which the eye is applied.

F.

Faç'ade (Fr.). The front view of a building.

Fa'cet (Fr. : a little face). A small face ; applied to the small terminal faces of crystals and cut gems.

Fa'cial (Lat. *fa'cies*, the face). Belonging to the face.

Fa'cial An'gle. In *anatomy*, the angle formed by a line drawn through the opening of the ear and the base of the nostrils, with another drawn from the most projecting part of the forehead through the front of the upper jaw ; regarded as a measure of intelligence in animals.

Fac'tor (Lat. *fac'io*, I make). A maker up or agent ; in *arithmetic* and *algebra*, the factors of a quantity are those by the multiplication of which into each other it is formed.

Fa'cules (Lat. *fa'cula*, a little torch). A term applied to varieties in the intensity of the brightness of different parts of the sun's disk.

Fæ'ces (Lat. *fæx*, dregs). Excrement or refuse matter.

Falcate (Lat. *falx*, a sickle). Bent like a sickle.

Fal'ciform (Lat. *falx*, a hook or sickle ; *forma*, shape). Shaped like a sickle.

Falx Cer'ebri (Lat. *falx*, a sickle). A curved projection downwards of the dura mater, which divides the brain into two hemispheres ; a similar structure also divides the cerebellum, or little brain.

Fari'na (Lat. *far*, corn). Meal or flour ; consisting of gluten, starch, and gum ; in *botany*, the pollen or fine dust of the anther.

Farina'ceous (Lat. *fari'na*, flour). Consisting of, or containing meal or flour.

Fas'cia (Lat. a band). A band ; in *architecture*, a band-like structure; a surgical bandage ; a membranous expansion.

Fas'ciate (Lat. *fas'cia*, a band). Bound, or apparently bound, with a band.

Fas'cicle or **Fascic'ulus** (Lat. *fascic'ulus*, a little bundle). A small bundle; in *anatomy*, a bundle of muscular fibres.

Fascic'ulate (Lat. *fascic'ulus*, a small bundle). Arranged in small bundles or clusters.

Fasci'ne (Lat. *fus'cia*, a band). A fagot used in military operations for raising batteries, filling ditches, &c.

Fau'ces (Lat. *faux*, the jaws). The opening by which the back part of the mouth communicates with the pharynx.

Fault (Lat. *fallo*, I deceive or fail). A failing ; in *geology*, an interruption of the continuity of strata.

Fauna (Lat. *Faunus*). The entire collection of animals peculiar to a country.

Favose (Lat. *favus*, a honey-comb). Resembling a honey-comb.

Favus (Lat. a honey-comb). A disease of the skin, popularly known as scaldhead.

Feather-edged. In *architecture*, made thin at one edge.

Febric'ula (Lat. *febris*, a fever ; *ula*, denoting smallness). A slight fever.

Feb'rifuge (Lat. *febris*, a fever ; *fugo*, I drive away). Diminishing or preventing fever.

Fe'brile (Lat. *febris*, a fever). Relating to, or indicating fever.

Fec'ula (Lat. *fœx*, dregs ; *ula*, denoting smallness). Starch.

Fec'ulent (Lat. *fœcula*, small dregs). Containing dregs or sediment.

Fe'cundate (Lat. *fecun'dus*, fruitful). To make fruitful.

Fecun'dity (Lat. *fecun'dus*, fruitful). Fruitfulness ; power of producing.

Feld'spar (Germ. *feld*, a field ; *spar*). The soft part of granite ; consisting of a mixture of alumina, lime, and potash or soda, with silicic acid.

Feldspath'ic (*Feld'spar*). Consisting of, or abounding in feldspar.

Fe'line (Lat. *fe'lis*, a cat). Belonging to cats, or to the cat tribe.

Fel'spar—Felspath'ic. *See* Feldspar and Feldspath'ic.

Fem'oral (Lat. *femur*, the thigh). Belonging to the thigh.

Femur (Lat.). In *anatomy*, the thighbone ; in *entomology*, the third joint of the leg in insects.

Fenes'tra (Lat. a window). A term applied in *anatomy* to two small openings in the bones of the ear.

Fenes'tral (Lat. *fenes'tra*, a window). Having openings like a window.

Fenes'trate (Lat. *fenes'tra*, a window). Belonging to, or resembling a window.

Feræ (Lat. *fera*, a wild beast). An order of mammalia in the Linnæan classification.

Ferment (Lat. *fer'veo*, I boil). That which causes fermentation.

Fermenta'tion (Lat. *fermentum*, leaven). A peculiar change or organic substances, by a rearrangement of their elements under the agency of an external disturbing force, different from ordinary chemical attraction.

Fer'reous (Lat. *ferrum*, iron). Relating to or consisting of iron.

Fer'ric (Lat. *ferrum*, iron). Derived from iron.

Ferrif'erous (Lat. *ferrum*, iron ; *fero*, I bear). Producing iron.

Ferro- (Lat. *ferrum*, iron). A prefix denoting that iron enters into the composition of the substance named.

Ferru'ginous (Lat. *ferrum*, iron ; *gigno*, I produce). Producing or yielding iron.

Fertilisa'tion (Lat. *fero*, I bear). In *botany*, the application of pollen to the stigma of a plant.

Fer'tilise (Lat. *fero*, I bear). To make fruitful or productive.

Fetal (Lat. *fœtus*, the young of a creature). Belonging to the fœtus.

Fetus or **Fœtus** (Lat). The young unborn animal, in which all the parts of the body are formed.

Fibre (Lat. *fibra*, a small sprout). A thread ; a minute slender structure entering into the composition of various parts of animals and vegetables.

Fi'bril (Lat. *fibra*, a small sprout ; *il*, denoting smallness). A minute fibre.

Fi'brin (*Fibre*). An organic substance found in the blood, which forms, on removal, long white elastic filaments.

Fi'bro-car'tilage (*Fibre ; cartilage*).

F

An animal tissue composed of fibrous tissue mixed with cartilage.

Fi'bro-se'rous (*Fibre ; serum*). Consisting of fibrous tissue covered by a serous membrane.

Fi'brous (Lat. *fibra*, a small sprout or fibre). Containing or consisting of fibres.

Fib'ula (Lat. a buckle). The outer or small bone of the leg.

Fib'ular (*Fib'ula*). Belonging to or situated near the fibula.

Fic'tile (Lat. *fingo*, I mould). Manufactured by the potter's art.

Fig'urate Numbers. A series of numbers, the units of which are capable of being placed in such order as to represent a geometrical figure.

Fil'ament (Lat. *filum*, a thread). A thread; in *anatomy*, a thread-like structure; in *botany*, the part of the stamen which supports the anther.

Fil'icoid (Lat. *filix*, fern ; Gr. εἶδος, *eidos*, form). Resembling fern.

Fi'liform (Lat. *filum*, a thread; *forma*, shape). Like a thread.

Filter (*Felt*, fulled wool). A strainer : to strain, in order to separate fluid from solid matter.

Filtrate. The liquid which has passed through a filter.

Filtra'tion. The act of filtering or straining.

Fim'briæ (Lat. *fim'bria*, a fringe). In *anatomy*, a structure resembling a fringe.

Fim'briated (Lat. *fim'bria*, a border or hem). Having a fringed edge.

Fi'nite (Lat. *finis*, an end). Having a limit.

Fire-damp. Light carburetted hydrogen : the explosive gas of coal-mines.

Firestone. A stone that stands heat ; in *geology*, a stone of lime and sand.

First Intention. In *surgery*, the process by which wounds heal by direct union.

Fis'sile (Lat. *findo*, I cleave). Capable of being split.

Fissip'arous (Lat. *findo*, I cleave ; *par'io*, I produce). Multiplying the species by the division of the individual into two parts, as in polygastric animalcules and polypes.

Fissiros'tres (Lat. *findo*, I cleave ; *rostrum*, a beak). A tribe of insessorial or perching birds, having the beak much depressed or flattened horizontally, so as to give a wide opening, as the swallows and kingfishers.

Fis'sure (Lat. *findo*, I cleave). A cleft ; in *anatomy*, an opening in a bone or other part resembling a cleft.

Fis'tula (Lat. a pipe). In *surgery*, a deep, narrow, callous ulcer.

Fis'tulous (Lat. *fis'tula*, a pipe). Like a pipe ; in *botany*, applied to cylindrical bodies which are hollow but closed at each end.

Fixed (Lat. *figo*, I fix). Firm ; fixed air, carbonic acid gas ; fixed stars.

Fixed Oils. Oils which are not capable of being distilled without decomposition.

Flabel'liform (Lat. *flabel'lum*, a fan ; *forma*, shape). Like a fan.

Flat'ulency (Lat. *flatus*, a blast). A generation of gases in the stomach and intestines.

Flex'ible (Lat. *flecto*, I bend). Capable of bending ; a changing form in obedience to a force exerted across the length of the material.

Flex'ion (Lat. *flecto*, I bend). A bending.

Flex'or (Lat. *flecto*, I bend). A bender ; applied to the muscles which bend the limbs.

Flex'ure (Lat. *flecto*, I bend). The bending or curve of a line or surface.

Flex'uose (Lat. *flecto*, I bend). Winding.

Floc'culent (Lat. *floccus*, a lock of wool). Consisting of or containing flocks, as of wool.

Flora (Lat. the Goddess of Flowers). The entire collection of plants belonging to a country.

Flo'ral (Lat. *flos*, a flower). Belonging to a flower.

Flower-bud. A bud which becomes developed into a flower.

Flu'ate (*Flu'orin*). A compound of fluoric acid with a base.

Fluid (Lat. *fluo*, I flow). Capable of

flowing ; not having sufficient force of adhesion in the component parts to prevent their separation by their mere weight readily changing their position.

Fluid'ity (Lat. *fluo*, I flow). The state of being fluid.

Flu'or, or Fluor-spar. A mineral consisting of fluoride of calcium, or the element fluorine with the metallic base of lime.

Fluor'ic. Relating to, or containing the element fluorine.

Flu'oride (*Flu'orine*). A compound of fluorine with another elementary body.

Flu'orine (*Fluor-spar*). An elementary substance which, in combination with calcium, forms fluor-spar.

Flu'vial (Lat. *flu'vius*, a river). Belonging to a river, or fresh water.

Flu'viatile (Lat. *flu'vius*, a river). Belonging to a river, or fresh water.

Flux (Lat. *fluo*, I flow). A flowing ; a substance used in chemical operations to promote the melting of metals or minerals.

Flux'ion (Lat. *fluo*, I flow). A flowing ; in *mathematics*, the finding of an infinitely small quantity, which, taken an infinite number of times, becomes equal to a given quantity.

Flywheel. A wheel used in machinery for the purpose of rendering motion equable and regular.

Focal. (Lat. *focus*, a fire-hearth). Relating to a focus.

Focal Distance. The distance of a focus from some fixed point ; in *optics*, the distance between the centre of a lens or mirror, and the point into which the rays are collected.

Focus (Lat. a hearth). A point in which rays meet.

Folia'ceous (Lat. *fo'lium*, a leaf). Consisting of, or resembling leaves.

Fo'liated (Lat. *fo'lium*, a leaf). Consisting of, or resembling a plate or leaf; arranged in layers like leaves.

Folia'tion (Lat. *fo'lium*, a leaf). The arrangement of leaves on a tree.

Fol'licle (Lat. *follis*, a bag). A little

bag ; in *botany*, a form of fruit with one suture.

Follic'ulated (Lat. *follic'ulus*, a little bag). Having follicles.

Fon'tanel (Lat. *fons*, a fountain). The opening in the skull of infants, between the bones, at each end of the sagittal suture.

Footstalk. The stem of a leaf.

For'alites (Lat. *foro*, I bore ; Gr. λιθος, *lithos*, a stone). Tube-like markings in sandstones and other geological strata, apparently the burrows of worms.

Fora'men (Lat. *foro*, I pierce). A hole or aperture.

Foraminif'erous (Lat. *fora'men*, a hole ; *fero*, I bear). Having a hole or holes ; applied to a class of marine animals, having shells consisting of chambers separated by partitions having in each a small hole.

Forma'tion (Lat. *formo*, I shape or build up). In *geology*, a term applied to any assemblage of rocks connected by geological position, by immediate succession in time, and by organic and mineral affinities.

For'miate. (*Formic* acid). A compound of formic acid with a base.

Formic (Lat. *formi'ca*, an ant). Belonging to or obtained from ants : applied to an organic acid procurable from ants, and also from the oxidation of wood-spirit under the influence of finely divided platinum.

Formica'tion (Lat. *formi'ca*, an ant). A sensation of ants or small insects creeping over the skin.

For'mula (Lat. *forma*, a form ; *ula*, denoting smallness). A form; in *mathematics*, a general expression by means of letters ; in *chemistry*, an expression denoting the composition of a substance ; in *medicine*, a prescription, or directions for making up medicines.

Fos'sil (Lat. *fo'dio*, I dig). Dug out of the earth ; in *geology*, applied generally to mineralised animal and vegetable remains, found in rocks or in the earth.

Fossilif'erous (Lat. *fo'dio*, I dig ; *fero*,

I bear). Producing or containing fossil remains of animals and vegetables.

Fos'silize (Lat. *fos'silis*, that which may be dug out). To convert into a fossil.

Fourchette (Fr. a fork). The bone in birds formed by the junction of the clavicles ; the merrythought.

Fovil'la (Lat. white ashes). The minute granular matter which exists in the interior of the pollen-grains in flowers.

Frac'tion (Lat. *frango*, I break). A broken part of an entire quantity or number.

Frac'ture (Lat. *frango*, I break). A break ; the manner or direction in which a break takes place.

Freezing Mixture. A mixture which produces cold sufficient to freeze other liquids.

Freezing Point. The point at which the mercury stands in the thermometer when immersed in a fluid in the act of freezing.

Frem'itus (Lat. *frem'o*, I roar or murmur). A vibrating sensation felt on applying the hand to the chest.

Fri'able (Lat. *frio*, I break or crumble). Easily crumbled.

Fric'tion (Lat. *frico*, I rub). The act of rubbing one body against another.

Frieze. The part of the entablature of a column which is between the architrave and cornice.

Frig'id (Lat. *fri'gus*, cold). Cold; wanting heat.

Frigorif'ic (Lat. *fri'gus*, cold ; *fa'cio*, I make). Producing cold ; freezing.

Frond (Lat. *frons*, a leaf, or bough with leaves). In *botany*, the flattened expansion produced by the spores of some acotyledonous or flowerless plants : leaf of a tree-fern.

Frondip'arous (Lat. *frons*, a leaf; *pa'rio*, I produce). In *botany*, applied to fruits which produce leaves from their upper part.

Fron'tal (Lat. *frons*, the forehead). Belonging to the forehead.

Fructifica'tion (Lat. *fruc'tus*, fruit ; *fac'io*, I make). The production of fruit.

Fruc'tify (Lat. *fruc'tus*, fruit; *fac'io*, I make). To make fruitful ; to fertilise.

Frugiv'orous (Lat. *fru'ges*, fruit; *voro*, I devour). Eating or living on fruits.

Frus'tum (Lat. a broken piece). A piece broken off; in *geometry*, the part of a solid body nearest the base, which remains after the top has been cut off by a plane parallel to the base.

Fuciv'orous (Lat. *fu'cus*, sea-weed ; *voro*, I devour). Eating or living on sea-weed.

Fu'coid (Lat. *fu'cus*, sea-weed ; Gr. εἶδος, *eidos*, form). Resembling sea-weed.

Ful'crum (Lat. *ful'cio*, I support). A support : the fixed point on which a lever turns.

Ful'minate (Lat. *ful'men*, thunder). To detonate : a compound of fulminic acid with a base, characterised by a tendency to explode violently.

Ful'minic Acid (Lat. *ful'men*, thunder). An acid produced by the action of nitric acid on alcohol in the presence of a salt of silver or mercury, and forming salts which have a tendency to explode violently.

Fu'maroiles (Lat. *fu'mus*, smoke). Crevices in the earth in volcanic districts from which steam and boiling fluids are emitted.

Fu'migate (Lat. *fu'mus*, smoke). To apply smoke or vapour.

Func'tion (Lat. *fungor*, I perform). In *physiology*, the use of a part or organ.

Fun'gi (Lat. *fun'gus*, a mushroom). An order of flowerless plants of which the mushroom is the type.

Fun'goid (Lat. *fun'gus*, a mushroom : Gr. εἶδος, *eidos*, form). Resembling a mushroom.

Fungos'ity (Lat. *fun'gus*, a mushroom). A soft excrescence, often of rapid growth.

Fun'gous (Lat. *fun'gus*, a mushroom). Consisting of, or resembling mushrooms.

Funic'ulus (Lat. *fu'nis*, a bundle). A little bundle : in *anatomy*, a bundle of fibrils of a nerve, enclosed in a sheath; in *botany*, the stalk by which the ovule is attached.

Furfura'ceous (Lat. *fur'fur*, bran). Resembling bran.

Fuse'e (Lat. *fu'sus*, a spindle). The conical part of a watch or clock which has the chain or cord wound round it.

Fusibil'ity (Lat. *fun'do*, I pour out). Capability of being melted, or converted from a solid to a liquid state by heat.

Fu'sible (Lat. *fun'do*, I pour out). Capable of being melted, or converted from a solid to a liquid state by heat.

Fu'siform (Lat. *fu'sus*, a spindle; *forma*, shape). Like a spindle : tapering at each end.

Fu'sion (Lat. *fun'do*, I pour out). A melting by heat.

G

Ga'bion (Fr.). A large cylindrical basket of wicker-work, filled with earth, used in fortifications.

Gable (Welsh, *gavael*, a hold). The upright triangular end of a house.

Galac'tagogue (Gr. γαλα, *gala*, milk ; ἀγω, *agō*, I drive). Increasing the secretion of milk.

Galac'tic Circle. In *astronomy*, the circle at right angles to the diameter forming the galactic poles.

Galac'tic Poles. In *astronomy*, the opposite points of the celestial sphere, round which the stars are most sparse.

Galactom'eter (Gr. γαλα, *gala*, milk : μετρον, *metron*, a measure). An instrument for ascertaining the purity of milk by means of its specific gravity.

Galactoph'agous (Gr. γαλα, *gala*, milk ; φαγω, *phag'ō*, I eat.) Living on milk.

Galactoph'orous (Gr. γαλα, *gala*, milk ; φερω, *phcrō*, I bear). Producing or conveying milk.

Gal'axy (Gr. γαλα, *gala*, milk). The milky way : a dense cluster of stars, giving to the naked eye an appearance of whitish nebulous light.

Gal'eated (Lat. *gal'ea*, a helmet). Covered as with a helmet : having a flower like a helmet.

Gale'na. Sulphuret of lead ; a compound of sulphur with lead.

Galen'ic (*Gale'nus*, an ancient physician). Relating to Galen : applied to medicines derived from the vegetable kingdom.

Gal'late. A compound of gallic acid with a base.

Gall-ducts. The ducts or canals which convey the bile from the liver.

Gal'lic (Lat. *galla*, a gall). Belonging to gall-nuts : applied to an organic acid derived from them.

Gallina'ceous (Lat. *galli'na*, a hen). Belonging to the order of birds of which the domestic fowl and the pheasant are examples.

Galli'næ (Lat. *galli'na*, a hen). An order of birds of which the common fowl is the type.

Galvan'ic. Relating to, containing, or exhibiting galvanism.

Gal'vanism. *See* Voltaic Electricity.

Gal'vanise. To affect with galvanism.

Galvanom'eter (*Gal'vanism ;* Gr. μετρον, *metron*, a measure). An instrument for measuring the intensity of galvanic or voltaic action.

Galvan'oscope (*Gal'vanism ;* Gr. σκοπεω, *skop'eō*, I view). An apparatus for ascertaining the direction in which the pole of a magnetic needle is moved by a galvanic current.

Gamopet'alous (Gr. γαμος, *gam'os*, marriage ; πεταλον, *pet'alon*, a petal). Having petals united by their margins.

Gamosep'alous (Gr. γαμος, *gam'os*, marriage; *sep'al*). Having sepals united by their margins.

Gan'gliated (*Ganglion*). Provided with ganglia.

Gan'glion (Gr. γαγγλιον, *gan'glion*, a knot). In *anatomy*, a small mass of nervous matter resembling a knot, found in the course of various parts of the nervous system; in *surgery*, a tumour consisting of a cyst filled with serous fluid, occurring generally at the wrist and ankle.

Ganglion'ic (Gr. γαγγλιον, *gan'glion*, a knot). Containing, or belonging to ganglia : applied especially to a part of the nervous system in which these structures abound, otherwise called the sympathetic nerve.

Gan'grene (Gr. γαγγραινα, *gangrai'na*, an eating sore). Death of a limited portion of the body, or of any of its tissues.

Ganoceph'ala (Gr. γανος, *gan'os*, splendour; κεφαλη, *keph'alē*, a head). An order of fossil reptiles having polished bony plates covering the head.

Gan'oid (Gr. γανος, *gan'os*, splendour; ειδος, *eidos*, appearance). Of splendid appearance; applied to an order of fishes, mostly extinct, with angular scales covered by a thick coat of shining enamel.

Gar'goyle (Lat. *gurgu'lio*, the throat-pipe). A spout in the cornice or parapet of a building for discharging water from the roof.

Gas (Saxon *gast*, German *geist*, a spirit). A body of which the component particles are not held together by mutual cohesion, and also have a disposition to separate from each other.

Gasholder. An apparatus for holding gases.

Gasom'eter (*Gas*; Gr. μετρον, *metron*, a measure). An apparatus for measuring, collecting, or mixing gases.

Gas'teropod (Gr. γαστηρ, *gastēr*, the stomach; πους, *pous*, a foot). Moving on the belly : applied to an order of molluscous invertebrate animals, of which the snail and slug are examples.

Gastral'gia (Gr. γαστηρ, *gastēr*, the stomach; αλγος, *algos*, pain). Pain in the stomach.

Gastric (Gr. γαστηρ, *gastēr*, the stomach). Pertaining to the stomach.

Gastri'tis (Gr. γαστηρ, *gastēr*, the stomach; *itis*, denoting inflammation). Inflammation of the stomach.

Gas'tro- (Gr. γαστηρ, *gastēr*, the stomach). In *anatomy* and *medicine*, a prefix in compound words signifying relation to, or connection with, the stomach.

Gastrocne'mius (Gr. γαστηρ, *gastēr*, the stomach; κνημη, *knēmē*, the leg). A muscle which forms the chief part of the calf of the leg.

Gastrodyn'ia (Gr. γαστηρ, *gastēr*, the stomach; οδυνη, *odunē*, pain). Pain in the stomach.

Gas'tro-enteri'tis (Gr. γαστηρ, *gastēr*, the stomach; εντερον, *en'teron*, an intestine; *itis*, denoting inflammation). Inflammation of the stomach and intestines.

Gastro-pul'monary (Gr. γαστηρ, *gastēr*, the stomach; Lat. *pulmo*, a lung). Connected with the lungs and intestinal canal : applied to a track of mucous membrane.

Gastro'raphy (Gr. γαστηρ, *gastēr*, the stomach; ραφη, *raphē*, a suture). Union of a wound of the stomach or abdomen by suture.

Gault. In *geology*, a common term for the chalky clays of the lower division of the chalk system.

Gel'atine (Lat. *gelo*, I congeal). The softish substance produced by dissolving animal membranes, skin, tendons, and bones, in water at a high temperature ; animal jelly.

Gelat'inize (*Gel'atine*). To change into gelatine.

Gelat'inous (*Gel'atine*). Belonging to or consisting of gelatine.

Gemina'tion (Lat. *gem'ini*, twins). A doubling.

Gemma'tion (Lat. *gemma*, a bud). Budding ; the construction of a leaf-bud ; multiplication by budding.

Gemmip'arous (Lat. *gemma*, a bud ;

par'io, I produce). Producing buds ; multiplying by a process of budding.

Gem'ucule (Lat. *gemma*, a bud ; *ule*, denoting smallness). The growing point of the embryo in plants.

Geneal'ogy (Gr. *γενος, genos*, a race ; *λογος, logos*, a description). A history of the descent of a person or family from an ancestor.

Gener'ic (Lat. *genus*, a kind). Pertaining to a genus; distinguishing a genus from a species or from another genus.

Gen'esis (Gr. *γενναω, gennaō*, I produce). A production or formation.

Genet'ic (Gr. *γενναω, gennaō*, I produce). Relating to the origin of a thing or its mode of production.

Ge'nio- (Gr. *γενειον, genei'on*, the chin). In *anatomy*, a prefix in several names of muscles, denoting attachment to the chin.

Gen'itive (Lat. *gigno*, I produce). In *grammar*, applied to that case which denotes the person or thing to which something else stands in the relation of descent, possession, or other connection.

Gen'us (Lat. a kind). An assemblage of species possessing certain common distinctive characters.

Geocen'tric (Gr. *γη, gē*, the earth ; *κεντρον, kentron*, a centre). Having the earth as a centre : applied to the position and motion of a heavenly body as viewed from the earth.

Ge'ode (Gr. *γη, gē*, the earth). In *geology*, a rounded nodule with internal cavities.

Geod'esy (Gr. *γη, gē*, the earth ; *δαιω, daiō*, I divide). The science which measures the earth and portions of it by mathematical observation.

Geognos'tic (Gr. *γη, gē*, the earth ; *γνωσις, gnōsis*, knowledge). Relating to a knowledge of the structure of the earth.

Geogno'sy (Gr. *γη, gē*, the earth ; *γνωσις, gnōsis*, knowledge). The knowledge of the earth.

Geograph'ical (Gr. *γη, gē*, the earth ; *γραφω, graphō*, I write). Relating to geography.

Geog'raphy (Gr. *γη, gē*, the earth ; *γραφω, graphō*, I write). The science which describes the surface of the earth, its divisions, their inhabitants, productions, &c. This is general or universal geography. Mathematical geography applies the knowledge of mathematics to the solution of problems connected with the earth's figure, the position of places, &c. Medical geography describes the distribution of diseases on the globe. Physical geography describes the various climates, the causes influencing them, and their bearing on animal and vegetable life. Political geography describes the political and social organisation of the various human inhabitants of the earth.

Geol'ogy (Gr. *γη, gē*, the earth ; *λογος, logos*, a description). The science which describes the condition or structure of those parts of the earth which lie beneath the surface.

Geomet'rical (*Geometry*). According to geometry.

Geomet'rical Progres'sion. A form of progression in which numbers increase or decrease by being multiplied or divided by the same number.

Geom'etry (Gr. *γη, gē*, the earth ; *μετρον, metron*, a measure). Literally and originally, the art of measuring the earth; but now denoting the science of the mensuration and relations of bodies, and their physical properties.

Geothermom'eter (Gr. *γη, gē*, the earth ; *θερμος, thermos*, warm ; *μετρον, metron*, a measure). An instrument for measuring the temperature of the earth at different points, as in mines, artesian wells, &c.

Ger'minal (Lat. *germen*, a bud). Belonging to a germ or bud.

Ger'minal Membrane. The membrane, formed of cells, which immediately surrounds the ovum or egg after segmentation.

Ger'minal Spot. The opaque spot on the germinal membrane, which is intended to be developed into the embryo.

Ger'minal Ves'icle. The small vesicular body within the yolk of the ovum or egg.

Ger'minate (Lat. *ger'men*, a sprout). To sprout or begin to grow.

Germina'tion (Lat. *ger'men*, a sprout). The act of sprouting.

Ger'und (Lat. *ger'o*, I bear). A part of a verb, partaking of the character of a noun.

Geyser. A boiling spring or fountain, of volcanic origin.

Gib'bous (Lat. *gibbus*, a bunch on the back). Humped; presenting one or more large elevations.

Gin'glymoid (Gr. γιγγλυμος, *gin'glumos*, a hinge or joint; ειδος, *eidos*, form). Resembling a hinge.

Gin'glymus (Gr. γιγγλυμος, *gin'glumos*, a hinge or joint). A joint allowing motion in two directions only, such as that of the elbow and lower jaw.

Gla'brous (Lat. *gla'ber*, smooth). Smooth; destitute of hair.

Glac'ial (Lat. *glac'ies*, ice). Resembling ice.

Glac'ier (Lat. *glac'ies*, ice). A mass of snow and ice, formed in the higher valleys, and descending into the lower valleys, carrying with them masses of rocks and stones.

Gland (Lat. *glans*, an acorn). A structure in animal and vegetable bodies, for the purpose of secreting or separating some peculiar material.

Gland'ula (*Gland*). In *anatomy*, a little gland.

Gland'ular. Consisting of or relating to glands; in *botany*, applied to hairs having glands at their tips containing some special secretion, or fixed on glands in the epidermis.

Glauco'ma (Gr. γλαυκος, *glaucos*, blue-grey). A disease of the eyes, attended with a greenish discoloration of the pupil.

Gle'noid (Gr. γληνη, *glēnē*, the pupil, or a shallow pit; ειδος, *eidos*, form). A term applied to a round shallow excavation in a bone, to receive the head of another bone.

Glo'bose (Lat. *globus*, a globe). In *botany*, forming nearly a true sphere.

Glob'ular (Lat. *globus*, a globe). A very small round body.

Glob'ular Projection. That projection of the sphere which so represents it as to present the appearance of a globe.

Glob'uline (*Glob'ule*). An organic substance, somewhat resembling albumen, found in the red corpuscles of the blood.

Glochid'iate (Gr. γλωχις, *glōchis*, a projecting point; the point of an arrow). In *botany*, applied to hairs, the divisions of which are barbed like a fish-hook.

Glom'erule (Lat. *glo'mus*, a clew of thread; *ule*, denoting smallness). In *botany*, a kind of dense tuft of flowers; also the powdering leaf lying on the thallus of lichens.

Glomer'ulus (Lat. *glo'mus*, a clew of thread). A name applied to small red bodies in the kidney, consisting of tufts of minute vessels, covered in by the dilate end of the secreting tubes of the organ.

Glos'sary (Gr. γλωσσα, *glōssa*, a tongue). A dictionary of difficult words; sometimes an ordinary dictionary.

Glossi'tis (Gr. γλωσσα, *glōssa*, a tongue; *itis*, denoting inflammation). Inflammation of the tongue.

Glos'so- (Gr. γλωσσα, *glōssa*, the tongue). In *anatomy*, a prefix in several compound words, signifying connection with the tongue.

Glosso-hyal (Gr. γλωσσα, *glōssa*, the tongue; *hyoid* bone). Connected with the tongue and the hyoid bone.

Glottis (Gr. γλωττα, *glōtta*, the tongue). The narrow opening at the top of the windpipe.

Glu'cose (Gr. γλυκυς, *glu'kus*, sweet). Grape-sugar, or the sugar of fruits.

Glume (Lat. *gluma*, chaff). The bracts covering the flower of grasses and corn.

Glumel'læ (Lat. *gluma*, chaff; *ella*, denoting smallness). The scales forming the flowers of grasses and corn.

Glu'teal (Gr. γλουτος, *glou'tos*, the hinder region). Belonging to the buttocks.

Gluten (Lat. glue). An insoluble substance obtained from wheat-flour by washing with water and straining.

Glyc'erine (Gr. γλυκυς, *glukus*, sweet). An organic substance existing in fats and oils, and obtained by saponifying them with an alkali or with oxide of lead.

Glycogen'esis (Gr. γλυκυς, *glukus*, sweet ; γενναω, *genna'ō*, I produce). The formation of sugar in the animal body.

Glyphog'raphy (Gr. γλυφω, *gluphō*, I engrave ; γραφω, *graphō*, I write). A process by which designs are engraved on a coating of wax or other soft substance spread on a metal, a sheet of other metal being then deposited on it by the electrotype process.

Glyptothe'ca (Gr. γλυφω, *gluphō*, I engrave ; τιθημι, *tithēmi*, I place). A building or room for preserving works of sculpture.

Gneiss. A hard tough crystalline rock, composed mostly of quartz, feldspar, mica, and hornblende, differing from granite in having its crystals broken, indistinct, and confusedly aggregated.

Gneiss'oid (*Gneiss ;* Gr. ειδος, *eidos*, form). Resembling gneiss ; applied to rocks intermediate between granite and gneiss, or between mica-slate and gneiss.

Gnomiomet'rical (Gr. γνωμων, *gnūmūn*, an index ; μετρον, *metron*, a measure). Relating to the measurement of angles by reflexion.

Gno'mon (Gr. γνωμων, *gnūmūn*, one that knows or interprets). The index of a dial.

Goitre (Fr). A large soft swelling in front of the neck.

Gompho'sis (Gr. γομφος, *gomphos*, a nail). A form of joint in which a conical body is fastened into a socket ; as the teeth.

Go'niodont (Gr. γωνια, *gōnia*, an angle ; οδους, *odous*, a tooth). Having angular teeth ; applied to certain fishes.

Goniom'eter (Gr. γωνια, *gōnia*, an angle ; μετρον, *metron*, a measure). An instrument for measuring angles.

Gorget (Fr. *gorge*, the throat). A piece of armour for defending the throat or neck ; in *surgery*, a certain cutting instrument.

Goth'ic. Belonging to the Goths : in *architecture*, applied to the architecture of the middle ages.

Gouty Concretions. Calculi or deposits of urate of soda in the joints, arising from gout.

Gov'ernor. A contrivance in machinery for maintaining uniform velocity with varying resistance.

Gra'dient (Lat. *grad'ior*, I step). The degree of slope of the ground over which a railway passes.

Grad'uate (Lat. *gradus*, a step). To receive a degree from an university ; to mark with regular divisions ; to change gradually.

Gradua'tion. The receiving a degree from an university ; the marking instruments with regular divisions.

Gral'læ or **Grallato'res** (Lat. *gralla'tor*, one who goes on stilts). An order of birds, remarkable for the length of the legs, as bustards, cranes, herons, and snipes.

Gramina'ceous or **Gramin'eous** (Lat. *gramen*, grass). Belonging to grasses, or the order of plants which includes grasses and corn.

Graminiv'orous (Lat. *gramen*, grass ; *voro*, I devour). Eating grass.

Gramme. A French weight ; the weight of a cubic *centimètre* of distilled water, or 15·438 grains Troy.

Gran'ite (Lat. *granum*, a grain, from its appearance). A stone or rock consisting of grains of quartz, felspar, and mica ; chemically composed for the most part of silica or flint-earth and-alumina.

Granit'ic (*Granite*). Relating to cr formed of granite.

Gran'itoid (*Granite ;* Gr. ειδος, *eidos*, form). Resembling granite.

Graniv'orous (Lat. *granum*, a grain or seed ; *voro*, I devour). Eating grains or seeds.

Gran'ular (Lat. *granum*, a grain). Consisting of or resembling grains.

Gran'ulate (Lat. *granum*, a grain). To form, or be formed, into grains or small masses.

Granula'tion (Lat. *granum*, a grain). The act of forming into grains; a small fleshy body springing up on the surface of wounds.

Graph'ite (Gr. γραφω, *graphō*, I write). Black-lead; a mineral consisting of carbon, generally with a small quantity of iron.

Grap'tolites (Gr. γραφω, *graphō*, I write; λιθος, *lith'os*, a stone). Fossil zoophytes or protozoa which give the appearance of writing or sculpture to the stone in which they are found.

Grauwac'ke or **Greywac'ke** (Germ. *grau*, grey; *wacke*, a kind of stone so called). A kind of sandstone consisting of different minerals.

Gravim'eter (Lat. *gravis*, heavy; Gr. μετρον, *metron*, a measure). An instrument for measuring specific gravities.

Grav'itate (Lat. *gravis*, heavy). To tend towards the centre of a body.

Gravita'tion (Lat. *gravis*, heavy). The act of tending towards a centre; the force by which bodies are drawn towards the centre of the earth or other centres.

Grav'ity (Lat. *gravis*, heavy). Weight; the force by which bodies tend towards the centre of the earth or another centre. Specific gravity is the weight of a body compared with the weight of an equal bulk of some other body, taken as unity.

Greensand. The lower group of the chalk system, in which many of the beds are coloured green.

Greenstone. A rock composed of feldspar and hornblende.

Grega'rious (Lat. *grex*, a herd). Living in flocks or herds.

Grego'rian Year. The year according to the ordinary reckoning, as reformed by Pope Gregory XIII.

Greywac'ke. *See* Grauwacke.

Grit. In *geology*, a term applied to any hard sandstone in which the grains are sharper than in ordinary sandstone.

Groined. In *architecture*, formed of vaults or arches which intersect and form angles with each other.

Gummif'erous (Lat. *gummi*, gum; *fero*, I bear). Producing gum.

Gun Cotton. An explosive material, formed by steeping cotton-wool or vegetable fibre in a mixture of nitric and sulphuric acids.

Gus'tatory (Lat. *gusto*, I taste). Belonging to taste.

Gutta Sere'na (Lat.). An old term for blindness from loss of power in the nervous system of the eye.

Guttif'erous (Lat. *gutta*, a drop; *fero*, I bear). Producing gum or resin.

Gut'tural (Lat. *guttur*, the throat). Belonging to, or formed by, the throat.

Gymna'sium (Gr. γυμνος, *gumnos*, naked). Originally, a place for athletic exercises; but also applied to schools for mental instruction.

Gymnas'tic (Gr. γυμνος, *gumnos*, naked). Pertaining to athletic exercises.

Gym'nodont (Gr. γυμνος, *gumnos*, naked; οδους, *odous*, a tooth). Having naked teeth: applied to some fishes in which the jaws are covered with an ivory-like substance in place of teeth.

Gym'nogens (Gr. γυμνος, *gumnos*, naked; γενναω, *gennaō*, I produce). Plants with naked seeds.

Gym'nosperms (Gr. γυμνος, *gumnos*, naked; σπερμα, *sperma*, seed). Plants having seeds apparently without a covering.

Gym'nospore (Gr. γυμνος, *gumnos*, naked; σπορα, *spora*, seed). A term applied to the spores of acotyledonous plants, when they are developed outside the cell in which they are produced.

Gynan'dria (Gr. γυνη, *gunē*, a female; ανηρ, *anēr*, a man). A class of plants in the Linnæan system, in which the stamens and pistils are consolidated.

Gy'nobase (Gr. γυνη, *gunē*, a female; βασις, *basis*, a base). In *botany*, a fleshy substance in the centre of a flower, bearing a single row of carpels.

Gynœ'ceum (Gr. γυνη, *gunē*, a female;

οἶκος, *oikos*, a house). The female apparatus of flowering plants ; the pistils.

Gy'nophore (Gr. γυνη, *gunē*, a female ; φερω, *pher'ō*, I bear). The stalk of a carpel in plants.

Gyp'seous (*Gypsum*). Containing or consisting of gypsum or sulphate of lime.

Gyp'sum (Gr. γυψος, *gupsos*, chalk or plaster of Paris). Sulphate of lime.

Gyra'tion (Gr. γυρος, *guros*, a whirling). A turning or whirling.

Gyrenceph'ala (Gr. γυροω, *guroō*, I wind; ἐγκεφαλος, *enkeph'alos*, the brain). Winding-brained ; applied by Professor Owen to a sub-class of mammalia in which the surface of the brain is convoluted, but not to the same extent as in man.

Gyri (Gr. γυρος, *guros*, a turning). In *anatomy*, a name given to the convolutions of the brain.

Gy'roscope (Gr. γυρος, *guros*, a whirling ; σκοπεω, *skop'eō*, I look at). An instrument for demonstrating the rotation of the earth by another apparent motion artificially produced.

H.

Hab'itat (Lat. *hab'ito*, I dwell). The natural abode or locality of an animal or plant.

Hæma- or **Hæmat-** (Gr. αἱμα, *haima*, blood). A part of some compound words, signifying blood.

Hæmadynamom'eter(Gr. αἱμα, *haima*, blood ; δυναμις, *du'namis*, force ; μετρον, *metron*, a measure). An instrument for measuring the force of the flow of blood in the vessels.

Hæmal (Gr. αἱμα, *haima*, blood). Relating to blood : applied to the arch proceeding from a vertebra, which encloses and protects the organs of circulation.

Hæmapoph'ysis (Gr. αἱμα, *haima*, blood ; *apoph'ysis*). A name given to the parts projecting from a vertebra which form the hæmal arch.

Hæmatem'esis (Gr. αἱμα, *haima*, blood ; ἐμεω, *em'eō*, I vomit). A vomiting of blood.

Hæ'matin (Gr. αἱμα, *haima*, blood). The colouring matter of the blood.

Hæ'matite (Gr. αἱμα, *haima*, blood). Blood-stone ; native sesquioxide of iron.

Hæmat'ocele (Gr. αἱμα, *haima*, blood ; κηλη, *kēlē*, a tumour). A tumour filled with blood.

Hæmatoc'rya (Gr. αἱμα, *haima*, blood ; κρυος, *kruos*, frost). Cold-blooded vertebrate animals.

Hæmatol'ogy(Gr. αἱμα, *haima*, blood ; λογος, *logos*, discourse). A description of the blood.

Hæmat'osin. *See* Hæmatin.

Hæmato'sis (Gr. αἱμα, *haima*, blood). The formation of blood.

Hæmatother'ma (Gr. αἱμα, *haima*, blood ; θερμος, *thermos*, warm). Warm-blooded vertebrate animals.

Hæmatu'ria (Gr. αἱμα, *haima*, blood ; οὑρον, *ouron*, urine). A discharge of blood with the urine.

Hæmop'tysis (Gr. αἱμα, *haima*, blood ; πτυω, *ptuō*, I spit). A spitting of blood.

Hæm'orrhage(Gr. αἱμα, *haima*, blood ; ῥηγνυμι, *rhēgnu'mi*, I burst forth). An escape of blood from its vessels.

Hæm'orrhoid (Gr. αἱμα, *haima*, blood ; ῥεω, *rheō*, I flow ; εἰδος, *eidos*, form). An enlargement of the veins of the lower bowel, commonly attended with loss of blood.

Hæmostat'ic (Gr. αἱμα, *haima*, blood ; ἱστημι, *histēmi*, I make to stand). Arresting the flow of blood.

Hagiog'rapha (Gr. ἁγιος, *hag'ios*, holy ; γραφω, *graphō*, I write). Sacred writings.

Hal'itus (Lat. *halo*, I breathe out). A breathing; the odour or vapour which escapes from blood.

Hallucina'tion (Lat. *hallu'cinor*, I blunder). An error of the senses.

Halo (Gr. ἁλως, *halōs*, a threshing-floor or area). A circle apparently round the sun or moon, sometimes white and sometimes coloured, produced by the passage of light through or near vapours in the atmosphere.

Hal'ogen (Gr. ἀλς, *hals*, salt ; γεννάω, *genna'ō*, I produce). Producing salts by combination with metals.

Ha'loid (Gr. ἀλς, *hals*, salt; εἶδος, *eidos*, form). Resembling salt : a name given to a class of saline substances constituted of a metal, and another element which is a salt radical ; after the type of common salt or chloride of sodium, where sodium is the metal, and chlorine the salt radical or halogen.

Ham'ite (Lat. *hamus*, a hook). A genus of fossil shells of cephalopods, with a hook at the e..d.

Harmo'nia (Gr. ἀρμοζω, *harmozō*, I fit together). A form of articulation in which the surfaces of bones are merely placed in apposition to each other, so as not to allow motion.

Harmon'ical (Gr. ἀρμοζω, *harmozō*, I fit together). Relating to harmony ;. concordant.

Harmonical Proportion. In *arithmetic*, that relation of four quantities to each other, in which the first is to the fourth as the difference between the first and second is to the difference between the third and fourth.

Har'mony (Gr. ἀρμοζω, *harmozō*, I fit together). A proper fitting of parts together ; agreement ; in *music*, the effect produced on the ear by the sounding of notes, the vibrations of which have a certain limit of coincidence.

Has'tate (Lat. *hasta*, a spear). Like a spear.

Haustel'late (Lat. *haustel'lum*, a sucker). Having a sucker for sucking or pumping up fluids ; applied to a large division of insects.

Haustel'lum (Lat. *hau'rio*, I draw). A sucker, such as some insects are provided with for taking their liquid food.

Haver'sian Canals (*Havers*, a physician, their discoverer). Small longi-

tudinal canals bone.

Heat. The se the contact quality of the sensation is the agent to due. Sensibl is perceptible heat is that ceives or lose increased or warmth. Spec required to r given degree

Hebdom'adal (G a period of s to a week.

Hectic (Gr. ἔξις, of fever arisin in a weakened

Hec'togramme (a hundred ; F weight of 10 3½ pounds av

Hec'tolitre (Gr. hundred ; Fr French measu

Hec'tometre (G hundred ; F measure of 1 328 British fe

Heli'acal (Gr. ἡ Emerging fro light of the s

Helianthoi'da (sun ; ἀνθος, a *eidos*, shape). resembling a ance ; of whi anemone is an

Hel'icoid (Gr. body ; εἶδος, e like the shell

Hel'ical (Gr. ἑ body). Spira

Helicotre'ma (G τρημα, *trēma*, in the apex of structure of t

Heliocen'tric ((sun : κεντρον, Having relati the sun.

Heliocen'tric L

formed at the sun's centre by the projection of the radius vector of a planet on the ecliptic with a line drawn from the sun's centre to the first point of Aries.

Heliograph'ic (Gr. ἡλιος, hēlios, the sun; γραφω, graphō, I write). Delineated by the rays of the sun.

He'liolites (Gr. ἡλιος, hēlios, the sun; λιθος, lith'os, a stone). A genus of fossil corals, distinguished by the central radiating or sun-like aspect of the pores.

Heliom'eter (Gr. ἡλιος, hēlios, the sun; μετρον, metron, a measure). An instrument for measuring the diameter of the heavenly bodies.

He'lioscope (Gr. ἡλιος, hēlios, the sun; σκοπεω, skop'ō, I view). A telescope fitted for viewing the sun without injury to the eyes.

He'liostat (Gr. ἡλιος, hēlios, the sun; ἱστημι, histēmi, I make to stand). An instrument for fixing (as it were) a sunbeam in an horizontal position.

Helisphe'rical (Gr. ἑλιξ, helix, a spire; σφαιρα, sphaira, a sphere). Applied to a course in navigation, which winds spirally round the globe.

Helix (Gr. ἑλιξ, helix, from ἑλισσω, helissō, I turn round). A spiral line or winding; the cartilaginous structure forming the external rim of the ear.

Hellen'ic (Gr. Ἑλλην, Hellēn, a Greek). Belonging to the Hellenes or inhabitants of Greece.

Hel'lenism (Gr. Ἑλλην, Hellēn, a Greek). The Grecian idiom used by the Jews living in countries where Greek was spoken.

Helminth'agogue (Gr. ἑλμινς, helmins, a worm; ἀγω, agō, I drive). Removing or expelling intestinal worms.

Helmin'thoid (Gr. ἑλμινς, helmins, a worm; εἰδος, shape). Like a worm.

Hema- or Hemat-. For words with this beginning, see the same words commencing with Haema- or Haemat-.

Hemeralo'pia (Gr. ἡμερα, hēmera, day; ἀλαομαι, ala'omai, I grope about; ὠψ, ōps, the eye). A defect of sight, in which the patients can see by night, but not by day.

Hemicra'nia (Gr. ἡμισυς, hēmisus, half; κρανιον, kra'nion, the skull). A painful affection of one side of the head and face.

Hemihed'ral (Gr. ἡμισυς, hēmisus, half; ἑδρα, hedra, a side). Half-sided; a form assumed by crystals from the excessive growth of some of their sides and the obliteration of others, so that they have only half the number of faces required by the laws of symmetry.

Hemily'tra (Gr. ἡμισυς, hēmisus, half; ἑλυτρον, elu'tron, a cover). Wing in insects, of which one half is firm, like an elytrum, and the other membranous.

Hemio'pia (Gr. ἡμισυς, hēmisus, half; ὠψ, ōps, the eye). A defect of sight in which only half of an object is seen.

Hemiple'gia (Gr. ἡμισυς, hēmisus, half; πλησσω, plēssō, I strike). Loss of power in one lateral half of the body.

Hemip'tera (Gr. ἡμισυς, hēmisus, half; πτερον, pteron, a wing). An order of insects which have the upper wings half hard and half membraneous; as the cock-roach and grasshopper.

Hem'isphere (Gr. ἡμισυς, hēmisus, half; σφαιρα, sphaira, a round body). A half sphere; the half of the earth, divided by the equator; a map of half the globe; in anatomy, applied to each lateral half of the brain.

Hemispher'ical (Gr. ἡμισυς, hēmisus, half; σφαιρα, sphaira, a round body). Having the shape of half a globe.

He'mitrope (Gr. ἡμισυς, hēmisus, half; τρεπω, trep'ō, I turn). Half turned.

Hemop'tysis. See Hæmop'tysis.

Hem'orrhage. See Hæm'orrhage.

Hendec'agon (Gr. ἑνδεκα, hen'deka, eleven; γωνια, gūnia, an angle). A figure of eleven sides and as many angles.

Hepat'ic (Gr. ἡπαρ, hēpar, the liver). Belonging to the liver; applied to a tube or duct conveying the bile from the liver.

Hepati'tis (Gr. ἧπαρ, *hēpar*, the liver; *itis*, denoting inflammation). Inflammation of the liver.

Hepatiza'tion (Gr. ἧπαρ, *hēpar*, the liver). A diseased condensation of parts of the body, or the lungs, so that they resemble liver.

Hepato- (Gr. ἧπαρ, *hēpar*, the liver). A prefix in compound words, signifying connection with, or relation to, the liver.

Hepatoga'stric (Gr. ἧπαρ, *hēpar*, the liver; γαστηρ, *gastēr*, the stomach). Belonging to the liver and stomach.

Hep'tagon (Gr. ἑπτα, *hepta*, seven; γωνια, *gōnia*, an angle). A figure of seven sides and seven angles.

Heptagyn'ia (Gr. ἑπτα, *hepta*, seven; γυνη, *gunē*, a female). A Linnæan order of plants, having seven pistils.

Heptan'dria (Gr. ἑπτα, *hepta*, seven; ἀνηρ, *anēr*, a man). A Linnæan class of plants, having seven stamens.

Heptas'tichous (Gr. ἑπτα, *hepta*, seven; στιχος, *stichos*, a row). In seven rows; in *botany*, applied to the arrangement of leaves in seven spiral rows, the eighth leaf in the series being placed above the first.

Herba'ceous (Lat. *herba*, a herb). Pertaining to herbs; applied to plants which perish yearly, at least as far as the root.

Herbiv'orous (Lat. *herba*, a herb; *voro*, I devour). Feeding on vegetables.

Her'borize (Lat. *herba*, a herb). To search for plants for scientific purposes.

Hered'itary (Lat. *hæres*, a heir). Acquired from ancestors; transmitted from parents to children.

Hermaph'rodite (Gr. Ἑρμης, *Hermēs*, Mercury; Αφροδιτη, *Aphrodi'tē*, Venus). Partaking of both male and female natures in the same individual.

Hermeneu'tic (Gr. ἑρμενευω, *hermeneu'ō*, I interpret; from Ἑρμης, *Hermēs*, Mercury). Relating to interpretation or explanation.

Hermeneu'tics (Gr. ἑρμενευω, *hermeneu'o*, I interpret). The art of explaining the meaning of a writing.

Hermet'ically (Gr. Ἑρμης, *Hermēs*, the supposed inventor of chemistry). Chemically; a vessel is hermetically sealed, when the neck is heated to melting, and closed by pincers until it is air-tight.

Her'nia (Gr. ἑρνος, *hernos*, a branch). A protrusion of any organ of the body from the cavity containing it.

Herpes (Gr' ἑρπω, *herpō*, I creep). Tetters or shingles; an eruptive spreading disease of the skin.

Herpet'ic (Gr. ἑρπω, *herpō*, I creep). Relating to, or of the nature of herpes.

Herpetol'ogy (Gr. ἑρπετον, *her'peton*, a reptile; λογος, *logos*, discourse). The description of reptiles.

Het'ero- (Gr. ἑτερος, *het'eros*, another). A prefix in many compound words, signifying another, or different.

Heterocer'cal (Gr. ἑτερος, *het'eros*, another; κερκος, *kerkos*, a tail). A term applied to fishes in which the caudal fin, or tail, is unsymmetrical; arising from the prolongation of the vertebral column into its upper lobe.

Het'eroclite (Gr. ἑτερος, *het'eros*, another; κλινω, *klinō*, I bend). Leaning another way; applied to words which depart from the ordinary form in declension or conjugation.

Heterod'romous (Gr. ἑτερος, *het'eros*, another; δρομος, *drom'os*, course). In *botany*, applied to the arrangement of leaves in branches in a different manner from the stem.

Heterog'amous (Gr. ἑτερος, *het'eros*, another; γαμος, *gamos*, marriage). Having florets of different sexes on the same flower-head.

Heterogan'gliate (Gr. ἑτερος, *het'eros*, another; γαγγλιον, *gan'glion*, a knot or nervous ganglion). Having the nervous ganglia scattered unsymmetrically; applied to the molluscous invertebrate animals.

Heteroge'neous (Gr. ἑτερος, *het'eros*, another; γενος, *genos*, kind). Unlike in kind ; consisting of elements of different nature.

Heterome'ra (Gr. *ἑτερος, het'eros,* another ; *μηρον, mēron,* a thigh). A section of coleopterous insects, having five joints in the four anterior tarsi, and one joint less in the hind tarsi.

Heteromor'phous (Gr. *ἑτερος, het'eros,* another ; *μορφη, morphē,* form). Having an irregular or unusual form ; applied to the larvæ of insects which differ in form from the imago.

Het'eropa (Gr. *ἑτερος, het'eros,* another ; *πους, pous,* a foot). A section of amphipodous crustacea, having fourteen legs, of which at least the four posterior are fitted only for swimming.

Heterophyl'lous (Gr. *ἑτερος, het'eros,* another ; *φυλλον, phullon,* a leaf). Having two different kinds of leaves on the same stem.

Het'eropods (Gr. *ἑτερος, het'eros,* another ; *πους, pous,* a foot). An order of gasteropodous molluscous animals, in which the foot forms a vertical plate, serving as a fin.

Heterop'tera (Gr. *ἑτερος, het'eros,* another ; *πτερον, pteron,* a wing). A section of hemipterous insects, having the wing-cases membranous at the end.

Heterorhi'zal (Gr. *ἑτερος, het'eros,* another ; *ριζα, rhiza,* a root). In *botany,* applied to acotyledonous plants, because their roots arise from every part of the cellular axis or spore.

Heteros'cian (Gr. *ἑτερος, het'eros,* one of two ; *σκια, skia,* a shadow). Having a shadow only in one direction ; applied to the inhabitants of the earth between the tropics and polar circles.

Heterot'ropous (Gr. *ἑτερος, het'eros,* another ; *τρεπω, trep'ō,* I turn). Turned another way ; applied to the embryo of seeds when it lies in an oblique position.

Hex'agon (Gr. *ἑξ, hex,* six ; *γωνια, gōnia,* an angle). A figure having six sides and six angles.

Hexagyn'ia (Gr. *ἑξ, hex,* six ; *γυνη,*

gunē, a female). A Linnæan order of plants, having six pistils.

Hexahed'ron (Gr. *ἑξ, hex,* six ; *ἑδρα, hedra,* a base). A regular solid body of six sides ; a cube.

Hexam'eter (*ἑξ, hex,* six ; *μετρον, metron,* a measure). A verse in ancient poetry consisting of six feet, as in the Iliad and Æneid.

Hexan'dria (Gr. *ἑξ, hex,* six ; *ανηρ, anēr,* a man). A Linnæan class of plants having six stamens.

Hexan'gular (Gr. *ἑξ, hex,* six ; Lat. *an'gulus,* an angle). Having six angles.

Hex'apod (Gr. *ἑξ, hex,* six ; *πους, pous,* a foot). Having six feet.

Hex'astyle (Gr. *ἑξ, hex,* six ; *στυλος, stulos,* a pillar). A building with six columns in front.

Hia'tus (Lat. *hio,* I gape). An opening or chasm ; the effect produced by the uttering of similar vowel sounds in succession.

Hiber'nate (Lat. *hiber'nus,* belonging to winter). To pass the winter in a torpid state, as some animals.

Hierat'ic (Gr. *ἱερος, hi'eros,* sacred). Sacred ; applied to the characters used in writing by the ancient Egyptian priests.

Hieroglyph'ic (Gr. *ἱερος, hi'eros,* sacred ; *γλυφω, gluphō,* I carve). A sacred character ; the representation of animals and other objects used by the ancient Egyptians to represent words and ideas.

High-pressure Engine. A steam-engine in which the direct power of steam is used, or that produced by the evaporation of water.

Hilum (Lat. the black of a bean). The scar marking the union of a seed with the fruit.

Hippocrat'ic (Gr. *Ἱπποκρατης, Hippocratēs,* an ancient physician). Pertaining to Hippocrates ; applied to the appearance of the face indicative of approaching death, as described by him.

Hippopathol'ogy (Gr. *ἱππος, hippos,* a horse ; *pathology*). The doctrine or description of the diseases of horses.

Hippu'ric (Gr. *ἱππος, hippos,* a horse ;

ουρον, ouron, urine). A term applied to an acid existing in the urine of horses.

Hippu'rites (Gr. *ιππος, hippos*, a horse; *ουρα, oura*, a tail). A genus of plants in the coal-formation, resembling the hippuris or mare's tail.

Hirsute (Lat. *hirsu'tus*, hairy). In *botany*, applied to plants having long, distinct, and tolerably soft hairs.

His'pid (Lat. *his'pidus*, rough). Shaggy or prickly; in *botany*, applied to plants having long soft hairs.

Histogen'esis or **Histog'eny** (Gr. *ιστος, histos*, a tissue; *γενναω, gennaō*, I produce). The formation of organic tissues.

Histolog'ical (Gr. *ιστος, histos*, a tissue; *λογος, logos*, discourse). Relating to histology or the description of tissues.

Histol'ogy (Gr. *ιστος, histos*, a tissue; *λογος, logos*, discourse). The description of the tissues which form an animal or plant.

His'tory (Gr. *ιστορεω, historeō*, I learn by inquiry). A narration of events; a description of things that exist.

Homo- (Gr. *ὁμος, homos*, the same). A prefix in compound words, signifying identity or exact similarity.

Homocen'tric (Gr. *ὁμος, homos*, the same; *κεντρον, kentron*, a centre). Having the same centre.

Homocer'cal (Gr. *ὁμος, homos*, the same; *κερκος, ker'kos*, a tail). Having a symmetrical tail; applied to fishes.

Homod'romous (Gr. *ὁμος, homos*, similar; *δρομος, drom'os*, a course). In *botany*, applied to the arrangement of leaves on branches in the same manner as on the stem.

Homœ'o- (Gr. *ὁμοιος, homoi'os*, similar). A prefix in compound words, implying similarity but not identity.

Homœomer'ic (Gr. *ὁμοιος, homoi'os*, similar; *μερος, meros*, a part). Having or relating to similarity of parts.

Homœop'athy (Gr. *ὁμοιος, homoi'os*,

similar; *παθος, pathos*, suffering). A system by which it is alleged that diseases can be cured by doses of substances capable of exciting similar diseased states in healthy persons.

Homogan'gliate (Gr. *ὁμος, homos*, the same; *γαγγλιον, gan'glion*, a knot or nervous ganglion). Having the nervous ganglia arranged symmetrically; applied to the articulated invertebrate animals.

Homogen'eous (Gr. *ὁμος, homos*, the same; *γενος, genos*, a kind). Of the same kind; consisting of elements of a like nature.

Homol'ogous (Gr. *ὁμος, homos*, the same; *λογος, logos*, reasoning). Constructed on the same plan, though differing in form and function.

Hom'ologue (Gr. *ὁμος, homos*, the same; *λογος, logos*, reasoning). The same part or organ, as far as its anatomical relation is concerned, although differing in form and functions; as the arms of man, the wings of birds, and the pectoral fins of fishes.

Homol'ogy (Gr. *ὁμος, homos*, the same; *λογος, logos*, reasoning). The doctrine of the corresponding relations of parts in different beings, having the same relations but different functions; affinity depending on structure, and not on similarity of form or use.

Homomor'phous (Gr. *ὁμος, homos*, the same; *μορφη, morphē*, form). Of similar form; applied to certain insects of which the larva is like the perfect insect, but without wings.

Homop'oda (Gr. *ὁμος, homos*, the same; *πους, pous*, a foot). A section of amphipodous crustaceans, having fourteen feet all terminated by a hook or point.

Homop'tera (Gr. *ὁμος, homos*, the same; *πτερον, pteron*, a wing). Having the four wings alike; restricted to a section of the hemipterous class of insects.

Hom'otype (Gr. *ὁμος, homos*, the same; *τυπος, tupos*, a type). A part homologous with another in a series.

Ho'rary (Lat. *hora*, an hour). Relating to, or denoting an hour.

Hori'zon (Gr. ὁριζω, *hori'zō*, I bound). The line in the celestial hemisphere which bounds the view on the surface of the earth.

Horizon'tal (*Horizon*). Parallel to the horizon.

Hornblende (Germ. *blenden*, to dazzle). A mineral, generally of a black or dark green colour, found frequently in granitic and trappean rocks.

Ho'rologe (Gr. ὡρα, *hōra*, an hour; λεγω, *legō*, I describe). An instrument for indicating the hours of the day.

Horol'ogy (Gr. ὡρα, *hōra*, an hour; λεγω, *legō*, I tell.) The art of constructing machines for indicating time.

Horom'etry (Gr. ὡρα, *hōra*, an hour; μετρον, *metron*, a measure). The art of measuring time by hours.

Horse-power. The power of a horse, estimated as equal to the raising of 33,000 pounds one foot high per minute, used in calculating the power of steam-engines.

Horse-shoe Magnet. An artificial magnet, in the form of a horse-shoe.

Horticul'ture (Lat. *hortus*, a garden; *colo*, I cultivate). The art of cultivating gardens.

Hortus Siccus (Lat. a dry garden). A collection of dried plants.

Hot Blast. A current of heated air thrown into a furnace.

Hu'mate (Lat. *humus*, the ground). A compound of humic acid with a base.

Humecta'tion (Lat. *humec'to*, I moisten). A making wet.

Hu'meral (Lat. *hu'merus*, the shoulder). Belonging to the humerus, or upper part of the arm above the elbow.

Hu'merus (Lat. the shoulder). The arm from the shoulder to the elbow; the bone of this part.

Hu'mic (*Humus*). Belonging to humus; applied to an acid produced from the decomposition of humus by alkalies.

Hu'moral (Lat. *humor*, moisture). Belonging to humours or fluids : *in medicine*, humoral pathology is the doctrine which attributes diseases to a disordered state of the fluids of the body.

Humour (Lat. *humor*, moisture). Moisture ; in *anatomy*, applied to certain parts of the eye which abound in fluid.

Hu'mus (Lat. soil). The common vegetable mould or soil, consisting of carbon, hydrogen, and oxygen, arising from the decay of vegetable matter.

Hy'ades (Gr. ὑω, *huō*, I rain). A cluster of five stars in the Bull's Head, supposed by the ancients to bring rain.

Hy'aline (Gr. ὑαλος, *hu'alos*, glass). Like glass ; transparent.

Hy'aloid (Gr. ὑαλος, *hu'alos*, glass ; εἰδος, *eidos*, form). Resembling glass ; transparent.

Hy'bodonts (Gr. ὑβος, *hu'bos*, humped; ὁδους, *odous*, a tooth). A family of fossil shark-like fishes with knobbed teeth.

Hy'brid (Gr. ὑβρις, *hubris*, force or injury). The offspring of two animals or plants of different varieties or species ; in *etymology*, applied to words compounded from different languages.

Hydat'id (Gr. ὑδωρ, *hudōr*, water). A transparent vesicle filled with water ; often applied to parasitic animal growth found in the liver and other organs.

Hydrac'id (*Hydrogen* ; Lat. *ac'idus*, acid). An acid containing hydrogen as one of its forming elements.

Hy'dragogue (Gr. ὑδωρ, *hudōr*, water ; αγω, *agō*, I lead). Producing a discharge of fluid ; applied to certain medicines.

Hy'drate (Gr. ὑδωρ, *hudōr*, water); A compound body in which water exists in chemical combination.

Hydrau'lic (Gr. ὑδωρ, *hudōr*, water ; αὐλος, *aulos*, a pipe). Relating to the conveyance of water through pipes.

G

Hydrau'lic Depth. The depth which a volume of flowing water would take in a channel, whose breadth is equal to the outline of the bottom and sides of the actual bed.

Hydrau'lic Head. The measure of a given hydraulic pressure, expressed in terms of the height of a barometrical column of the fluid.

Hydrau'lic Press. A machine in which powerful pressure is produced by water forced into a cylinder, and therein acting on a piston which raises a table on which the material to be pressed is placed.

Hydrau'lic Pressure. The pressure which a liquid moving in a closed channel, exerts on the surfaces by which it is confined.

Hydrau'lics (Gr. ὕδωρ, *hudōr*, water; αὐλος, *aulos*, a pipe). The science which teaches the application of the knowledge of the forces influencing the motion of fluids, to their conveyance through pipes and canals.

Hydrenceph'alocele (Gr. ὕδωρ, *hudōr*, water; ἐγκεφαλον, *enkeph'alon*, the contents of the skull; κηλη, *kēlē*, a tumour). A hernial protrusion from the head containing water.

Hydrenceph'aloid (Gr. ὕδωρ, *hudōr*, water; ἐγκεφαλον, *enkeph'alon*, the brain; εἰδος, *eidos*, from). Resembling hydrocephalus or dropsy of the brain.

Hydri'odate (*Hy'drogen* and *Iodine*). A compound of hydriodic acid with a base; now described by chemists as an iodide, or compound of iodine with a metal, together with an equivalent of water.

Hydriod'ic (*Hy'drogen* and *I'odine*). Consisting of hydrogen and iodine.

Hydro- (Gr. ὕδωρ, *hudōr*, water). A prefix implying the existence of water; but, in chemical terms, implying that hydrogen is a component part of the substance.

Hydrocar'bon (*Hy'drogen* and *Carbon*). A compound of carbon and hydrogen.

Hydrocar'buret (*Hy'drogen* and *Carbon*). A compound of carbon and hydrogen.

Hydroceph'alus (Gr. ὕδωρ, *hūdōr*, water; κεφαλη, *keph'alē*, the head). A disease characterised by the presence of water within the head; a dropsy of the membranes covering the brain.

Hydrochlo'rate (*Hy'drogen* and *Chlorine*). A compound of hydrochloric acid with a base: now described by chemists as a compound of chlorine with a metal, together with an equivalent of water.

Hydrochlo'ric (*Hy'drogen* and *Chlorine*). Consisting of hydrogen and chlorine.

Hy'drocy'anate (*Hy'drogen* and *Cyan'ogen*). A compound of hydrocyanic acid with a base: now described by chemists as a compound of cyanogen and a metal, together with an equivalent of water.

Hydrocyan'ic (*Hy'drogen* and *Cyanogen*). Consisting of hydrogen and cyanogen.

Hydrodynam'ics (Gr. ὕδωρ, *hudōr*, water; δυναμις, *du'namis*, force). The science which treats of the motion of liquids and the causes influencing it.

Hydro-elec'tric (Gr. ὕδωρ, *hudōr*, water; *electric*). A term applied to a machine in which electricity is developed by the action of the steam of water; also to the voltaic current into the combination of which a liquid element enters.

Hydrofiu'ate (*Hy'drogen* and *Flu'orine*). A compound of hydrofluoric acid with a base.

Hydrofluor'ic (*Hy'drogen* and *Flu'orine*). Consisting of hydrogen and fluorine.

Hy'drogen (Gr. ὕδωρ, *hudōr*, water; γεννaω, *gennaō*, I produce). The lightest of elementary bodies; a colourless combustible gas, which, with oxygen, forms water.

Hydrog'rapher (Gr. ὕδωρ, *hudōr*, water; γραφω, *graphō*, I write). A person who describes the physical or geographical conformation of seas or other bodies of water.

Hydrog'raphy (Gr. ὕδωρ, *hudōr*, water; γραφω, *graphō*, I write). The science of describing the physical or geographical conformation of seas, lakes, and other bodies of water.

Hydrol'ogy (Gr. ὕδωρ, *hudōr*, water; λογος, *logos*, discourse). The science which describes water.

Hydrom'eter (Gr. ὕδωρ, *hudōr*, water; μετρον, *metron*, a measure). An instrument for measuring the specific gravity of fluids.

Hydromet'rograph (Gr. ὕδωρ, *hudōr*, water; μετρον, *metron*, a measure; γραφω, *graphō*, I write). An instrument for recording the quantity of water discharged from a pipe or orifice in a given time.

Hydropericar'dium (Gr. ὕδωρ, *hudōr*, water; *pericar'dium*). Dropsy of the pericardium or covering membrane of the heart.

Hydropho'bia (Gr. ὕδωρ, *hudōr*, water; φοβος, *phob'os*, fear). A disease characterised by a dread of water.

Hy'drophyte (Gr. ὕδωρ, *hudōr*, water; φυω, *phuō*, I grow). A plant which grows in the water.

Hydro-salts (Gr. ὕδωρ, *hudōr*, water). A name given to salts, the acid or base of which contains hydrogen.

Hydrostat'ic (Gr. ὕδωρ, *hudōr*, water; ἱστημι, *histēmi*, I place). Relating to the pressure of fluids at rest.

Hydrostat'ic Pressure. The pressure of water or any fluid, at rest, on a given surface.

Hydrostat'ics (Gr. ὕδωρ, *hudōr*, water; ἱστημι, *histēmi*, I make to stand). The science which treats of the properties of fluids at rest.

Hydrosul'phuret (*Hydrogen* and *Sulphur*). A compound of hydro-sulphuric acid with a base: now described by chemists as a sulphide, or compound of sulphur with a metal, together with an equivalent of water.

Hydrotho'rax (Gr. ὕδωρ, *hudōr*, water; θωραξ, *thōrax*, the chest). A disease characterised by the presence of water in the chest; dropsy of the chest.

Hydrous (Gr. ὕδωρ, *hudōr*, water). Containing water; watery.

Hydrozo'a (Gr. ὕδρα, *hudra*, a water-serpent; ζωον, *zōon*, an animal).

The polypes which are organised like the hydra.

Hyetog'raphy (Gr. ὑετος, *hu'etos*, rain; γραφω, *graph'ō*, I write). The science of rain; the knowledge of the quantities and localities in which rain has fallen in a given time.

Hyg'iene (Gr. ὑγιης, *hu'giēs*, healthy). The science which treats of the preservation of health.

Hygien'ic (Gr. ὑγιης, *hu'giēs*, healthy). Relating to the health and its preservation.

Hygro- (Gr. ὑγρος, *hu'gros*, moist). A prefix in compound words, implying moisture.

Hygrom'eter (Gr. ὑγρος, *hugros*, moist; μετρον, *metron*, a measure). An instrument for measuring the amount of moisture in the atmosphere.

Hygromet'ric (Gr. ὑγρος, *hugros*, moist : μετρον, *metron*, a measure). Relating to the measurement of the moisture in the air; readily absorbing moisture from the air.

Hygrom'etry (Gr. ὑγρος, *hu'gros*, moist; μετρον, *metron*, a measure). The branch of meteorological science which treats of the measuring the pressure, quantity, and effects, of watery vapour in the atmosphere.

Hy'groscope (Gr. ὑγρος, *hugros*, moist; σκοπεω, *skop'eō*, I view). An instrument for ascertaining approximatively the moisture of the atmosphere.

Hygroscop'ic (Gr. ὑγρος, *hugros*, moist; σκοπεω, *skop'eō*, I view). Liable to absorb moisture from the air.

Hyme'nium (Gr. ὑμην, *humēn*, a membrane). The mass formed by the union of the organs of fructification in the mushroom tribe.

Hymenop'tera (Gr. ὑμην, *humēn*, a membrane; πτερον, *pter'on*, a wing). An order of insects having fine membranous wings, as bees and wasps.

Hy'o- (The Greek letter υ, or *upsilon*). In *anatomy*, a prefix in compound words, implying connection with the hyoid bone.

Hy'oid (The Greek letter υ, or *upsilon*; εἶδος, shape). Resembling the letter υ; applied to the bone which supports the tongue, from its shape.

Hypæ'thral (Gr. ὑπο, *hupo*, under ; αἰθηρ. *aithēr*, the air). Exposed to the open air ; without a roof.

Hypal'lage' (Gr. ὑπο, *hupo*, under : ἀλλασσω, *allas'sō*, I exchange). In *grammar*, an interchange of cases ; as an accusative of the thing given and a dative of the recipient, for an accusative of the recipient and a dative of the thing given.

Hypapoph'ysis (Gr. ὑπο, *hupo*. under ; *apoph'ysis*). An apophysis of a vertebra growing downwards.

Hyper- (Gr. ὑπερ, *huper*, above). A preposition signifying excess in compound words.

Hyperæ'mia (Gr. ὑπερ, *huper*, beyond ; αἱμα, *haima*, blood). An excessive supply of blood.

Hyperæm'ic (Gr. ὑπερ, *huper*, beyond ; αἱμα, *haima*, blood). Relating to, or having an excessive supply of blood.

Hyperæsthe'sia (Gr. ὑπερ, *huper*, beyond ; αἰσθανομαι, *aisthan'omai*, I feel). Excessive sensibility.

Hyper'baton (Gr. ὑπερ, *huper*, beyond ; βαινω, *bainō*, I go). A figure in *grammar*, in which the natural order of words or sentences is inverted.

Hyper'bola (Gr. ὑπερ, *huper*, beyond ; βαλλω, *ballō*, I throw). A curve formed by the section of a cone by a plane passing parallel to its axis.

Hyper'bole' (Gr. ὑπερ, *huper*, beyond ; βαλλω, *ballō*, I throw). A figure of speech, characterised by exaggeration, or the representation of the qualities of an object as greater or less than they really are.

Hyper'boloid (*Hyper'bola ;* Gr. εἶδος, *eidos*, form). A solid formed by the revolution of an hyperbola about its axis.

Hyperbor'ean (Gr. ὑπερ, *huper*, beyond ; βορεας, *bor'eas*, the north wind). Dwelling far to the north.

Hypercathar'sis (Gr. ὑπερ, *huper*, beyond ; καθαιρω, *kathai'rō*, I cleanse). Excessive purgation.

Hyperino'sis (Gr. ὑπερ, *huper*, beyond ; ἰς, *is*, force or fibre). A state characterised by an excessive formation of fibrine in the blood.

Hyper'trophy (Gr. ὑπερ, *huper*, beyond ; τρεφω, *trephō*, I nourish). Excessive growth of a part.

Hypo- (Gr. ὑπο, *hupo*, under). A preposition implying diminution or inferiority, in quality or situation.

Hypocarpoge'an (Gr. ὑπο, *hupo*, under ; καρπος, *karpos*, fruit ; γη, *gē*, the earth). Producing fruit under ground.

Hypochon'drium (Gr. ὑπο, *hupo*, under ; χονδρος, *chondros*, a cartilage). The part of the abdomen which lies under the cartilages of the lower ribs.

Hypochondri'asis (Gr. ὑποχονδρια, *hupochon'dria*, the *hypochondria*, because formerly supposed to be connected with this region). A form of insanity, in which the patient converts an idea of purely mental origin into what appears to him to be a real material change.

Hypocrater'iform (Gr. ὑπο, *hupo*, under ; κρατηρ, *kratēr*, a cup ; Lat. *forma*, shape). Shaped like a saucer or salver.

Hypogas'tric (Gr. ὑπο, *hupo*, below ; γαστηρ, *gastēr*, the stomach). Relating to the middle part of the abdomen.

Hypoge'al (Gr. ὑπο, *hupo*, under ; γη, *gē*, the earth). Under the earth.

Hy'pogene (Gr. ὑπο, *hupo*, under ; γενναω, *genna'ō*, I produce). A term proposed to be applied to the primary strata in geology, to denote their formation from below.

Hypoglos'sal (Gr. ὑπο, *hupo*, under ; γλωσσα, *glōssa*, the tongue). Under the tongue.

Hypog'ynous (Gr. ὑπο, *hupo*, under ; γυνη, *gunē*, a female). Inserted beneath the pistil.

Hypophos'phite (*Hypophos'phorous*). A compound of hypophosphorous acid with a base.

Hypophos'phorous (Gr. ὑπο, *hupo*, under ; *phos'phorus*). A name applied to an acid which contains less oxygen than phosphorous acid.

Hypo'pion (Gr. ὑπο, *hupo*, under; ὠψ, *ōps*, the eye). A collection of pus in the anterior part of the eye.

Hyposul'phate (Gr. ὑπο, *hupo*, under; *sulphate*). A compound of hyposulphuric acid with a base.

Hyposul'phite (Gr. ὑπο, *hupo*, under; *sulphite*). A compound of hyposulphurous acid with a base.

Hyposulphu'ric (Gr. ὑπο, *hupo*, under; *sulphu'ric*). Applied to an acid containing less oxygen than sulphuric and more than sulphurous acid.

Hyposul'phurous (Gr. ὑπο, *hupo*, under : *sul'phurous*). Applied to an acid containing less oxygen than sulphurous acid.

Hypoth'enuse, or, more correctly, **Hypot'enuse** (Gr. ὑπο, *hupo*, under; τεινω, *teinō*, I stretch). The side of a right-angled triangle which subtends or is opposite to the right angle.

Hypoth'esis (Gr. ὑπο, *hupo*, under; τιθημι, *tithēmi*, I place). An explanation of phenomena, not founded on the actual observation of facts, but assumed in order to demonstrate a point in question.

Hypozo'ic (Gr. ὑπο, *hupo*, under; ζωον, *zōon*, an animal). A term applied in *geology* to the rocks in which no organic remains have been found.

Hypsom'etry (Gr. ὑψος, *hup'sos*, height; μετρον, *metron*, measure). The art of measuring the heights of places on the earth, by the barometer or by trigonometrical observations.

Hysteran'thous (Gr. ὑστερος, *hus'teros*, later; ἀνθος, *anthos*, a flower). In *botany*, applied to plants of which the leaves expand after the flowers have opened.

Hyste'ria. A diseased state, consisting in a morbid condition of the nervous centres, giving rise to paroxysmal symptoms, and to the imitation of various diseases.

I.

Iam'bic (Gr. ιαμβος, *iam'bos*). Relating to or consisting of the iambus.

Iam'bus (Gr. ιαμβος, *iam'bos*). A foot in verse consisting of a short syllable followed by a long one.

Ia'tro- (Gr. ιατρος, *ia'tros*, a physician). A part of some compound words, signifying a connection with medicine or physicians.

-Ic. In *chemistry*, a termination denoting the acid containing most oxygen, when more than one is formed from the same element.

Iceberg (*Ice*; Germ. *berg*, a mountain). A mountain or hill of ice.

Ich'nites (Gr. ιχνος, *ichnos*, a footstep). In *geology*, fossil foot-prints.

Ich'nolite (Gr. ιχνος, *ichnos*, a footstep; λιθος, *lithos*, a stone). A stone retaining the impression of the foot mark of a fossil animal.

Ichnol'ogy (Gr. ιχνος, *ichnos*, a footstep; λογος, *logos*, a discourse). The science of fossil foot-prints.

Ichor (Gr. ιχωρ, *ichēr*). A thin watery humour.

Ich'thyic (Gr. ιχθυς, *ichthus*, a fish). Relating to fishes.

Ichthyodor'ulites (Gr. ιχθυς, *ichthus*, a fish; δορυ, *doru*, a spear; λιθος, *lithos*, a stone). Fossil spines of fishes.

Ich'thyoid (Gr. ιχθυς, *ichthus*, a fish; εἰδος, *eidos*, shape). Like a fish; applied to certain saurian reptiles.

Ich'thyolite (Gr. ιχθυς, *ichthus*, a fish; λιθος, *lithos*, a stone). A fossil fish, or portion of a fish.

Ichthyol'ogy (Gr. ιχθυς, *ichthus*, a fish; λογος, *logos*, a discourse). The description of fishes.

Ichthyoph'agous (Gr. ιχθυς, *ichthus*, a fish; φαγω, *phagō*, I eat). Living on fishes as food.

Ichthyopteryg'ia (Gr. ιχθυς, *ichthus*, a fish; πτερυγιον, *pteru'gion*, a fin). An order of fossil reptiles with limbs formed for swimming, like fins.

Ichthyosau'rus (Gr. ἰχθυς, *ichthus*, a fish ; σαυρος, *sauros*, a lizard). A fossil animal, having a structure between that of a lizard and a fish.

Ichthyo'sis (Gr. ἰχθυς, *ichthus*, a fish). A disease in which the body, or parts of it, are covered by scales overlapping each other like those of a fish.

Icosahed'ron (Gr. εἰκοσι, *ei'kosi*, twenty ; ἑδρα, *hedra*, a base). A figure having twenty sides or faces.

Icosan'dria(Gr. εἰκοσι, *ei'kosi*, twenty; ανηρ, *anēr*, a man). A class of plants having twenty or more stamens on the calyx.

Icter'ic (Lat. *ic'terus*, jaundice). Relating to, or affected with jaundice.

Ic'terus (Lat.). The jaundice.

-Idæ (Greek termination -ιδης, -*idēs*, signifying descent). A termination employed in *zoology*, signifying some degree of likeness to the animal to the name of which the termination is affixed.

-Ide. A termination applied in *chemistry*, to denote combinations of non-metallic elements with metals, or with other non-metallic elements.

Idea (Gr. εἰδω, *eidō*, I see). An image or model formed in the mind.

Ide'alism (*Idea*). A system of philosophy, according to which what we call external objects are mere conceptions of the mind.

Ideograph'ic (Gr. ἰδεα, *idea* ; γραφω, *graphō*, I write). Expressing ideas.

Id'iocy (Gr. ἰδιωτης, *idiōtēs*, a private or ignorant person). A state of defective intellect existing from birth.

Idioelec'tric (Gr. ἰδιος, *id'ios*, peculiar or separate ; *electric*). Having the property of manifesting electricity on friction.

Id'iom (Gr. ἰδιος, *id'ios*, proper or peculiar). The form of speech peculiar to a country.

Idiomat'ic (Gr. ἰδιος, *id'ios*, proper or peculiar). Pertaining to the particular modes of expression belonging to a language.

Idiopath'ic (Gr. ἰδιος, *id'ios*, peculiar ; παθος, *path'os*, suffering). Applied to diseases which arise without any apparent exciting cause.

Idiosyn'crasy (Gr. ἰδιος, *idios*, peculiar ; συγκρασις, *sunkra'sis*, a mixing together). A peculiarity of constitution and susceptibility. The disposition or habit of body characteristic, belonging to, and distinguishing an individual. Idiocrasy.

Id'iot (Gr. ἰδιωτης, *idiōtēs*, a private or ill-informed person). A person whose intellect is altogether deficient from birth.

Idol (Gr. εἰδωλον, *eidōlon*, an image, phantom, or fancy). A term used by Bacon to denote fallacies of the mind.

Idols of the Den. The mental fallacies arising from the nature of the mind and body of the individual.

Idols of the Market. The fallacies arising from reciprocal intercourse, and the popular application of words and names.

Idols of the Theatre. The fallacies arising from false theories or perverted laws of demonstration.

Idols of the Tribe. The fallacies inherent in human nature.

Ig'neous (Lat. *ignis*, fire). Arising from, or connected with fire ; in *geology*, applied to the apparent results of subterraneous heat.

Ignis Fat'uus (Lat. foolish fire). A luminous appearance sometimes seen at night, and produced by the combustion of phosphorus which has escaped from organic matter.

Ignit'ion (Lat. *ignis*, fire.) A setting on fire.

I'leo- (*Ileum*). In *anatomy*, a prefix denoting connection with, or relation to, the intestine called ileum.

Ileo-cœ'cal (*Ileum* ; *cœcum*). Belonging to, or lying between, the ileum and cœcum.

Il'eum (Gr. εἰλεω, *ei'leō*, I roll). The lower portion of the small intestines.

Il'iac (Lat. *i'lia*, the flank). Belonging to the ileum ; or to the bone called ilium.

Il'io- (*Ilium*). In *anatomy*, a prefix denoting connection with, or relation to, the iliac bone.

Il'ium (Lat. *i'lia*, the flank). The large partly flattened bone which forms the principal part of the pelvis, and enters into the composition of the hip-joint.

Illa'tive (Lat. *in*, on ; *la'tus*, borne). Denoting an inference ; applied in *logic*, where the truth of the converse follows from the truth of the proposition itself.

Ima'go (Lat. an image). A name given to the perfect state of an insect.

Imbecil'ity (Lat. *in*, on ; *bacil'lus*, a staff). Weakness : a defective state of intellect, not amounting to idiocy.

Im'bricate (Lat. *imbrex*, a tile). Lying over each other like tiles ; in *botany*, applied to the arrangement in the bud in which the outer leaves successively overlap the inner.

Immer'sion (Lat. *in*, in ; *mergo*, I dip). A putting beneath the surface, as of a fluid ; in *astronomy*, the entrance of one body into such a position with regard to another, as to apparently sink into it, and become invisible.

Im'pact (Lat. *in*, on ; *pango*, I drive). A stroke ; the action of two bodies on each other in coming together.

Impal'pable (Lat. *in*, not ; *palpo*, I feel). Incapable of being felt.

Imparisyllab'ic (Lat. *in*, not : *par*, equal ; *syl'laba*, a syllable). Not having the same number of syllables ; applied to nouns which have not the same number of syllables in all their cases.

Impenetrabil'ity (Lat. *in*, not ; *pen'etro*, I pierce). In *physics*, the property in virtue of which a body occupies a certain space, which cannot at the same time be occupied by another body.

Imper'ative (Lat. *im'pero*, I command). Commanding ; in *grammar*, implying a command or entreaty.

Imper'meable (Lat. *in*, not ; *per*, through ; *meo*, I pass). Incapable of being passed through by a fluid.

Imper'sonal (Lat. *in*, not ; *perso'na*, a person). Without persons ; applied to verbs which have only the third person singular.

Imper'vious (Lat. *in*, not ; *per*, through ; *via*, a way). Incapable of being passed through.

Impeti'go (Lat. *im'peto*, I attack). A disease of the skin characterised by clusters of pustules which run together into a crust ; a running tetter.

Im'petus (Lat. from *in*, against ; *peto*, I urge). The force with which a body is driven.

Imping'e (Lat *impin'go*, I strike against). To strike or dash against.

Implu'vium (Lat. *in ;* plu'via, rain). A basin to receive rain, in the middle of the *atrium* or courtyard of ancient Roman houses.

Impon'derable (Lat. *in*, not ; *pondus*, weight). Without perceptible weight.

Impulse (Lat. *in*, on or against ; *pello*, I drive). The effect of one body striking on another, being the result of the motion of the striking body.

Impul'sion (Lat. *in*, against ; *pello*, I drive). The act of driving against : the process by which a moving body changes the motion of another by striking it.

Inan'imate (Lat. *in*, not ; *an'ima*, animal life). Without animal life.

Inanit'ion (Lat. *ina'nis*, empty). Emptiness ; want of nutrition ; starvation.

Inartic'ulate (Lat. *in*, not ; *artic'ulus*, a joint). Not having the power of articulation or speech ; in *botany*, without joints.

Incandes'cence (Lat. *in ;* candes'co, I grow white). A white heat ; the luminous appearance which bodies assume when heated to a certain point.

Incandes'cent (Lat. *in ;* candes'co, I grow white). White or glowing from heat.

Incep'tive (Lat. *incip'io*, I begin).

Beginning ; applied to verbs which imply a commencement of action.

In'cidence (Lat. *in*, on ; *cado*, I fall). A falling on ; in *dynamics* and *optics*, the angle of incidence is the angle made by a body or ray of light falling on an object, with a line drawn perpendicularly to the surface struck.

In'cident (Lat. *in*, on ; *cado*, I fall). Falling on.

Incin'erate (Lat *in*, into ; *cinis*, ashes). To burn to ashes.

Incinera'tion (Lat. *in*, into ; *cinis*, ashes). A burning to ashes.

Incis'ion (Lat. *in*, into ; *cædo*, I cut). A cutting into ; a cut.

Inci'sor (Lat. *in*, into ; *cædo*, I cut). A cutter ; applied to the fore teeth, which cut the food.

Inclina'tion (Lat. *in* ; *clino*, or Gr. κλινω, *klinō*, I lean). A leaning ; in *physics*, the direction of one body with respect to another, as measured by the angle formed at their point of meeting.

Incline (Lat. *in*, towards ; *clino*, I bend). A slope ; the direction of a surface, as of a road, with respect to the horizon.

Inclined Plane. A plane forming an angle, less than a right angle, with the horizon.

Inclu'ded (Lat. *in*, in ; *claudo*, I shut). In *botany*, applied to stamens when they do not project beyond the corolla.

Incombus'tible (Lat. *in*, not; *combu'ro*, I burn up). Incapable of being burned.

Incommen'surable (Lat. *in*, not ; *con*, with ; *mensu'ra*, a measure). Not capable of being measured together ; applied to quantities and magnitudes which do not exactly measure each other, or of which one is not contained a definite number of times in the other ; or which cannot be divided without a remainder by some other number.

Incommis'cible (Lat. *in*, not ; *con*, together; *mis'ceo*, I mix). Incapable of being mixed together.

Incompat'ible (Lat. *in*, not ; *con*, with ; *pat'ior*, I suffer). Not capable of subsisting with something else ; applied to substances which chemically decompose each other when brought into contact in a solution.

Incompressibil'ity (Lat. *in*, not ; *con*, together ; *prem'o*, I press). The property of resisting forcible reduction into a smaller space.

Incompres'sible (Lat. *in*, not ; *con*, together ; *prem'o*, I press). Resisting compression into a smaller space.

Incor'porate (Lat. *in*, into ; *corpus*, a body). To mix into one body or mass.

Incorpora'tion (Lat. *in*, into ; *corpus*, a body). A mixing into one body or mass.

In'crement (Lat. *in* ; *cresco*, I grow). An increase ; in *mathematics*, the quantity by which a variable quantity increases.

Incrusta'tion (Lat. *in*, in ; *crusta*, a crust). The covering of a body with a rough coating, as with a crust.

Incuba'tion (Lat. *in*, on ; *cumbo*, I lie). The act of sitting on eggs for the purpose of hatching young.

Incum'bent (Lat. *in*, on ; *cumbo*, I lie). In *botany*, applied when the radicle lies on the back of the cotyledons.

Incurva'tion (Lat. *in*, towards ; *curvus*, bent). A bending, or turning out of a straight course.

Indecli'nable (Lat. *in*, not ; *de*, from ; *clino*, I bend). Not declinable ; applied to words incapable of being varied by terminations.

Indef'inite (Lat. *in*, not ; *de*, down ; *finis*, an end). Not definite or limited ; in *botany*, applied to inflorescence, in which the central or terminal flower is the last to expand.

Indehis'cent (Lat. *in*, not ; *dehis'co*, I gape). Not gaping ; applied to fruits which do not split open, as the apple.

Indent' (Lat. *in*, in ; *dens*, a tooth). To notch, as if by the teeth, or into inequalities like teeth.

Indent'ed (Lat. *in*, in ; *dens*, a tooth). Notched, as if bitten by teeth, or into margins like teeth.

Indent'ure (Lat. *in*, in ; *dens*, a

tooth). A deed of agreement between two persons, of which the upper edge of the first page has a waving line like a row of teeth.

Indeter'minate (Lat. *in*, not; *de*, down; *ter'minus*, a limit). Not limited; in *mathematics*, applied to problems which admit an unlimited number of solutions; in *botany*, applied to inflorescence with the same meaning as indefinite.

In'dicator (Lat. *in'dico*, I point out). A pointer: applied to the muscle which extends the fore-finger.

Indig'enous (Lat. *in*, in; *gigno*, I produce). Native; produced naturally in a country.

Induc'tion (Lat. *in*, into; *duco*, I lead). A bringing in: the leading an inference or general conclusion from a number of particular instances; in *electricity* and *magnetism*, the process by which an electrified or magnetic body produces an electrical or magnetic state in surrounding bodies.

Inductom'eter(*Induction;* Gr. μετρον, *metron*, a measure). An instrument for measuring differences of electrical induction.

Induc'tive (Lat. *in*, into; *duco*, I I lead). Leading to inferences: applied to those sciences which are based on the observation of facts and the conclusions drawn from them.

Indu'plicate (Lat. *in*, in; *duplex*, double). Doubled inwards: in *botany*, applied to the arrangement of a flower-bud in which the edges of the petals are slightly turned inwards.

Indura'tion (Lat. *in*, into; *durus*, hard). Hardening.

Indu'sium (Lat. *in'duo*, I put on). A covering: in *botany*, the epidermic covering which encloses the spores or analogues of seeds in some ferns.

Inen'chyma (Gr. *ls, is*, fibre; ἐγχυμα, *en'chuma*, a tissue). In *botany*, a tissue consisting of cells with spiral fibres in them.

Iner'tia (Lat. *iners*, inactive). The quality in virtue of which matter is incapable of spontaneous change, whether from motion to rest, or from rest to motion; inactivity.

In'fantile (Lat. *infans*, an infant). Belonging to or occurring in infants.

Infec't (Lat. *infic'io*, I taint). To introduce into a healthy body the emanation or miasma proceeding from one which is diseased, so as to propagate the disease.

Infec'tion (Lat. *infic'io*, I taint). The communication of disease by means of the miasm or emanation proceeding from a diseased body.

Infec'tious (Lat. *infic'io*, I taint). Capable of being communicated by infection.

Infe'rior (Lat. below). In *botany*, applied to the ovary when it is adherent to the calyx, or to the calyx when it is not adherent to the ovary.

Inferobran'chiate (Lat. *in'ferus*, below; Gr. βραγχια, *bran'chia*, gills). Having the gills arranged along the sides of the body under the margin of the mantle: applied to an order of gasteropods.

Infiltra'tion (Lat. *in*, into; *filter*). The process of entering a body through pores; the substance which has so entered.

In'finite (Lat. *in*, not; *finis*, an end). Without a limit; an infinite decimal or series is one which cannot be brought to an end.

Infinites'imal (Lat. *in*, not; *finis*, an end). Indefinitely small: having relation to indefinitely small numbers or quantities.

Infin'itive (Lat. *in*, not; *finis*, I limit). Placing no limit: in *grammar*, applied to that part of the verb which expresses its name.

Inflam'mable (Lat. *in*, into; *flamma*, flame). Capable of being set on fire.

Inflamma'tion (Lat. *in*, into; *flamma*, flame). A getting on fire: in *medicine*, a diseased state, characterised by redness, heat, pain, swelling, and disturbance of the function of a part.

Inflect'ed (Lat. *in*, on; *flecto*, I bend). Bent or turned out of a straight course; curved inwards.

Inflec′tion (Lat. *in*, towards ; *flecto*, I bend). A turning from a straight course : in *optics*, the effect produced by the edges of an opaque body on the light passing in contact with them, by which the rays are bent out of their course either inwards or outwards ; in *grammar*, the variation of words by changes of termination.

Inflex′ible (Lat. *in*, not ; *flecto*, I bend). Incapable of being bent.

Inflores′cence (Lat *in*, in ; *flos*, a flower). The arrangement of flowers on the flowering stem or branch.

Influen′za (Italian, *influenza*, influence). An epidemic catarrh or cold, attended with great loss of strength and severe fever.

Influx (Lat. *in*, into ; *fluo*, I flow). A flowing into.

Infracos′tal (Lat. *infra*, beneath ; *costa*, a rib). Beneath ribs.

Inframaxil′lary (Lat. *infra*, beneath ; *maxil′la*, a jaw). Beneath the jaw.

Infraor′bital (Lat. *infra*, beneath ; *or′bita*, an orbit). Beneath the orbit.

Infraspi′nous (Lat. *infra*, beneath ; *spina*, a spine). Beneath a spine or spinous process.

Infundib′uliform (Lat. *infundib′u-lum*, a funnel ; *forma*, shape). Shaped like a funnel.

Infu′sion (Lat. *in*, on ; *fundo*, I pour). The process of steeping substances in liquid, so as to extract certain qualities from them ; the liquid thus prepared.

Infuso′ria (*Infusion*). A term given to microscopic animals of several orders, found in water in which organic matter has been infused.

Inges′ta (Lat. *in*, in ; *gero*, I carry). Things taken in ; applied to food.

Inglu′vies (Lat. a crop). A crop or partial dilatation of the œsophagus.

In′guinal (Lat. *in′guen*, the groin). Relating or belonging to the groin.

Inhala′tion (Lat. *in*, into ; *halo*, I breathe). A breathing in ; the act of drawing in fumes or vapours with the breath.

Inha′le (Lat. *in*, into ; *halo*, I breathe). To draw in air or vapours by means of the breathing organs.

Inject′ (Lat. *in*, into ; *jac′io*, I throw). To throw into.

Injec′tion (Lat. *in*, into ; *jac′io*, I throw). A throwing in ; a medicine thrown into the body : the act of filling the vessels of a body with some coloured substance, so as to render them distinct ; also the substance thrown in.

Inna′te (Lat. *in*, into or on ; *nascor*, I am born). Natural ; applied to ideas supposed to exist in the mind from birth ; in *botany*, applied to anthers when attached to the top of the filaments.

Innerva′tion (Lat. *in*, into ; *nervus*, a nerve). The properties or functions of the nervous system.

Innom′inate (Lat. *in*, not ; *nomen*, a name). Without a name ; applied to a bone forming the pelvis, constituted of three bones which grow together ; also to a large arterial trunk arising from the aorta.

Inoc′ulate (Lat. *in*, into ; *oc′ulus*, an eye). To engraft buds ; to communicate disease to a person by inserting infectious matter into his skin.

Inoper′cular (Lat. *in*, not ; *oper′culum*, a lid). Without an operculum or lid.

Inor′dinate (Lat. *in*, not ; *or′dino*, I put in order). Irregular : in *mathematics*, applied to two ranks of quantities, which are proportionate in a cross order.

Inorgan′ic (Lat. *in*, not ; *organ′ic*). Without the organs or instruments of life : in *medicine*, not apparently connected with change in structure.

Inos′culate (Lat. *in*, into ; *os′culum*, a little mouth). To open into, as by little mouths.

Insal′ivation (Lat. *in*, into ; *saliva*). The blending of the saliva with the food.

Insa′ne (Lat. *in*, not ; *sanus*, sound or healthy). Unsound in mind.

Insan′ity (Lat. *in*, not ; *sanus*, sound or healthy). A term used to express in general derangements of the mind, except the temporary delirium occasioned by fever.

In'sect (Lat. *in*, into ; *seco*, I cut). A class of invertebrate animals, having a body composed of three distinct parts jointed together, with three pairs of feet, and generally wings.

Insectiv'orous (Lat. *insec'ta*, insects ; *voro*, I devour). Living on insects.

Insensibil'ity (Lat. *in*, not ; *sentio*, I perceive). Loss of the power of feeling or sensation.

Insesso'res (Lat. *in*, on ; *sed'eo*, I sit). An order of birds, including those which habitually perch on trees, excepting the rapacious and the climbing birds ; as the crow, starling, finch, and swallow.

In situ (Lat. *in*, in ; *situs*, a situation). In the place where it was originally formed or deposited.

Insola'tion (Lat. *in*, in ; *sol*, the sun). Exposure to the rays of the sun ; or the effects of such exposure.

Insol'uble (Lat. *in*, not ; *solvo*, I melt). Incapable of being melted.

Inspira'tion (Lat. *in*, into ; *spiro*, I breathe). The act of drawing in air by the lungs.

Inspi'ratory (Lat. *in*, into ; *spiro*, I breathe). Relating to the act of inspiration.

Inspi're (Lat. *in*, into ; *spiro*, I breathe). To draw in air by the breathing organs.

Inspiss'ate (Lat. *in*, in ; *spissus*, thick). To thicken.

In'stinct (Lat. *instin'guo*, I urge on). The power by which, independently of instruction or experience, animals are unerringly directed to do whatever is necessary for their preservation and the continuance of their species, in a manner incapable of modification or improvement by experience.

Instinc'tive (Lat. *instin'guo*, I urge on). Arising from instinct.

In'sulate (Lat. *in'sula*, an island). To separate ; to surround a body with substances incapable of carrying off the electricity or caloric accumulated in it.

Insula'tion (Lat. *in'sula*, an island). The state of being separated or insulated.

In'sulator (Lat. *in'sula*, an island). The substance which prevents the passage of electricity from a body.

In'teger (Lat. entire). The whole : applied especially to whole numbers, in contradistinction from fractions.

In'tegral (Lat. *in'teger*, entire). Entire ; making part of a whole.

Integral Calculus. A branch of mathematical analysis, in which the primitive function is derived from its differentiate, or its differential co-efficient.

In'tegrant (Lat. *in'teger*, entire). Making part of a whole ; applied to parts which are of the same nature as the whole.

In'tellect (Lat. *intel'ligo*, I understand). The faculty of the human mind which receives and comprehends the idea enunciated by the senses or by other means.

Intel'ligence (Lat. *intel'ligo*, I understand). The faculty which leads to the performance of operations as the result of experience, and capable of improvement by exercise.

Interambula'cra (Lat. *inter*, between ; *ambula'crum*). The plates between the perforated plates, or ambulacra, in the echinoderms.

Interartic'ular (Lat. *inter*, between ; *artic'ulus*, a joint). Between joints.

Interauric'ular (Lat. *inter*, between ; *auric'ula*, an auricle). Between the auricles of the heart.

Intercal'ary (Lat. *inter*, between ; *calo* (Gr. καλεω, *kaleō*), I call). Inserted : applied to the day inserted in the calendar every fourth year to compensate for the deficiency in the three preceding years : also to a month inserted in the old Roman calendar to make up a deficiency.

Intercel'lular (Lat. *inter*, between ; *cel'lula*, a cell). Between cells.

Intercep'ted (Lat. *inter*, between ; *cap'io*, I take). Included or comprehended between.

Interclavic'ular (Lat. *inter*, between ; *clav'icle*). Between clavicles.

Intercon'dyloid (Lat. *inter*, between; Gr. κονδυλος, *kon'dulos*, a condyle). Between condyles.

Intercos'tal (Lat. *inter*, between; *costa*, a rib). Between ribs.

Intercur'rent (Lat. *inter*, between; *curro*, I run). Running between; in *medicine*, applied to diseases which occur in a scattered manner during the prevalence of epidemic disorders.

Interdig'ital (Lat. *inter*, between; *dig'itus*, a finger). Between the fingers.

Interfa'cial (Lat. *inter*, between; *fac'ies*, a face). Included between two faces or planes.

Interfe'rence (Lat. *inter*, between; *fero*, I bear). A term applied to the phenomenon of the effacement of an undulation by the meeting of two waves; and in *optics* especially, to the mutual intersection of rays of light under certain conditions, so that they extinguish each other.

Interfo'liar (Lat. *inter*, between; *fo'lium*, a leaf). Between two opposite leaves.

Interganglion'ic (Lat. *inter*, between; Gr. γαγγλιον, *gan'glion*, a knot). Lying or extending between ganglions.

Interhæ'mal (Lat. *inter*, between; Gr. αἱμα, *haima*, blood). Between the hæmal processes in vertebræ.

Interlob'ular (Lat. *inter*, between; *lo'bulus*, a little lobe). Between lobules or little lobes.

Intermaxil'lary (Lat. *inter*, between, *maxil'la*, a jaw). Between the maxillary or jaw bone.

Intermis'sion (Lat. *inter*, between; *mitto*, I send). Temporary cessation as applied to fevers; complete cessation for a time.

Intermit'tent (Lat. *inter*, between; *mitto*, I send). Ceasing for a time; applied to diseases in which the symptoms leave the patient entirely for a time, and then return.

Intermus'cular (Lat. *inter*, between; *mus'culus*, a muscle). Between muscles.

Interneu'ral (Lat. *inter*, between, (Gr. νευρον, *neuron*, a nerve). Between the neural processes in vertebræ.

In'ternode (Lat. *inter*, between; *nodus*, a knot). The space in a stem between the nodes, or parts where the leaves are formed.

Interos'seous (Lat. *inter*, between; *os*, a bone). Between bones.

Interpedun'cular (Lat. *inter*, between; *ped'uncle*). Between peduncles.

Interpet'iolar (Lat. *inter*, between; *pet'iole*). Between petioles of opposite sides.

Interpola'tion (Lat. *inter'polo*, I place between). The insertion of words, passages, or numbers between others.

Interposit'ion (Lat. *inter*, between; *pono*, I put). A placing or coming between.

Intersect' (Lat. *inter*, between; *seco*, I cut). To cut or cross mutually.

Intersec'tion (Lat. *inter*, between; *seco*, I cut). A mutual cutting or crossing.

Interspi'nal or **Interspi'nous** (Lat. *inter*, between; *spina*, a spine). Inserted between the spinous processes of the vertebræ.

Interstel'lar (Lat. *inter*, between; *stella*, a star). Between the stars, beyond the limits of our solar system.

Inter'stice (Lat. *inter*, between; *sto*, I stand). A small space between the parts which compose a body.

Interstit'ial (Lat. *inter*, between; *sto*, I stand). Relating to or occupying interstices; taking place gradually throughout a body.

Interstrat'ified (Lat. *inter*, between; *stratum*, a layer; *fac'io*, I make). Interposed in strata between other bodies.

Intertu'bular (Lat. *inter*, between; *tubule*). Between tubules or small tubes.

Interver'tebral (Lat. *inter*, between; *ver'tebra*, a bone of the spine). Between vertebræ.

Intes'tines (Lat. *intus*, within). The alimentary canal from the stomach to its termination.

Intine (Lat. *intus*, within). The inner covering of the pollen-grain.

Intona'tion (Lat. *in*, in; *tonus*, a

tone). The manner of sounding the notes of a musical scale.

Intracel'lular (Lat. *intra*, within ; *cell'ula*, a cell). Within cells ; applied in *histology* to the formation of cells within cells.

Intralob'ular (Lat. *intra*, within ; *lo'bulus*, a lobule). Within lobules or little lobes.

Intran'sitive (Lat. *in*, not ; *trans*, over ; *eo*, I go). Not passing on : applied to verbs in which the action does not pass to or act on an object.

Intrau'terine (Lat. *intra*, within ; *u'terus*, the womb). Within the uterus or womb.

In'trorse (Lat. *intror'sum*, within). Turned inwards ; in *botany*, applied to anthers which open on the side next the pistil.

Intuit'ion (Lat. *in*, on ; *tu'eor*, I look). The process by which the mind perceives a fact at once, without the intervention of other ideas, or of reasoning.

Intu'itive (Lat. *in*, on ; *tu'eor*, I look). Perceived immediately by the mind, without a process of reasoning.

Intumes'cence (Lat. *in*, in ; *tu'meo*, I swell). A swelling.

Intussuscep'tion (Lat. *intus*, within ; *suscip'io*, I take up). A drawing of one part of a tube or canal into another.

Inven'tion (Lat. *inven'io*, I find). A finding ; the production of some combination or contrivance that did not before exist.

Inver'se (Lat. *in ; verto*, I turn). Placed in a contrary order ; as in an arithmetical proportion, when the ratio of the numbers to each other appears to be reversed.

Inver'sion (Lat. *in ; verto*, I turn). A placing in a contrary order ; a mutual changing of position.

Inver'tebrate (Lat. *in*, not ; *ver'tebra*, a bone of the spine). Without vertebræ or spinal bones.

Involu'cel (*Involu'crum ; cel*, denoting smallness). In *botany*, the collection of bractlets which surrounds a secondary or partial umbel.

Involu'crum (Lat. *in*, in ; *volvo*, I roll). A covering membrane : in

botany, a collection of bracts round a cluster of flowers : the layer of epidermis covering in the spore-cases on ferns.

Invol'untary (Lat. *in*, not ; *volun'tas*, will). Not dependent on or proceeding from the will.

In'volute (Lat. *in*, in ; *volvo*, I roll). Rolled inwards ; in *botany*, applied to a leaf which has each of its edges rolled inwards towards the midrib.

Involu'tion (Lat. *in*, into ; *volvo*, I roll). A folding or rolling in ; in *arithmetic* and *algebra*, the raising a number from its root to a power, as if it were folded or rolled on itself.

I'odate (*I'odine*). A compound of iodic acid with a base.

Iod'ic (*I'odine*). Containing iodine.

I'odide (*I'odine*). A compound of iodine with a metal or other substance.

I'odine (Gr. *lov, i'on*, a violet). A solid elementary body, the vapour of which has a violet colour.

I'odism (*I'odine*). In *medicine*, a morbid condition sometimes produced by the use of iodine.

I'on (Gr. *lων, iōn*, going). A name applied to the elements of substances capable of decomposition by the voltaic current, and which are evolved at the poles of the battery.

Iris (Gr. *lρις, iris*, the rainbow). The ring-shaped diaphragm which surrounds the pupil of the eye ; so called from being coloured.

Irides'cence (Gr. *lρις, iris*, the rainbow). A play of colours like a rainbow.

Irides'cent (Gr. *lρις, iris*, the rainbow). Marked with colours like the rainbow.

Ironstone. A term for the carbonates of iron found in nodules or thin layers in secondary rocks.

Irra'diation (Lat. *in ; ra'dius*, a ray). Emission of light ; illumination.

Ir'rigate (Lat. *in*, on ; *ri'go*, I moisten). To moisten.

Irritabil'ity (Lat. *irrito*, I excite). Excitability : the property of muscles by which they contract on the application of an exciting cause.

Irrup'tion (Lat. *in*, in; *rumpo*, I break). A breaking in.

I'sagon (Gr. ἰσος, *isos*, equal; γωνια, *gōnia*, an angle). A figure with equal angles.

Ischiat'ic (Gr. ἰσχιον, *is'chion*, the hip). Belonging to the hip.

Iso- (Gr. ἰσος, *isos*, equal). A prefix in compound words, denoting equality.

Isobaromet'ric (Gr. ἰσος, *isos*, equal; *barom'eter*). Applied to lines connecting places on the earth's surface which present the same mean difference between the monthly extremes of the barometer.

Isochei'mal (Gr. ἰσος, *isos*, equal; χειμα, *cheima*, winter). Having the same winter temperature.

Isochromat'ic (Gr. ἰσος, *isos*, equal; χρωμα, *chrōma*, colour). Having the same colour.

Isoch'ronal (Gr. ἰσος, *isos*, equal; χρονος, *chronos*, time). Uniform in time; occurring in equal times.

Isoclin'ic (Gr. ἰσος, *isos*, equal; κλινω, *klinō*, I bend). Bending equally; applied to curves in the earth's surface in which the dip of the magnetic parallels is equal.

Isodynam'ic (Gr. ἰσος, *isos*, equal; δυναμις, *du'namis*, power). Of equal power; applied to lines on the earth where the magnetic intensities are equal.

Isogeother'mal (Gr. ἰσος, *isos*, equal; γη, *gē*, the earth; θερμος, *thermos*, warm). *See* Isothermal.

Isogo'nic (Gr. ἰσος, *isos*, equal; γωνια, *gōnia*, an angle). Having equal angles; applied to lines on the earth's surface in which the magnetic needle has the same declinations.

Isohyeto'ses (Gr. ἰσος, *isos*, equal; ὑετος, *hu'etos*, rain). Lines connecting places on the surface of the globe where the quantity of rain which falls annually is the same.

Isomer'ic (Gr. ἰσος, *isos*, equal; μερος, *meros*, a part). Consisting of the same elements in the same proportions, but possessing different physical and chemical properties.

Isom'erism (Gr. ἰσος, *isos*, equal; μερος, *meros*, a part). The state of compounds which contain the same elements in the same proportions, but have different properties.

Isomor'phism (Gr. ἰσος, *isos*, equal; μορφη, *morphē*, form). The property which certain substances have of replacing each other in crystallised compounds without change of form.

Isomor'phous (Gr. ἰσος, *isos*, equal; μορφη, *morphē*, form). Of equal form; applied to substances capable of replacing each other in crystalline compounds without alteration of form.

Isop'odous (Gr. ἰσος, *isos*, equal, πους, *pous*, a foot). Applied to an order of crustaceans with fourteen legs, not having the respiratory organs attached to them.

Isos'celes (Gr. ἰσος, *isos*, equal; σκελος, *skel'os*, a leg). Having two equal legs or sides.

Isoste'monous (Gr. ἰσος, *isos*, equal; στημων, *stēmōn*, a stamen). In *botany*, applied when the stamens are equal in number to the sepals or petals.

Isoth'eral (Gr. ἰσος, *isos*, equal; θερος, *theros*, summer). Having the same mean summer temperature.

Isother'mal (Gr. ἰσος, *isos*, equal; θερμως, *thermos*, hot). Having equal heat: applied to lines drawn round the globe, and passing over points where the mean temperature is equal.

-Ite. A termination in chemistry, denoting a salt formed of an acid in a lower state of oxygenation.

-Itis. A termination denoting inflammation.

J.

Jacob's Membrane. A layer of the retina in the eye, described by Dr. Jacob as a serous membrane, but consisting of numerous rod-like bodies placed vertically together.

Jactita'tion (Lat. *jac'tito*, I throw about). A tossing about of the body; restlessness.

Jeju'num (Lat. *jeju'nus*, empty; because often found empty). A part of the small intestines, reaching from the duodenum to the ileum.

Jo'vian (Lat. *Jovis*, the genitive case of *Jupiter*). Belonging to the planet Jupiter.

Ju'ga (Lat. *jugum*, a yoke). The elevated portions traversing the carpels of umbelliferous plants.

Ju'gate (Lat. *jugum*, a yoke). In *botany*, applied to the pairs of leaflets in compound leaves.

Ju'gular (Lat. *ju'gulum*, the throat). Belonging to or connected with the neck or throat.

Ju'lian (*Julius Cæsar*). A term applied to the system of reckoning the year promulgated by Julius Cæsar, and which continued until the adoption of the new style.

Juras'sic (*Jura*, Mont Blanc in Switzerland). A name given in geology to the oolitic system, from its occurrence in the Jura mountains.

Jurispru'dence (Lat. *jus*, law ; *pruden'tia*, knowledge). The science of law.

Juxtaposit'ion (Lat. *juxta*, near ; *pono*, I put). A placing side by side.

K.

Kalei'dophone (Gr. καλος, *kalos*, beautiful ; ειδος, *eidos*, form ; φωνη, *phōnē*, sound). An instrument consisting of an elastic rod, with a polished knob at the free end, which exhibits beautiful curves of vibration when put in motion.

Kalei'doscope (Gr. καλος, *kalos*, beautiful ; ειδος, *eidos*, shape ; σκοπεω, *skop'eō*, I look at). An optical instrument, formed on the principle of multiplied reflection of light, for the purpose of exhibiting a variety of beautiful colours and symmetrical forms.

Ka'olin. A very fine earth or clay consisting of decomposed feldspar, used in the manufacture of porcelain.

Kathetom'eter (Gr. καθετος, *kath'etos*, perpendicular height; μετρον, *metron*, a measure). An instrument for measuring small differences of perpendicular height.

Kelænone'sian (Gr. κελαινος, *kelai'nos*, black ; νησος, *nēsos*, an island). A term applied to the inhabitants of the islands in the Pacific, whose skin is of a dark colour.

Kelp. The ashes of seaweed, from which carbonate of soda was produced.

Kepler's Laws. The laws of the courses of the planets, according to Kepler: viz., that a line drawn from the sun to the planets describes equal areas in equal times ; that the planets move in elliptic orbits ; and that the squares of the periods of revolution of the planets are very nearly in the ratio of the cubes of their mean distances.

Kil'ogramme (Gr. χιλιοι, *chil'ioi*, a thousand; Fr. *gramme*). A French weight equal to a thousand *grammes*, or 2·205 pounds avoirdupois.

Kil'olitre (Gr. χιλιοι, *chil'ioi*, a thousand ; *litre*). A French measure of a thousand *litres*, or 220 gallons.

Kil'ometre (Gr. χιλιοι, *chil'ioi*, a thousand; *metre*). A French measure of a thousand *mètres*, or about 1094 English yards.

Kim'meridge Clay. A blue and greyish yellow clay of the oolite formation.

Kleptoma'nia (Gr. κλεπτω, *kleptō*, I steal; μανια, *ma'nia*, madness). An irresistable desire to steal.

Kinet'ics (Gr. κινεω, *kineō*, I move). The part of mechanical science which treats of motion without reference to the forces producing it.

Kreasote. *See* Cre'asote.

Kre'atin and Kreat'inin. *See* Cre'atin and Creat'iniu.

Ky'anize (Mr. *Kyan*, the inventor of the process). To steep timber in a solution of corrosive sublimate in order to preserve it from dry rot.

L.

Label'lum (Lat. *la'bium*, a lip). A little lip.

La'bial (Lat. *la'bium*, a lip). Belonging to the lips; produced by the lips.

La'biate (Lat. *la'bium*, a lip). Having lips; applied in *botany* to a form of flower in which the corolla presents two portions resembling lips.

Labioden'tal (Lat. *la'bium*, a lip; *dens*, a tooth). Formed by the action of the lips and teeth.

La'bium (Lat. a lip). The lower lip of insects; the inner lip of a shell.

Labor'atory (Lat. *labo'ro*, I work). A place where operations or experiments are carried on.

Lab'radorite (*Labrado'r*). A mineral, consisting of a species of feldspar; consists chiefly of silica, alumina, and lime, with some oxide of iron.

La'brum (Lat., the brim of a vessel). The upper lip of insects; the outer lip of a shell.

Lab'yrinth (Gr. λαβυρινθος, *laburin'thos*, a maze). A name given to the internal ear, from its complex structure.

Labyrinth'iform (Gr. λαβυρινθος, *laburin'thos*, a maze; Lat. *forma*, shape). Having the form of a labyrinth; applied to a family of fishes in which there are a number of cells for containing water, formed by the plates of the pharyngeal bones above the gills.

Labyrinth'odonts (Gr. λαβυρινθος, *laburin'thos*, a labyrinth; ὁδους,

odous, a tooth). An order of fossil reptiles, so called from the complex undulating structure of the teeth as seen in section.

Lacer'tian (Lat. *lacer'tus*, a lizard). Relating to the lizard tribe.

Lacertil'ia (Lat. *lacertus*, a lizard). An order of reptiles of which the lizard is the type.

Lach'rymal (Lat. *lach'ryma*, a tear). Relating to the tears.

Lach'rymal Canals. The canals which convey the tears from the eye to the nasal ducts.

Lach'rymal Ducts. The ducts or small tubes which convey the tears from the lachrymal gland to the eyes.

Lach'rymal Gland. The gland which secretes the tears.

Lacin'iated (Lat. *lacin'ia*, fringe). Irregularly cut into narrow segments.

Lac'tate (Lat. *lac*, milk). A salt of lactic acid with a base.

Lacta'tion (Lat. *lac*, milk). The act of giving milk; suckling.

Lac'teal (Lat. *lac*, milk). Conveying milk, or a fluid like milk; applied to the vessels which take up the chyle from the alimentary canal and convey it to the thoracic duct.

Lactes'cence (Lat. *lac*, milk). A state resembling milk.

Lactes'cent (Lat. *lac*, milk). Yielding milky juice.

Lac'tic (Lat. *lac*, milk). Belonging to milk; applied to an acid obtained from milk.

Lactif'erous (Lat. *lac*, milk ; *fero*, I carry). Conveying milk.

Lac'tin (Lat. *lac*, milk). Sugar of milk; a sweetish substance existing in milk.

Lactom'eter (Lat. *lac*, milk ; Gr. μετρον, *metron*, a measure). An instrument for ascertaining the specific gravity of milk.

Lacu'na (Lat. a ditch). A little pit or depression, or hollow cavity.

Lacus'trine (Lat. *lacus*, a lake). Belonging to or produced in lakes.

Læmodip'oda (Gr. λαιμος, *laimos*, a throat; πους, *pous*, a foot). An order of crustacea in which the two fore-legs form part of the head.

Lamb'doid (The Greek letter Λ, λαμβδα, *lambda*; ειδος, *eidos*, shape). Resembling the Greek letter Λ or lambda.

Lamel'la (Lat.). A little plate or scale.

Lamel'lar (Lat. *lamel'la*, a little plate). Arranged in thin scales or plates.

Lamellibran'chiate (Lat. *lamel'la*, a little plate; Gr. βραγχια, *bran'chia*, gills). Having gills in symmetrical semicircular layers.

Lamel'liform (Lat. *lamel'la*, a small plate; *forma*, shape). Having the form of a small plate.

Lamelliros'tral (Lat. *lamel'la*, a small plate ; *rostrum*, a beak). Having the margins of the beak furnished with plates, as the duck and goose.

Lam'ina (Lat. a plate). A plate or scale ; in *botany*, the blade of a leaf, or the broad part of a sepal or petal.

Lam'inar or Lam'inated (Lat. *la'mina*, a plate). Arranged in plates or scales.

Lamina'tion (Lat. *la'mina*, a plate). An arrangement in plates or scales.

Lanate (Lat. *lana*, wool). Covered with a curly hair like wool.

Lan'ceolate (Lat. *lan'cea*, a lance). Gradually tapering to the outer end.

Lania'riform (Lat. *lanio*, I tear ; *forma*, shape). Shaped like the canine teeth of carnivorous animals.

La'niary (Lat. *lanio*, I tear). Formed for tearing.

Lanig'erous (Lat. *lana*, wool ; *gero*, I bear). Bearing or producing wool.

Lanu'ginous (Lat. *lanu'go*, down). In *botany*, woolly ; covered with interlaced hairs.

Lanu'go (Lat. down, or fine hair). The first hair which is produced in the fœtus.

Lapidif'ication (Lat. *lap'is*, a stone ; *fac'io*, I make). Conversion into stone.

Lapid'ify (Lat. *lap'is*, a stone; *fac'io*, I make). To convert into stone.

Lapil'li (Lat. *lapil'lus*, a little stone). A variety of volcanic cinders.

Lap'is (Lat. a stone). A term applied to various mineral substances.

Larda'ceous (Lat. *lardum*, lard or bacon). Resembling lard or bacon.

Larva (Lat. a mask). An insect in the caterpillar or grub state.

Lar'viform (Lat. *larva* ; *forma*, shape). Like a larva.

Larvip'arous (Lat. *larva* ; *par'io*, I bring forth). Producing young in the state of larvæ or grubs.

Larynge'al (Gr. λαρυγξ, *larunx*, the larynx). Belonging to the larynx or windpipe.

Laryngis'mus (Gr. λαρυγξ, *larunx*, the larynx). Spasmodic action of the larynx.

Laryngi'tis (Gr. λαρυγξ, *larunx*, the larynx ; *itis*, denoting inflammation). Inflammation of the larynx.

Laryngot'omy (Gr. λαρυγξ, *larunx*, the larynx ; τεμνω, *temnō*, I cut). The operation of opening the larynx.

Larynx (Gr. λαρυγξ, *larunx*). The enlarged upper part of the windpipe, projecting in the neck.

La'tency (Lat. *lat'eo*, I lie hid). A lying hid.

Latent (Lat. *lat'eo*, I lie hid). Hidden; not apparent to the senses.

Lat'eral (Lat. *latus*, a side). Belonging to or placed at a side.

Lat'erigrade (Lat. *latus*, a side ; *gradus*, a step). Able to walk sideways.

H

Laterit'ious (Lat. *later*, a brick). Like bricks or brick-dust.

Latex (Lat. a liquor or juice). The elaborated sap of plants.

Laticif'erous (Lat. *latex; fero*, I carry). Conveying latex or elaborated sap.

Lat'itude (Lat. *latus*, wide). Width. Terrestrial latitude is the position of a place on the surface of the earth north or south of the equator. Celestial latitude is the distance of a heavenly body from the ecliptic, measured in a direction perpendicular to the ecliptic.

Lava. The general name for melted rocky matter discharged from volcanoes.

Lax'ative (Lat. *laxo*, I loosen). Loosening; mildly purgative.

Laxa'tor (Lat. *laxo*, I loosen). That which relaxes or makes loose; applied to certain muscles.

Leaf-bud. A bud which produces leaves.

Leg'ume (Lat. *legu'men*, pulse). In *botany*, a pod opening at the front and back, as in the pea.

Legu'minous (Lat. *legu'men*, pulse). Belonging to the bean tribe, the fruit of which is a legume or pod.

Lemma (Gr. λαμβανω, *lam'banō*, I receive). A proposition laid down to demonstrate for the purpose of rendering more plain another that is to follow.

Lens (Lat. a lentil). A transparent substance, with two curved surfaces, or with a curved surface and a plane surface, for the purpose of altering the direction of rays of light passing through it.

Lentic'ular (Lat. *lentic'ulus*, a little lentil). Having the form of a double convex lens, or the form or size of a lentil.

Lentor (Lat. *lentus*, slow). Slowness; viscidity or thickness of fluids.

Lepidoden'dron (Gr. λεπις, *lep'is*, a scale; δενδρον, *dendron*, a tree). A family of fossil plants in the coal formation, so called from the scale-like arrangement of the scars of their leaves.

Lepidogan'oid (Gr. λεπις, *lep'is*, a scale; γανος, *ganos*, splendour; ειδος, *eidos*, form). A sub-order of fossil fishes.

Lep'idoid (Gr. λεπις, *lep'is*, a scale; ειδος, *eidos*, shape). Resembling scales.

Lep'idote (Gr. λεπις, *lep'is*, a scale). Covered with scales.

Lepidop'tera (Gr. λεπις, *lep'is*, a scale; πτερον, *pter'on*, a wing). An order of insects having four membranous wings covered with fine scales, as butterflies and moths.

Lepra (Gr. λεπις, *lep'is*, a scale). The leprosy; a disease of the skin characterised by the formation of whitish opaque scales.

Le'sion (Lat. *lædo*, I hurt). An injury.

Leth'argy (Gr. ληθη, *lēthē*, oblivion; αργος, *argos*, idle). Preternatural drowsiness.

Leucæ'mia (Gr. λευκος, *leukos*, white; αιμα, *haima*, blood). White blood.

Leucin (Gr. λευκος, *leukos*, white). A white crystallisable organic substance obtained from muscular fibre, and from the compounds of protein.

Leucocythæ'mia (Gr. λευκος, *leukos*, white; κυτος, *kutos*, a cell; αιμα, *haima*, blood). A diseased state characterised by an excess of white corpuscles in the blood.

Leucophlegma'sia (Gr. λευκος, *leukos*, white; φλεγμα, *phlegma*, phlegm). A condition of body characterised by paleness and flabbiness, with an excess of serum in the blood.

Leva'tor (Lat. *levo*, I raise). That which raises: applied to certain muscles.

Lever (Lat. *levo*, I raise). A solid bar turning on an axis or fulcrum, employed for the purpose of raising weights.

Lev'igate (Lat. *lævis*, smooth). To make smooth; to rub to a fine impalpable powder.

Lex'icon (Gr. λεγω, *legō*, I speak). A dictionary: applied generally to dictionaries of the Greek or Hebrew languages.

Leyden Jar. A glass jar coated on both sides with tinfoil to within several inches of the top, for the purpose of accumulating electricity.

Lias (said to be from *liers* or *layers*, from its occurrence in thin beds). The lowest portion of the oolitic system in *geology*, composed of clayey limestones, bluish clays, and bituminous and pyritous shales.

Liber (Lat. bark). The inner portion of the bark of a tree.

Libra'tion (Lat. *libra*, a balance). A state of balancing: in *astronomy*, a variation in the appearance of portions of the edge of the moon, whereby, under certain circumstances, they become alternately visible and invisible, as if the motion of the moon were subject to oscillations.

Li'chen (Gr. λειχην, *leichēn*, a tree-moss). A division of cryptogamic plants covering trees and rocks : a disease of the skin.

Lien'tery (Gr. λειος, *leios*, smooth ; ἐντερον, *en'teron*, an intestine). A disease in which food is discharged undigested from the bowels.

Lig'ament (Lat. *ligo*, I bind). That which binds together ; a fibrous structure connecting bones.

Ligamen'tous (Lat. *ligo*, I bind). Having the nature of or acting as a ligament.

Lig'ature (Lat. *ligo*, I bind). A band ; the act of binding ; a cord or string used in surgery for tying blood-vessels.

Lig'neous (Lat. *lignum*, wood). Consisting of or resembling wood.

Lignifica'tion (Lat. *lignum*, wood ; *facio*, I make. A making wood, or converting into wood.

Lignin (Lat. *lignum*, wood). Vegetable fibre ; the substance which constitutes the essential part of the structure of plants.

Lignite (Lat. *lignum*, wood). Brown coal : a variety of coal of recent formation, in which the woody structure is distinctly apparent.

Lig'ulate (Lat. *lig'ula*, a strap). Like a bandage or strap.

Lilia'ceous (Lat. *lil'ium*, a lily). Belonging to or resembling a lily.

Limb (Lat. *limbus*, an edge or bor-

der). In *astronomy*, the border or outer edge of the sun or moon.

Limestone. A mineral composed of carbonate of lime, and of which there are several varieties.

Linctus (Lat. *lingo*, I lick). A medicine of the consistence of honey or treacle.

Lin'eal (Lat. *lin'ea*, a line). Belonging to a line or length ; like a line.

Lin'ear Numbers. In *mathematics*, numbers which have relation to length only.

Lin'ear Perspective. That perspective which regards only the positions, forms, and sizes of objects.

Lin'eate (Lat. *lin'ea*, a line). Marked longitudinally, with parallel depressions.

Lin'gual (Lat. *lingua*, the tongue). Belonging to the tongue.

Linguis'tic (Lat. *lingua*, tongue or language). Relating to language or the affinities of languages.

Lin'iment (Lat. *lin'io*, I anoint). An oily composition for rubbing into external parts of the body.

Liqua'tion (Lat. *liquo*, I melt). The art of melting ; the process of melting out from an alloy an easily fusible metal from one less capable of fusion.

Liquefac'tion (Lat. *liquefac'io*, I make liquid). A melting.

Liq'uefy (Lat. *liquefac'io*, I make liquid). To melt or dissolve by heat.

Liq'uid (Lat. *liq'ueo*, I melt). A substance of which the component parts are not held together with sufficient force to prevent their separation by their own weight, but have not a mutual repulsion like gases.

Liquor San'guinis (Lat. the liquor of the blood). The transparent colourless fluid part of the blood, in which the corpuscles float.

Lissenceph'ala (Gr. λισσος, *lissos*, smooth ; ἐγκεφαλος, *enkeph'alos*, the brain). Smooth-brained animals ; a term applied by Owen to a sub-class of mammalia in which the brain is more connected than in lyencephala, but has few or no convolutions, as in the rodents and insectivorous animals.

Lit′eral (Lat. *lit′era*, a letter). According to the letter or exact expression; consisting of letters : in *algebra*, applied to equations in which the known quantities as well as the unknown are represented by letters.

Lith′ate (Gr. λιθος, *lith′os*, a stone). A salt of lithic acid with a base.

Lith′ic (Gr. λιθος, *lith′os*, a stone). Belonging to a stone or calculus ; applied to an acid formed in the animal body, and often forming a part of calculi.

Lith′o- (Gk. λιθος, *lith′os*, a stone). A prefix in compound words, signifying stone.

Lith′ocarp (Gr. λιθος, *lith′os*, a stone ; καρπος, *karpos*, fruit). Fossil fruit.

Lithog′raphy (Gr. λιθος, *lith′os*, a stone ; γραφω, *graphō*, I write). The art of tracing letters or figures on stone and transferring them to paper.

Lithol′ogy (Gr. λιθος, *lith′os*, a stone ; λογος, *logos*, discourse). The department of geology which describes the rocks and strata, without reference to fossils.

Lith′ophyte (Gr. λιθος, *lith′os*, a stone ; φυτον, *phuton*, a plant). Stone plants ; a tribe of polypi having a fixed internal axis of stony consistency.

Lithot′omy (Gr. λιθος, *lith′os*, a stone ; τεμνω, *temnō*, I cut). An operation for the removal of stones from the bladder.

Litmus. A blue colouring matter obtained by the action of ammonia on certain lichens, and used in chemistry to detect the presence of acids, which turn it red.

Litre (Fr.). The French standard measure of capacity, equal to a cubic *decimètre*, or about 1¾ English pint.

Lit′toral (Lat. *littus*, the shore). Belonging to the shore.

Lixiv′iate (Lat. *lixa*, ley of ashes). To impregnate with salts from wood ashes, as by passing water through them.

Llandei′lo Formation. In *geology*, the lowest series of the Silurian system.

Llanos (Spanish *llano*, flat, from Lat. *planus*). A name given to the plains extending along the banks of the Orinoco in South America.

Loadstone (*Lead* and *Stone*). The magnet ; an ore consisting of protoxide and peroxide of iron.

Loam. Any soil composed of clay and sand, containing neither in a distinct form.

Lobe (Gr. λοβος, *lobos*). A part or division of an organ, as of the brain, lungs, or liver ; or of a leaf.

Lob′ular (*Lobule*). Belonging to or affecting a lobule.

Lob′ule (Gr. λοβος, *lobos*, a lobe). A little lobe, or sub-division of a lobe.

Local (Lat. *locus*, a place). Belonging or confined to a part.

Locomo′tion (Lat. *locus*, a place ; *mov′eo*, I move). Motion from place to place.

Locomo′tive (Lat. *locus*, a place ; *mov′eo*, I move). Moving from place to place.

Loc′ulament (Lat. *loc′ulus*, a cell). In *botany*, a cavity in an ovary.

Loc′ular (Lat. *loc′ulus*, a cell). Having one or more cells.

Loculici′dal (Lat. *loc′ulus*, a cell ; *cædo*, I cut). In *botany*, applied to that form of opening of fruits in which the cells are split open at the back.

Loc′ulose (Lat. *loc′ulus*, a cell). Divided by one or more partitions into cells.

Locus (Lat. a place). In *geometry*, a term applied to a line by which a local or indeterminate problem is solved.

Lode (Sax. *lædan*, to lead). In *geology*, a vein or course, whether containing metal or not.

Log′arithm (Gr. λογος, *logos*, a ratio ; αριθμος, *arith′mos*, a number). The index or power to which any number, taken as a base, is to be raised so that the result may be equal to a given number.

Logic (Gr. λογος, *logos*, a word, reason). The science of the operations of the understanding which are subservient to the estimation of evidence; pointing out the relations between given facts and the conclusions to be drawn from them.

Logog'raphy (Gr. λογος, *logos*, a word; γραφω, *graphō*, I write). A system of printing by words instead of letters.

Logom'eter (Gr. λογος, *logos*, proportion; μετρον, *metron*, a measure). A scale for measuring chemical equivalents.

Logomet'ric (Gr. λογος, *logos*, a proportion; μετρον, *metron*, a measure). Measuring proportionate spaces.

Lomenta'ceous (Lat. *lomen'tum*, bean-meal). In *botany*, applied to legumes or pods with transverse partitions, each division containing one seed.

Longi- (Lat. *longus*, long). A prefix in compound words implying length.

Lon'gitude (Lat. *longus*, long). Length; the distance, eastward or westward, of any meridian on the earth's surface from some fixed meridian arbitrarily selected. The longitude of a celestial body is the arc of the ecliptic between the first point of Aries and the circle which measures its latitude.

Loph'iodon (Gr. λοφος, *loph'os*, a crest or ridge; οδους, *odous*, a tooth). An extinct pachydermatous or thick-skinned animal found in the tertiary strata; so called from the eminences on its teeth.

Lophobran'chiate (Gr. λοφος, *loph'os*, a tuft; βραγχια, *bran'chia*, gills). Having gills arranged in tufts: applied to an order of fishes.

Lo'ricate (Lat. *lori'ca*, a coat of mail). Covered as with a coat of mail or plate armour, as crocodiles, alligators, &c.

Loxodrom'ic (Gr. λοξος, *loxos*, oblique; δρομος, *drom'os*, a course). Having an oblique course; applied to a course in sailing, in which the ship is directed constantly towards the same point of the compass in an oblique direction.

Lu'bricate (Lat. *lu'bricus*, slippery). To make smooth or slippery.

Lu'cules (Lat. *lux*, light; *ule*, denoting smallness). A name given to the variations in the intensity of the brightness of the sun's disk.

Lumba'go (Lat. *lumbus*, the loin). A rheumatic affection of the region of the loins.

Lumbar (Lat. *lumbus*, the loin). Belonging to the loins.

Lumbrica'les (Lat. *lumbri'cus*, an earth-worm; from their shape). A name given to certain small long muscles of the fingers and toes.

Luminif'erous (Lat. *lumen*, light; *fero*, I bear). Producing or conveying light.

Lu'minous (Lat. *lumen*, light). Shining; applied to bodies which are original sources of light.

Lu'nacy (Lat. *luna*, the moon; because formerly supposed to be influenced by the moon). Insanity or madness; strictly, that form of insanity which is accompanied by intervals of reason, but commonly applied to all states of unsound mind.

Lunar (Lat. *luna*, the moon). Relating to the moon; measured by the revolutions of the moon.

Lu'nate (Lat. *luna*, the moon). Shaped like a crescent.

Lu'natic (Lat. *luna*, the moon). Affected with lunacy.

Luna'tion (Lat. *luna*, the moon). The period of the monthly revolution of the moon, or the time from one new moon to another.

Luniso'lar (Lat. *luna*, the moon; *sol*, the sun). Compounded of the periods of revolution of the sun and moon.

Lu'nula (Lat. a little moon). The portion of the human nail near the root, which is whiter than the rest; also the narrow portion at the margins of the semilunar valves of the heart.

Lupus (Lat. a wolf). In *medicine*, a disease characterised by its tendency to destructive ulceration of the parts which it attacks.

Luxate (Lat. *luxo*, I loosen). To put out of joint.

Luxa'tion (Lat. *luxo*, I loosen). A putting out of joint ; a dislocation.

Lyenceph'ala (Gr. λυω, *luō*, I loosen; ἐγκέφαλος, *enkeph'alos*, the brain). Loose-brained : a term proposed by Professor Owen to denote the lowest group of mammalia, in which the hemispheres of the brain are comparatively loose and disconnected, as in the monotremes and marsupials.

Lymph (Lat. *lympha*, water). A transparent and nearly colourless fluid, which is conveyed into the blood by the lymphatic vessels.

Lymphat'ic (Lat. *lympha*, water). Belonging to lymph : applied to the vessels which convey lymph.

Lyrate (Lat. *lyra*, a lyre). In *botany*, applied to leaves of which the apex consists of a large rounded lobe, and the divisions become gradually smaller towards the base.

M.

Mac'erate (Lat. *macer*, lean). To make lean or thin ; to soften and dissolve away by steeping in a fluid.

Macera'tion (Lat. *macer*, lean). The act of softening and dissolving away by steeping in a fluid.

Macro- (Gr. μακρος, *makros*, long). A prefix in compound words signifying length.

Macroceph'alous (Gr. μακρος, *makros*, long ; κεφαλη, *keph'alē*, the head). Having a long head ; applied in *botany* to embryos of which the two cotyledons grow together.

Macrodactyl'ic (Gr. μακρος, *makros*, long ; δακτυλος, *dak'tulos*, a finger or toe). Having long toes.

Macrom'eter (Gr. μακρος, *makros*, long ; μετρον, *metron*, a measure). An instrument for measuring inaccessible heights and objects.

Macrop'odous (Gr. μακρος, *makros*, long ; πους, *pous*, a foot). Having long feet ; applied to a family of crustacean invertebrate animals.

Macrou'rous (Gr. μακρος, *makros*, long ; ουρα, *oura*, a tail). Long-tailed ; applied to a tribe of crustaceans of which the lobster and shrimp are examples.

Mac'ula (Lat. a spot). A spot : the name is given in the plural (*maculæ*) to an order of diseases of the skin.

Mad'repore (Fr. *madre*, spotted ; *pore*) A kind of coral.

Maestricht Beds (*Maestricht*, a town in the Netherlands). In *geology*, the upper layers of the chalk formation, consisting of a soft yellowish limestone.

Mag'deburg Hemispheres. An apparatus for illustrating atmospheric pressure, consisting of two hollow brass hemispheres fitting together, which, when the air is withdrawn from their interior, cannot be separated.

Magellan'ic Clouds (*Magal'haens* or *Magel'lan*, a Portuguese navigator). A name given to two nebulous patches of stars in the southern hemisphere.

Magma (Gr. μασσω, I knead). A mass of matter worked up into a paste.

Magne'sian Limestone. A limestone containing magnesia ; in *geology*, the term characterises a portion of the Permian system, or new red sandstone.

Magnet (Gr. μαγνης, *magnēs*; from *Magnesia* in Asia Minor, where first observed). The loadstone ; an ore consisting of protoxide or sesquioxide of iron, which has the property of attracting small pieces of iron and of pointing to the poles; a piece of iron to which these properties have been imparted.

Magnet'ic (Gr. μαγνης, *magnēs*, a magnet). Belonging to or having the properties of the magnet.

Magnet'ic Bat'tery. A battery formed

of several magnets with all their poles similarly disposed.

Magnet'ic Equa'tor. A line on the earth traced through the points at which the magnetic needle rests horizontal.

Magnet'ic Merid'ian. A line on the earth's surface, bearing the same analogy to the magnetic equator as the terrestrial meridian to the terrestrial equator.

Magnet'ic Poles. The two regions of attraction separated by the equator of a magnet.

Mag'netism (Gr. μαγνης, *magnēs*, a magnet). The science which describes the properties of the magnet; the property which is possessed by the magnet.

Mag'netise (Gr. μαγνης, *magnēs*, a magnet). To impart magnetic properties : to become magnetic.

Mag'neto-electric'ity (*Magnet ; electricity*). The phenomena of electricity called into existence by magnetism.

Magnetom'eter (*Magnet ;* Gr. μετρον, *metron*, a measure). A magnetised bar of steel for the purpose of determining the absolute amount of magnetic declination, or the intensities of terrestrial magnetism in horizontal or vertical directions.

Mag'nitude (Lat. *magnus*, large). Size. Linear magnitude is length or distance. Superficial magnitude or area is the space included in length and breadth expressed in squares. Solid magnitude or volume is the bulk expressed by the length, breadth, and thickness of a body, or the space which it fills, expressed in cubes. Apparent magnitude, in *optics*, is the size of the picture formed on the retina, as measured by the angle formed between the object seen and the centre of the eye.

Mal'achite (Gr. μαλαχη, *mal'achē*, mallows ; from its appearance). A mineral, consisting of green carbonate of copper.

Mal'aco- (Gr. μαλακος, *mal'akos*, soft). A prefix in compound words, signifying softness.

Malacol'ogy (Gr. μαλακος, *mal'akos*,

soft ; λογος, *logos*, a description). The description of molluscous or soft-bodied animals.

Malacopteryg'ian (Gr. μαλακος, *mal'akos*, soft ; πτερυγιον, *pteru'-gion*, a little wing, or fin). Having soft fins ; applied to an order of fishes, of which the rays of the fins are cartilaginous.

Malacopteryg'ii abdomina'les. Abdominal malacopterygians ; soft-finned fishes, with the ventral fins situated under the abdomen behind the pectoral fins.

Malacopteryg'ii subbranchia'ti. Subbranchiate malacopterygians ; soft-finned fishes, with the ventral fins placed under the pectorals.

Malacopteryg'ii ap'odes. Apodal or footless malacopterygians ; soft-finned fishes, without ventral fins, the homologues of feet.

Malacos'teon (Gr. μαλακος, *mal'akos*, soft ; οστεον, *os'teon*, a bone). Softness of bones ; the disease otherwise called mollities ossium.

Malacos'tracous (Gr. μαλακος, *mal'akos*, soft ; οστρακον, *os'trakon*, a shell). A section of crustacea, of which the shell is generally solid ; named from the relative softness of the shell as compared with that of mollusca.

Malar (Lat. *mala*, the cheek). Belonging to the cheek.

Mala'ria (Italian, *mal*, bad ; *a'ria*, air). Bad air ; an exhalation, as from marshes, tending to produce disease.

Mala'rial (*Mala'ria*). Produced by malaria.

Mala'rious (*Mala'ria*). Containing or of the nature of malaria.

Ma'late (Lat. *malum*, an apple). A compound of malic acid, or acid of apples, with a base.

Ma'lic (Lat. *malum*, an apple). Belonging to apples : applied to an organic acid, found principally in apples.

Malleabil'ity (Lat. *mal'leus*, a hammer). The property of being reduced to thin plates or leaves by hammering or rolling.

Mal'leable (Lat. *mal'leus*, a hammer).

Capable of being beaten or rolled into thin plates.

Malle'olar (Lat. *mal'leolus*). Belonging to the ankle; applied to certain small arteries.

Malle'olus (Diminutive of Lat. *mal'leus*, a hammer). An ankle, or the joint formed with the legs on each side of the foot.

Mammal (Lat. *mamma*, the breast). A name given to those vertebrate animals which suckle their young.

Mammalif'erous (*Mammalia* or *mammals; fero*, I bear). Producing mammalian animals; applied to the geological strata which contain remains of mammals.

Mam'mary (Lat. *mamma*, the breast). Belonging to the breast.

Mam'mifer (Lat. *mamma*, the breast; *fero*, I carry). *See* Mammal.

Mammil'lary (Lat. *mammil'la*, a teat). Belonging to or resembling teats.

Mam'millated (Lat. *mammil'la*, a teat). Having protuberances like nipples.

Man'dible (Lat. *mando*, I chew). The upper jaw of an insect.

Mandib'ulate (Lat. *mando*, I chew). Provided with an upper jaw.

Manduca'tory (Lat. *mandu'co*, I chew). Relating to or employed in chewing.

Man'ganate (*Mangane'se*). A compound of manganic acid with a base.

Mangan'ic (*Mangane'se*). An acid consisting of an atom of manganese with three of oxygen.

Manipula'tion (Lat. *manip'ulus*, a handful). Work by hand; applied to the manual and mechanical operations in science.

Mannite. A variety of sugar obtained from manna.

Manom'eter (Gr. μανος, *manos*, thin; μετρον, *metron*, a measure). An instrument for measuring the rarity or density, or the elastic force of any gaseous substance.

Man'oscope (Gr. μανος, *manos*, thin; σκοπεω, *skop'eō*, I view). *See* Manometer.

Mantis'sa (Lat. over-measure). A name given to the decimal part of a logarithm.

Mantle. In *zoology*, the skin of molluscous animals, which covers in the viscera and a large part of the body.

Manu'brium (Lat. a handle). A name sometimes given to the upper part of the sternum or breast-bone.

Maras'mus (Gr. μαραινω, *marai'nō*, I cause to waste away). Atrophy; a wasting of the body.

Marces'cent (Lat. *marces'co*, I pine away). Withering or fading.

Mar'garate (Gr. μαργαριτης, *margari'tēs*, a pearl). A compound of margaric acid, with a base.

Margar'ic (Gr. μαργαριτης, *margari'tēs*, a pearl). Belonging to pearl, or to the pearl-like substance called margarine; applied to one of the acids existing in oils.

Mar'garine (Gr. μαργαριτης, *margari'tēs*, a pearl). A pearl-like substance obtained from oils by exposure to cold.

Mar'ginate (Lat. *margo*, a rim or edge). In *botany*, applied to the calyx when it is reduced to a mere rim.

Marine (Lat. *mare*, the sea). Belonging to or produced in the sea.

Marl. A general term for all friable or crumbly compounds of lime and clay.

Marlstone. A layer of calcareous, sandy, and irony beds, forming one of the strata of the lowest or liassic group in the oolitic system in geology.

Marsu'pial (Lat. *marsu'pium*, a pouch or bag). Having or belonging to a pouch; applied to an order of mammalia which bring forth their young in an imperfect state, and keep them, until developed, in a pouch formed by a peculiar arrangement of the skin on the abdominal surface of the animal.

Marsupia'ta (Lat. *marsu'pium*, a pouch or bag). *See* Marsupial.

Marsu'pium (Lat. a pouch). A dark coloured membrane in the vitreous body of the eyes of birds.

Mas'sicot. Yellow oxide of lead,

Mas'ticate (Gr. μαστος, *mastos*, the jaws or mouth). To chew.

Masti'tis (Gr. μαστος, *mastos*, the breast; *itis*, denoting inflammation). Inflammation of the breast.

Mas'todon (Gr. μαστος, *mastos*, a nipple; ὀδους, *odous*, a tooth.) A fossil animal of the elephant kind, so called from the nipple-like prominences on its teeth.

Mastodyn'ia (Gr. μαστος, *mastos*, the breast; ὀδυνη, *od'une*, pain). Pain of the breast.

Mas'toid (Gr. μαστος, *mastos*, a nipple; εἰδος, *eidos*, shape). Resembling a nipple.

Mater (Lat. a mother). A name given to two of the membranes covering the brain, because formerly supposed to be the source of all the other membranes.

Mate'ria Med'ica (Lat. medical material). The collective name for the substances used in medicine; the science which describes these substances, their properties, modes of preparation, &c.

Mathemat'ics (Gr. μαθημα, *mathēma*, learning; from μανθανω, *man'thanō*, I learn). The science which treats of whatever can be measured or numbered. Pure mathematics considers quantity and number without reference to matter. Mixed mathematics treats of magnitude in connection with material bodies.

Matrix (Lat. *mater*, a mother). The cavity or substance in which anything is formed or imbedded. A mould: as the matrix of a type, &c

Mat'urate (Lat. *matu'rus*, ripe). To ripen.

Maxil'la (Lat. a jaw). A jaw; the lower pairs of horizontal jaws in invertebrate animals.

Maxil'lary (Lat. *maxil'la*, a jaw). Belonging to the jaws.

Maxil'liped (Lat. *maxil'la*, a jaw; *pes*, a foot). A jaw-foot; applied to the foot-like organs covering the mouth in crustacea.

Max'imum (Lat. greatest). The greatest quantity or degree attainable.

Mean (Fr. *moyen*, from Lat. *me'dius*, middle). Having an intermediate or average value between two or more quantities.

Mea'tus (Lat. *meo*, I pass). A passage.

Mechan'ics (Gr. μηχανη, *mēchanē*, an artificial contrivance). The science which investigates the action of bodies on one another, either directly or by means of machinery.

Mec'onate (Gr. μηκων, *mēkōn*, a poppy). A salt of meconic acid with a base.

Mecon'ic (Gr. μηκων, *mēkōn*, a poppy). Belonging to the poppy; applied to an acid found in opium.

Mediæ'val (Lat. *me'dius*, middle; *ævum*, an age). Belonging to the middle ages.

Me'dian Plane (Lat. *me'dius*, middle). A plane or flat surface supposed to pass down through a body from before backwards, so as to leave equal parts on both sides.

Mediasti'num. The partition formed by the meeting of the pleuræ, dividing the chest into two lateral parts.

Med'ical Jurispru'dence. The science which treats of subjects in which both law and medicine are applied.

Med'icate (Lat. *med'icus*, a physician). To impregnate with medicinal substances.

Med'icine (Lat. *med'eor*, I cure; from Gr. μεδομαι, *med'omai*, I attend to). The science of relieving, curing, or preventing diseases; any substance used with these objects.

Medie'val. *See* Mediæval.

Me'dium (Lat. *me'dius*, the middle). The space, substance, or matter in which bodies exist, or in which they move; the agent through which a cause or power acts in producing its effect.

Medul'la (Lat.). Marrow; in *botany*, the pith of plants.

Medul'la Oblonga'ta (Lat.). The lengthened or prolonged marrow; the continuation of the spinal cord within the skull.

Medul'la Spina'lis (Lat.) The spinal marrow or cord.

Med'ullary (Lat. *medul'la*, marrow). Relating to marrow; in *botany*, belonging to or connected with pith.

Med'ullary Rays. In *botany*, masses of cells connecting the pith with the bark.

Med'ullary Sheath. The sheath which surrounds the pith in exogenous plants.

Mega- or **Megal-** (Gr. μεγας, *meg'as*, large). A prefix in compound words, denoting large size.

Megac'eros (Gr. μεγας, *meg'as*, great; κερας, *ker'as*, horn.) The fossil or sub-fossil deer of the British Isles, commonly named the Irish elk.

Megalich'thys (Gr. μεγας, *meg'as*, great; ιχθυς, *ichthus*, a fish). A large fossil fish.

Megalon'yx (Gr. μεγας, *meg'as*, great; ονυξ, *on'ux*, a nail). An extinct animal allied to the sloth; named from the large size of its claw-bones.

Megalosau'rus (Gr. μεγας, *meg'as*, great; σαυρος, *sauros*, a lizard). A large fossil land reptile.

Megathe'rioids (Gr. μεγας, *meg'as*, great; θηριον, *thērion*, a wild beast; ειδος, *eidos*, form). A family of fossil mammalia allied to the megatherium.

Megathe'rium (Gr. μεγας, *meg'as*, great; θηρ, *thēr*, a beast). A large extinct animal, allied to the sloth.

Melæ'na (Gr. μελας, *mel'as*, black). A discharge of dark blood from the bowels.

Melano'sis (Gr. μελας, *mel'as*, black). A diseased formation of a black or dark colour.

Melanot'ic (Gr. μελας, *mel'as*, black). Having or of the nature of melanosis.

Melas'ma (Gr. μελας, *mel'as*, black). A blackening or darkening.

Mellif'erous (Lat. *mel*, honey; *fero*, I bear). Producing honey.

Melliv'orous (Lat. *mel*, honey; *voro*, I devour). Feeding on honey.

Mel'ody (Gr. μελος, *mel'os*, a tune; ωδη, *ōdē*, an ode). An agreeable succession of sounds.

Membrana'ceous (Lat. *membra'na*, a membrane). Consisting of membrane.

Membra'na Nic'titans (Lat.) The winking membrane; a moveable fold of skin with which birds cover their eyes.

Mem'branous. *See* Membranaceous.

Menin'ges (Gr. μηνιγξ, *mēninx*, a membrane). The membranes covering the brain and spinal cord.

Meningi'tis (*Meninges; itis*, denoting inflammation). Inflammation of the membranes covering the brain.

Menis'cus (Gr. μηνισκος, *mēniskos*, a crescent; from μηνη, *mēnē*, the moon). A lens convex on one side and concave on the other, with a sharp edge.

Mensura'tion (Lat. *mensu'ra*, a measure). The art of measuring.

Mentag'ra (Lat. *mentum*, the chin; Gr. αγρα, *agra*, a seizing). An eruptive disease affecting the chin and upper lip.

Mephit'ic (Lat. *mephi'tis*, an ill smell). Offensive; pestilential; destructive to life.

Merca'tor's Chart (Gerrard *Merca'tor*, a Flemish geographer). A representation of the earth on a plane surface.

Mercu'rial (Lat. *Mercu'rius*, Mercury, also quicksilver). Belonging to or formed of mercury or quicksilver.

Mer'icarp (Gr. μερος, *mer'os*, a part; καρπος, *karpos*, fruit). The half of the fruit of an umbelliferous plant.

Merid'ian (Lat. *merid'ies*, mid-day). A great circle supposed to be drawn through the poles of the earth at right angles to the equator, dividing the hemisphere into eastern and western: when this circle arrives opposite the sun, it is midday at the place. Celestial meridian is the vertical circle which passes through the celestial pole. Magnetic meridian. *See* Magnetic.

Merid'ional (*Merid'ian*). Belonging to the meridian.

Merismat'ic (Gr. μεριζω, *meri'zō*, I

divide). Fissiparous ; multiplying by division.

Mesenceph'alic (Gr. μεσος, *mes'os*, middle ; ἐγκεφαλον, *enceph'alon*, the contents of the skull). Belonging to the middle part of the brain.

Mesenter'ic (Gr. μεσος, *mes'os*, midst ; ἐντερον, *en'teron*, the intestine). Belonging to the mesentery.

Mes'entery (Gr. μεσος, *mes'os*, middle ; ἐντερον, *en'teron*, an intestine). The fold of membrane which attaches the intestines to the spine.

Mes'o- (Gr. μεσος, *mes'os*, middle). A prefix in compound words, signifying middle.

Mesocæ'cum (Gr. μεσος, *mes'os*, middle ; Lat. *cæcum*, a portion of the large intestines). The part of the peritoneum which attaches the cæcum.

Mes'ocarp (Gr. μεσος, *mes'os*, middle ; καρπος, *karpos*, fruit). The middle of the three layers in fruits.

Mesoceph'alon (Gr. μεσος, *mes'os*, middle ; κεφαλη, *keph'alē*, a head). A name sometimes given to the pons Varolii of the brain, from its position.

Mesoco'lon (Gr. μεσος, *mes'os*, middle ; *colon*, a part of the intestines so called). The portion of mesentery which attaches the colon.

Mesogas'tric (Gr. μεσος, *mes'os*, middle ; γαστηρ, *gastēr*, the stomach). Attaching the stomach to the walls of the abdomen.

Mesono'tum (Gr. μεσος, *mes'os*, middle ; νωτος, *nōtos*, the back). The upper half of the middle segment of the thorax in insects, covering in the back.

Mesophlœ'um (Gr. μεσος, *mes'os*, middle ; φλοιος, *phloi'os*, bark). In *botany*, the middle layer of the bark of a tree.

Mesophyll'um (Gr. μεσος, *mes'os*, middle ; φυλλον, *phullon*, a leaf). The cellular substance of a leaf.

Mes'osperm (Gr. μεσος, *mes'os*, middle ; σπερμα, *sperma*, a seed). The middle coat of a seed.

Mesoster'num (Gr. μεσος, *mes'os*, middle ; στερνον, *sternon*, the breast).

The lower half of the middle segment of the thorax in insects.

Mesotho'rax (Gr. μεσος, *mes'os*, middle ; θωραξ, *thōrax*, a breast-plate). The middle part of the thorax of insects, bearing the anterior pair of wings and the middle pair of legs.

Mesozo'ic (Gr. μεσος, *mes'os*, middle ; ζωη, *zōē*, life). A name given in *geology* to the middle period, as regards animal remains ; comprehending the cretaceous, oolitic and triassic epochs.

Met'a- (Gr. μετα, *met'a*, beyond). A prefix in compound words, signifying beyond.

Metacar'pal (Gr. μετα, *met'a*, beyond ; καρπος, *karpos*, the wrist). Belonging to the metacarpos.

Metacar'pus (Gr. μετα, *met'a*, beyond ; καρπος, *karpos*, the wrist). The hand between the wrist and the fingers.

Metach'ronism (Gr. μετα, *met'a*, beyond ; χρονος, *chron'os*, time). The placing an event in chronology after its real time.

Metagen'esis (Gr. μετα, *met'a*, implying change ; γενναω, *gennaō*, I produce). Alternating generation ; the succession of individuals, which present the same form only at every alternate generation ; the changes of form which the representative of a species undergoes in passing from the egg to a perfect or more complete state.

Metagenet'ic (Gr. μετα, *met'a*, implying change ; γενναω, *gennaō*, I produce). Referring to the changes of form undergone in passing from the egg to a perfect state.

Metallif'erous (Lat. *metal'lum*, a metal ; *fero*, I bear). Producing or yielding metals.

Metal'loid (Gr. μεταλλον, *metal'lon*, a metal ; εἰδος, *eidos*, form). Like metal ; a name sometimes given to the non-metallic elements.

Met'allurgy (Gr. μεταλλον, *metal'lon*, a metal ; ἐργον, *ergon*, work). The art of working metals ; especially separating them from their ores.

Metamor'phic (Gr. μετα, *met'a*, implying change ; μορφη, *morphē*, form). Changing form ; a name given in *geology* to those rocks which have undergone a change in their original structure and texture; in *medicine*, applied to diseases having their seat in the processes of development and nutrition.

Metamor'phism (Gr. μετα, *met'a*, implying change ; μορφη, *morphē*, form). Change in form ; a term applied in *geology* to the change in structure and texture which has been undergone by some rocks.

Metamorph'osis (Gr. μετα, *met'a*, implying change ; μορφη, *morphē*, form). A change in shape ; the change undergone by some animals, such as insects and reptiles.

Metano'tum (Gr. μετα, *met'a*, behind ; νωτος, *nōtos*, the back). The upper half of the hinder division of the thorax in insects.

Met'aphor (Gr. μετα, *met'a*, beyond ; φερω, *pher'ō*, I bear). A similitude expressed without the sign of comparison.

Metaphys'ics (Gr. μετα, *met'a*, beyond ; φυσικη, *phu'sikē*, physics, or the science of nature). The science of mind or intelligence.

Metapoph'ysis (Gr. μετα, *met'a*, between ; *apoph'ysis*). A part growing between apophyses.

Metas'tasis (Gr. μετα, *met'a*, beyond ; ιστημι, *histēmi*, I place). A transference of diseases from one place to another.

Metaster'num (Gr. μετα, *met'a*, behind ; στερνον, *sternon*, the breast). The lower part of the posterior division of the thorax in insects.

Metatar'sal (Gr. μετα, *met'a*, beyond ; ταρσος, *tarsos*, the instep). Belonging to the metatarsus.

Metatar'sus (Gr. μετα, *met'a*, beyond ; ταρσος, *tarsos*, the instep). The foot from the ankles to the toes.

Metath'esis (Gr. μετα, *met'a*, implying change ; τιθημι, *tithēmi*, I place). A transposition of the letters or syllables of a word.

Metatho'rax (Gr. μετα, *met'a*, beyond; θωραξ, *thorax*, a breast-plate). The hinder part of the thorax of insects, bearing the posterior pair of wings and legs.

Me'teor (Gr. μετεωρος, *meteū'ros*, lifted up ; from μετα, *met'a*, beyond ; αιρω, *airō*, I raise up). Any atmospheric appearance or phenomenon of a transitory nature.

Meteor'ic (*Me'teor*). Relating to meteors.

Meteoric Stones. Aërolites, or masses of hard matter, containing metallic iron, nickel, and other bodies, occasionally falling on the earth.

Me'teorite (*Me'teor*). A solid substance falling on the earth from the higher regions of the atmosphere.

Me'teorolite (*Me'teor* ; Gr. λιθος, *lith'os*, a stone). *See* Meteorite.

Meteorol'ogy (*Me'teor* ; λογος, *logos*, a description). The science which describes atmospherical phenomena, whether accidental or permanent.

Meth'yl (Gr. μεθυ, *meth'u*, wine; ὑλη, *hulē*, material). An hypothetical compound of carbon and hydrogen, forming the base of certain compounds, as wood-spirit and chloroform, analogous to the alcohol series.

Meton'ic Cycle (Μητων, *Mētōn*, an Athenian astronomer). A cycle or period of nineteen years, at the end of which the lunations of the moon return to the same days of the month as at first.

Meton'ymy (Gr. μετα, *met'a*, implying change ; ονομα, *on'oma*, a name). A putting one word for another which has some relation to it ; as an effect for a cause ; an author's name for his writings ; &c.

Me'tre (Gr. μετρον, *metron*, a measure). A French measure of length, being the ten-millionth part of the distance from the equator to the north pole, equal to 39·37 English inches.

Met'ronome (Gr. μετρον, *metron*, a measure ; νομος, *nom'os*, a law). An instrument consisting of a pendulum suspended by a point between the extremities, used for measuring by its vibrations the quickness or slowness of musical compositions.

Mezzotin'to (Italian *mezzo*, middle or half; *tinto*, painted). A manner of engraving on copper, in which the lights of the figure represented are obtained by the erasure of dents and furrows previously scratched on the plate.

Mias'ma (Gr. μιαινω, *miaī'nō*, I taint or pollute). Effluvia floating in the air, often injurious to health.

Miasmat'ic (Gr. μιασμα, *mias'ma*). Pertaining to or characterised by miasma.

Mi'ca (Lat. *mī'co*, I glitter). A soft glistening mineral, chiefly composed of silica, potash, and magnesia; it forms the glistening scaly appearance in granite.

Mica-schist. A slaty rock, of which mica is the principal ingredient, together with quartz.

Mica'ceous (*Mica*). Belonging to or resembling mica, or chiefly consisting of mica.

Micro- (Gr. μικρος, *mikros*, small.) A prefix in compound words, signifying smallness.

Microm'eter (Gr. μικρος, *mikros*, small; μετρον, *met'ron*, a measure). An instrument for measuring small bodies or spheres, or small visual angles formed by remote objects, by means of which the magnitude of bodies seen through the telescope or microscope may be ascertained.

Mi'cropyle (Gr. μικρος, *mikros*, small; πυλη, *pulē*, a gate). The opening or foramen in a seed, towards which the radicle is always pointed.

Mi'croscope (Gr. μικρος, *mikros*, small; σκοπεω, *skop'eō*, I look at). An optical instrument formed of lenses which magnify the image of small objects placed in their focus, so as to render them visible or more distinct than before.

Microscop'ical (*Mi'croscope*). Relating to the microscope; visible by means of the microscope.

Midrib (*Mid* and *rib*). The principal vein of a leaf, which runs from the stem to the point.

Mil'iary (Lat. *mil'ium*, millet). Like millet-seeds; applied to an eruptive disease characterised by the presence of innumerable white pimples.

Milky Way. An appearance of nebulous light extending over a large extent of the celestial sphere, and found by the telescope to consist of countless multitudes of stars, so crowded as to give the place they occupy a whitish appearance.

Mil'ligramme (Lat. *mil'le*, a thousand; Fr. *gramme*). A French weight of a thousandth part of a *gramme*, or ·015 English grain.

Millime'tre (Lat. *mil'le*, a thousand; Fr. *mètre*). A French measure, equal to the thousandth part of a *mètre*, or ·03937 English inch.

Mimet'ic (Gr. μιμεομαι, *mim'eomai*, I imitate). Imitative.

Min'eral (*Mine*). A body destitute of organisation, existing naturally within the earth or at its surface.

Mineral'ogy (*Mineral;* Gr. λογος, *logos*, a description). The science which describes the properties and relations of simple mineral substances.

Min'imum (Lat. *min'imus*, least). The least quantity assignable in a given case.

Min'ium (Lat.) A compound of protoxide and deutoxide of lead, of a red colour.

Min'uend (Lat. *min'uo*, I diminish). That which is to be diminished; in *arithmetic*, the number from which another is to be subtracted or taken.

Min'ute (Lat. *minu'tus*, diminished). A sixtieth part of an hour or degree.

Mi'ocene (Gr. μειων, *meiōn*, less; καινος, *kainos*, new). A name given in geology to the middle group of the tertiary strata, from its containing a less number of shells identical with existing species than the upper or pliocene group.

Mira'ge (Fr.) The name given to an atmospheric phenomenon, consisting in the appearance in the air of inverted images of distant objects, produced by the rays of light proceeding from them through a dense stratum of air falling on the surface of a rarer stratum, and being, under certain conditions, reflected downwards.

Mi'tral (Lat. *mi'tra*, a head-dress, or mitre). Resembling a mitre; applied to the valve at the orifice of the left ventricle of the heart.

Mi'triform (Lat. *mi'tra*, a mitre; *forma*, shape). Shaped like a mitre.

Mnemon'ics (Gr. *μναομαι, mna'omai,* I remember). The art of assisting the memory.

Mobil'ity (Lat. *mo'bilis*, moveable). Capability of being moved.

Mo'dal (Lat. *mo'dus*, manner). Relating to manner or form; in *logic*, applied to propositions which show the manner in which the predicate is connected with the subject.

Mod'ule (Lat. *mod'ulus*, a measure). A model: in *architecture*, a measure taken to regulate the proportions of an edifice; generally the semi-diameter of the column at the bottom of the shaft.

Mo'lar (Lat. *mo'la*, a mill). Grinding; applied to the large double teeth by which the food is ground.

Molec'ular (*Mol'ecule*). Consisting of or relating to molecules.

Molec'ular Attraction. That form of attraction which operates on the molecules or particles of a body.

Molec'ular Forces. The attractive and repulsive forces existing between the molecules of a body.

Mol'ecules (Lat. *mo'les*, a mass; *ule*, denoting smallness). A very minute particle of a mass.

Mollit'ies (Lat. softness). In *medicine*, a diseased softening of various parts.

Mollus'ca (Lat. *mol'lis*, soft). A division of invertebrate animals, so called from the softness of their bodies; comprising cephalopods, pteropods, gasteropods, acephala, and brachiopods.

Mollus'coid (*Mollus'ca*; Gr. *εἶδος, eidos,* form). A subdivision of the molluscous division, including tunicata and bryozoaria.

Momen'tum (Lat. *moveo,* I move). The force which a moving mass of matter exercises against an object with which it comes into contact, being the product of its quantity of matter and its velocity.

Mon- or **Mon'o-** (Gr. *μονος, mon'os,* alone). A prefix in compound words signifying single.

Mon'ad (Gr. *μονος, mon'os, single*). An ultimate atom; a name given to the smallest of visible animalcules.

Monadel'phia (Gr. *μονος, mon'os,* single; *ἀδελφος, adelphos,* a brother). A class of plants in the Linnean system, in which all the stamens are united in a cylindrical body, through the midst of which the pistil passes.

Monan'dria (Gr. *μονος, mon'os,* single; *ἀνηρ, anēr,* a man). A class of plants in the Linnæan system, having only one stamen.

Mongo'lian (*Mongol*). A term applied to a class of mankind having the Mongols and Chinese as the type.

Monil'iform (Lat. *moni'le,* a necklace; *for'ma,* shape). Like a necklace; beaded.

Monoba'sic (Gr. *μονος, mon'os,* single; *βασις, ba'sis,* a foundation). Having a single atom of base.

Monocar'pous (Gr. *μονος, mon'os,* single; *καρπος, kar'pos,* fruit). Bearing a single fruit.

Monochlamyd'eous (Gr. *μονος, mon'os,* single; *χλαμυς, chlamus,* a tunic). Applied to flowers having a single envelope.

Monocli'nate (Gr. *μονος, mon'os,* single; *κλινω, klinō,* I bend). Having one of the axes turned obliquely; applied in *mineralogy* to certain crystals.

Mon'ochord (Gr. *μονος, mon'os,* single; *χορδη, chordē,* a chord or string). A musical instrument or apparatus of one string, used for the purpose of determining the rates of vibration of musical notes.

Monochromat'ic (Gr. *μονος, mon'os,* single; *χρωμα, chrōma,* colour). Of one colour only.

Monocotyle′donous(Gr. μονος, mon′os, single; cotyle′don). Having one cotyledon or seed-lobe.

Monoc′ular (Gr. μονος, mon′os, one; Lat. oc′ulus, an eye). Having one eye only.

Monœ′cia (Gr. μονος, mon′os, single; οικος, oikos, a house). A class of plants in the Linnæan system, having the stamens and pistils in different flowers, but on the same plant.

Monogam′ia (Gr. μονος, mon′os, single; γαμος, gamos, marriage). An order of plants in the Linnæan system having the anthers united.

Mon′ogram (Gr. μονος, mon′os, single; γραμμα, gramma, a letter). A character composed of two or more letters interwoven.

Mon′ograph (Gr. μονος, mon′os, single; γραφω, graphō, I write). A treatise or book on one subject or class of subjects.

Monogyn′ia (Gr. μονος, mon′os, single; γυνη, gunē, a female). An order of plants in the Linnæan system, consisting of plants having one pistil.

Monoma′nia (Gr. μονος, mon′os, single; μανια, ma′nia, madness). A form of insanity in which the mind is deranged with regard to one idea.

Monome′ra (Gr. μονος, mon′os, single; μηρον, mēron, a thigh). A section of hemipterous insects having only one joint in the tarsi.

Monomor′phous (Gr. μονος, mon′os, single; μορφη, morphē, form). Of a single form.

Monomy′ary (Gr. μονος, mon′os, single; μυς, mus, a muscle). Having one muscle; applied to certain bivalve mollusca, of which the shell is closed by a single muscle.

Monopet′alous (Gr. μονος, mon′os, single; πεταλον, pet′alon, a petal). Having petals united by their margins.

Mon′optote (Gr. μονος, mon′os, single; πτωσις, ptōsis, a case). A noun having only one case besides the nominative.

Monorgan′ic (Gr. μονος, mon′os, single; ὁργανον, or′ganon, an organ). Belonging to or affecting one organ or set of organs.

Monosep′alous (Gr. μονος, mon′os, single; sepal). Having sepals united by their margins.

Monosper′mous (Gr. μονος, mon′os, single; σπερμα, sper′ma, a seed). Having a single seed.

Monosyllab′ic (Gr. μονος, mon′os, single; συλλαβη, sul′labē, a syllable). Having one syllable only.

Monothal′amous (Gr. μονος, mon′os, single; θαλαμος, thal′amos, a chamber). Having one chamber only; not divided by partitions.

Monotre′matous (Gr. μονος, mon′os, single; τρημα, tre′ma, a hole or opening). Having only one external opening for the passage of excreted matter; applied to a small class of mammalia.

Monsoon. A name given to a modification of the course of the trade-winds in the eastern seas.

Moraine. A name given to the longitudinal mounds of stony detritus found at the bases and along the edges of glaciers.

Morbid (Lat. mor′bus, disease). Relating to disease; diseased.

Morbid Anatomy. The study of the alterations produced in the structure of the body by disease.

Morbif′ic (Lat. morbus, disease; fac′io, I make). Causing disease.

Morbil′li (Lat.). The measles.

Mordant (Fr. biting; from Lat. mor′deo, I bite). Any substance employed in dyeing for the purpose of fixing the colours.

Mor′phia (Gr. Μορφευς, Morpheus, the god of sleep). A vegetable alkaloid obtained from opium.

Morpholog′ical (Gr. μορφη, morphē, form; λογος, logos, description). Relating to modifications of form.

Morphol′ogy (Gr. μορφη, morphē, form; λογος, logos, a description). The study of the forms which different organs or parts assume, and of the laws that regulate their changes.

Mortifica′tion (Lat. mors, death; fac′io, I make). Loss of vitality or life in some part of a living body.

Mortise. A cavity cut in a piece of wood or other material, to receive a corresponding projecting piece called a tenon.

Mososau'rus (*Mo'sa*, the river Meuse; Gr. σαυρος, *sauros*, a lizard). A large fossil reptile found in the cretaceous formation.

Motor (Lat. *mov'eo*, I move). Producing or regulating motion; applied to certain nerves and muscles.

Mouldings. A term applied to all the varieties of outline or contour given to the surfaces or edges of the various subordinate parts of buildings, whether projections or depressions.

Mu'cilage (Lat. *mu'cus*). A kind of gum found in vegetables; a solution of gum in water.

Mucilag'inous (*Mu'cilage*). Pertaining to or of the nature of mucilage.

Mucor (Lat.). Mouldiness.

Mu'cous (Lat. *mu'cus*, slime). Pertaining to or of the nature of mucus; secreting mucus.

Mucous Membrane. A membrane secreting mucus, and lining internal passages and other cavities which open on the surface of the body, as well as the cavities which open into these passages.

Mu'cronate (Lat. *mu'cro*, the sharp point of a weapon). Ending in a stiff point.

Mucus (Lat.). The slimy substance effused on the surface of the membranes covering the inner surface of the body, as the alimentary canal, nose, lungs, &c.

Multi- (Lat. *mul'tus*, many or much). A prefix in compound words, signifying many.

Multan'gular (Lat. *mul'tus*, many; *an'gulus*, an angle). Having many angles.

Multiartic'ulate (Lat. *mul'tus*, many; *artic'ulus*, a joint). Having many joints.

Multicus'pidate (Lat. *mul'tus*, many; *cus'pis*, the point of a weapon). Having several points or tubercles; applied to the molar teeth.

Multicos'tate (Lat. *mul'tus*, many; *cos'ta*, a rib). Having many ribs.

Mul'tifid (Lat. *mul'tus*, many; *fin'do*, I cleave). Having many divisions; in *botany*, applied to leaves divided laterally about the middle between the edge and the midrib into numerous divisions.

Mul'tiform (Lat. *mul'tus*, many; *forma*, shape). Having many shapes.

Multilat'eral (Lat. *mul'tus*, many; *latus*, a side). Having many sides.

Multilin'ear (Lat. *mul'tus*, many; *li'nea*, a line). Having many lines.

Multiloc'ular (Lat. *mul'tus*, many; *loc'ulus*, a little place). Having many cells or chambers.

Multino'date (Lat. *mul'tus*, many; *nodus*, a knot). Having many knots.

Multino'mial (Lat. *mul'tus*, many; *nomen*, a name). Having many names or terms; applied in *algebra* to quantities consisting of several names or terms.

Multip'arous (Lat. *mul'tus*, many: *par'io*, I bring forth). Producing many young at a birth.

Multipar'tite (Lat. *mul'tus*, many: *par'tio*, I divide). Divided into many parts; applied in *botany* to leaves having numerous and deep divisions.

Mul'tiple (Lat. *mul'tus*, many; *plic'o*, I fold). Containing many times; a common multiple of two or more numbers is a number which can be divided by each of them without leaving a remainder.

Mul'tiplicand (Lat. *multip'lico*, I multiply). The number which is to be multiplied.

Multiplica'tion (Lat. *mul'tus*, many; *pli'co*, I fold). The process of repeating a quantity a certain number of times, as though it were repeatedly folded on itself.

Mul'tiplier (*Multiply*). That which multiplies; an instrument for indicating the deflecting influence of a weak electric current: so called because the influence of the current is multiplied by being conducted several times round a magnetic needle.

Multiply (Lat. *multus*, many; *pli'co*, I fold). To increase a quantity a given number of times.

Multiplying Glass. A kind of lens presenting a number of plane surfaces, so that the rays of light from an object enter the eye in different directions, and make the object appear as if increased in number.

Multispi'ral (Lat. *multus*, many; *spira*, a spire). Having many spiral turns.

Mul'tivalve (Lat. *multus*, many; *valvæ*, folding doors). Having many valves.

Multoc'ular (Lat. *multus*, many : *oc'ulus*, an eye). Having many eyes.

Multiun'gulate (Lat. *multus*, many ; *un'gula*, a hoof). Having the hoof divided into more than two parts.

Mural Circle (Lat. *murus*, a wall). An astronomical instrument, consisting of a large graduated metal circle, carried on an axis placed horizontally in the face of a stone wall or pier; it has a telescope fixed on it, and is so arranged that the whole instrument, including the telescope, moves on its axis in the plane of the meridian; it is used to determine with precision the instant at which an object passes the meridian.

Murex'ide (Lat. *murex*, a shell-fish yielding a purple dye). Purpurate of ammonia; an organic compound, which forms a purple colour with solution of potash.

Mu'riate (Lat. *mu'ria*, salt water). A term formerly applied to chlorides, on the supposition that they were compounds of muriatic acid with a base.

Muriat'ic (Lat. *mu'ria*, salt water). Relating to brine or salt-water, an old name for hydrochloric.

Mu'riform (Lat. *murus*, a wall; *forma*, shape). Like a wall; arranged like bricks on a wall.

Musch'elkalk (Germ. *muschel*, a shell; *kalk*, lime). Shell-lime-stone; a series of the Triassic

system in *geology* found in Germany, consisting of a compact greyish limestone, abounding in fossil remains.

Mus'cites (Lat. *muscus*, moss). Fossil plants of the moss family.

Muscle (Lat. *mus'culus*, a little mouse). An organ by which the active movements of the body are produced ; the name is derived probably from the shape of some of the muscles.

Mus'cular (Lat. *mus'culus*, a muscle). Relating to or performed by muscles ; provided with muscles.

Muscular Tissue. The tissue which forms the substance of muscles.

Mute (Lat. *mutus*, dumb). In *grammar*, applied to consonants which intercept the voice, as *k*, *p*, and *t*.

Myal'gia (Gr. μυς, *mus*, a muscle ; ἀλγος, *algos*, pain). Pain in muscles.

Myce'lia (Gr. μυκης, *mukēs*, a fungus). The flocculent filaments of fungi.

Mycol'ogy (Gr. μυκης, *mukēs*, a fungus; λογος, *logos*, a discourse). A description of fungi.

Myelenceph'ala (Gr. μυελος, *mu'elos*, marrow ; ἐγκεφαλον, *enkeph'alon*, brain). Animals possessing a brain and spinal chord ; vertebrate animals.

Myeli'tis (Gr. μυελος, *mu'elos*, marrow : *itis*, denoting inflammation). Inflammation of the spinal cord.

My'lodon (Gr. μυλος, *mulos*, a mill ; ὀδους, *odous*, a tooth). An extinct animal ; so named from the flat grinding surfaces of its molar teeth.

Myol'ogy (Gr. μυς, *mus*, a muscle ; λογος, *logos*, a discourse). A description of muscles.

Myo'pia (Gr. μυω, *muō*, I shut ; ὠψ, *ōps*, the eye). Near-sightedness.

Myosi'tis (Gr. μυς, *mus*, a muscle ; *itis*, denoting inflammation). Inflammation of muscles.

Myos'tici (Gr. μυς, *mus*, muscle ; ὀστεον, *os'teon*, bone) A name proposed to be given to diseases affecting bones and muscles.

Myot'omy (Gr. μυς, *mus*, a muscle ; τεμνω, *temnō*, I cut). The anatomy of the muscles.

Myr'iagramme (Gr. μυριοι, *mu'rioi*, ten thousand ; Fr. *gramme*). A French weight of ten thousand *grammes*, or about twenty-two pounds avoirdupois.

Myr'iametre (Gr. μυριοι, *mu'rioi*, ten thousand ; Fr. *mètre*). A French measure of ten thousand *mètres*, or 6·21 English miles.

Myr'iapods (Gr. μυριοι, *mu'rioi*, ten thousand ; τους, *pous*, a foot). A class of invertebrate animals, generally resembling insects, but with numerous legs ; as the centipede.

N.

Na'creous (Fr. *nacre*, mother-of-pearl). Having a pearly lustre.

Nadir (Arabic *natara*, to be like, or correspond). The point in the heavens of the opposite or invisible hemisphere, which would be reached by a perpendicular line drawn from an observer on the surface of the earth, and reaching at the other end a point in the visible hemisphere, called the zenith.

Nævus (Lat.). A tumour consisting essentially in an excessive growth of the vascular tissue of a part.

Na'piform (Lat. *napus*, a turnip ; *forma*, shape). Shaped like a turnip.

Narcot'ic (Gr. ναρκοω, *nar'koō*, I render torpid). Producing insensibility to pain and external impressions, with sleep.

Nar'cotism (Gr. ναρκοω, *nar'koō*, I render torpid). The effect of a narcotic medicine or poison.

Na'res (Lat. *naris*, a nostril). The nostrils.

Na'sal (Lat. *nasus*, the nose). Belonging to the nose ; formed by the nose.

Nascent (Lat. *nascor*, I am born). Beginning to exist : the nascent state of a gas is the condition in which it is at the moment when it is liberated from combination.

Nata'tion (Lat. *nato*, I swim). The act of swimming.

Natato'res (Lat. *nato*, I swim). Swimmers ; an order of birds with feet provided with webs for swimming, as ducks, geese, swans, and gulls.

Na'tatory (Lat. *nato*, I swim). Enabling or assisting to swim ; formed for swimming.

Nat'ural History. The science which describes the natural products of the earth, animal, vegetable, and mineral ; their characters, relations, arrangement, &c.

Nat'ural Philos'ophy. The science which describes the material world, the bodies which compose it, and their qualities and properties.

Nat'ural Sys'tem. The classification of animals or plants into orders, genera, and species, according to their alliances in points of structure which are regarded as essential.

Nau'sea (Gr. ναυς, *naus*, a ship). A disgust for food, with inclination to vomit ; probably at first applied to sea-sickness.

Nau'tical (Gr. ναυτης, *nautēs*, a sailor). Pertaining to seamen or navigation.

Nau'tilites (Lat. *nau'tilus*). Fossil shells apparently allied to the nautilus.

Navic'ular (Lat. *navic'ula*, a boat ; from *naris*, a ship). Belonging to or like a boat ; applied to one of the bones of the wrist, from its shape.

Neb'ula (Lat., a mist). A little cloud or mist : in *astronomy*, an object resembling stars seen through a mist, or a cloudy speck, but found by the telescope to consist of a cluster of stars.

Neb'ular (Lat. *neb'ula*, a mist). Relating to nebulæ ; the nebular hypothesis was a belief that the appearances called nebulæ were the results of the aggregation of a sort of luminous fluid diffused through different parts of the universe.

Neb'ulous (Lat. *neb'ula*, a mist). Misty ; having the appearance of a mist.

Necroph'agous (Gr. νεκρος, *nek'ros*, dead : φαγω, *phag'o*, I eat). Eating dead bodies of animals.

Necrop'olis (Gr. νεκρος, *nek'ros*, dead; πολις, *pol'is*, a city). A city of the dead ; a cemetery.

Nec'ropsy (Gr. νεκρος, *nek'ros*, dead ; ὀψις, *opsis*, sight). The examination of a dead body.

Necroscop'ic (Gr. νεκρος, *nek'ros*, dead ; σκοπεω, *skop'eo*, I view). Relating to the examination of bodies after death.

Necro'sis (Gr. νεκροω, *nek'roo*, I kill). A disease of bone terminating in its death ; a state analogous to mortification or gangrene in soft parts.

Nectariferous (Lat. *nectar; fer'o*, I produce). Having a honey-like secretion : in *botany*, applied to petals having furrows at their base yielding a sweet secretion.

Nec'tary (*Nectar*). In *botany*, any abnormal part of a flower ; but properly any organ secreting sweet matter.

Neg'ative (Lat. *nego*, I deny). Implying denial or absence ; in *physical science*, applied to one of the forms of electricity which a body is capable of assuming ; in *algebra*, applied to quantities which have the sign — (*minus*) prefixed.

Ne'matoid (Gr. νημα, *nema*, a thread; εἰδος, *eidos*, form). Like a thread ; applied to a class of parasitic worms.

Nematoneu'ra (Gr. νημα, *nema*, a thread : νευρον, *neuron*, a nerve). Having the nervous system arranged in filaments or threads.

Nemoc'era (Gr. νημα, *nema*, a thread; κερας, *ker'as*, a horn). A section of dipterous insects with filiform or thread-like antennæ, of six joints.

Neoco'mian (Lat. *Neocomum*, Neufchâtel). A term applied in *geology* to the green sand formation, which is especially developed in the vicinity of Neufchâtel.

Neol'ogy (Gr. νεος, *neos*, new ; λογος, *logos*, discourse). The introduction of new words or doctrines.

Neoter'ic (Gr. νεωτερος, *neoteros*, younger). Recent in origin ; modern.

Neozo'ic (Gr. νεος, *neos*, new ; ζωον, *zoon*, an animal). Having new animals ; a term applied in *geology* to a division of the fossiliferous strata, including the cainozoic and mesozoic of some geologists.

Nephral'gia (Gr. νεφρος, *neph'ros*, a kidney ; ἀλγος, *algos*, pain). Pain in the kidney.

Nephrit'ic (Gr. νεφρος, *neph'ros*, a kidney). Relating to the kidneys.

Nephri'tis (Gr. νεφρος, *neph'ros*, a kidney ; *itis*, denoting inflammation). Inflammation of the kidneys.

Neptu'nian (*Neptune*, the god of the sea). A term applied to stratified rocks, or those which have been deposited by water.

Nep'tunist (*Neptune*, the god of the sea). A name given to the geologists of the school of Werner, who believed all old rocks to have been of aqueous origin.

Nerve (Lat. *nervus*). A bundle of white fibres, forming an organ for the conveyance of impressions between any part of the body and the brain or spinal cord.

Nervine (Lat. *nervus*, a nerve). Acting on the nerves.

Nervous System. The collection of organs, comprising the brain, spinal cord, and nerves, the office of which is to receive and convey impressions.

Ner'vures (Lat. *nervus*, a sinew). The frame-work of the wings of insects : also applied sometimes, in *botany*, to the frame-work of leaves.

Neural (Gr. νευρον, *neuron*, a nerve). Belonging or having relation to the nervous system.

Neural'gia (Gr. νευρον, *neuron*, a nerve ; ἀλγος, *algos*, pain). Pain having its origin especially in the nerves.

Neurapoph'ysis (Gr. νευρον, *neuron*, a nerve ; *apoph'ysis*). The part projecting from a vertebra. which aids in forming the canal that protects the spinal cord.

Neurilem'ma (Gr. νευρον, *neuron*, a nerve ; λεμμα, *lemma*, a peel or skin). The sheath of a nerve.

Neurine (Gr. νευρον, *neuron*, a nerve). Nervous substance.

Neurol'ogy (Gr. νευρον, *neuron*, a nerve ; λογος, *logos*, discourse). A description of the nerves.

Neuro'ma (Gr. νευρον, *neuron*, a nerve). A swelling or tumour in the course of a nerve.

Neurop'athy (Gr. νευρον, *neuron*, a nerve ; παθος, *path'os*, suffering). Disease of a nerve.

Neurop'tera (Gr. νευρον, *neuron*, a nerve ; πτερον, *pter'on*, a wing). An order of insects with four membranous transparent wings, with a net-work of veins or nervures ; as the dragon-fly.

Neuro'ses (Gr. νευρον, *neuron*, a nerve). A term appled to nervous affectjons or diseases.

Neuroskel'eton (Gr. νευρον, *neuron*, a nerve ; σκελετον, *skel'eton*). The deep-seated bones of the vertebral skeleton which have relation to the nervous system and to locomotion.

Neutral (Lat. *ne*, not ; *uter*, which of the two). In *chemistry*, applied to salts composed of an acid and a base in such proportions that they exactly destroy each other's properties ; in *botany*, applied to flowers having neither stamens nor pistils.

Neutralisa'tion (Lat. *neuter*, neither). In *chemistry*, the process by which an acid is combined with a base in such proportion as to render inert the properties of both.

Neu'tralise (Lat. *neuter*, neither). To render neutral or inert ; to destroy the properties of a body by combining with it another body of different properties.

Nic'otin (*Nicotia'na*, the tobacco plant). A principle obtained from tobacco.

Nic'titate (Lat. *nic'tito*, I wink). To wink.

Nic'titating Membrane. A fold of skin with which birds cover their eyes.

Nidamen'tal (Lat. *nidamen'tum*, the material of which birds make their nests). Relating to the protection of the egg and young ; secreting material for constructing nests.

Nilom'eter (Gr. Νειλος, *Neilos*, the Nile ; μετρον, *met'ron*, a measure). An instrument for measuring the rise of the waters of the Nile.

Ni'trate (*Nitric*). A salt consisting of nitric acid with a base.

Ni'tric (*Nitre*). Produced from nitre or saltpetre ; applied to an acid obtained from nitre or nitrate of potash.

Ni'trite. A salt consisting of nitrous acid and a base.

Ni'trogen (*Nitre ;* Gr. γενναω, *genna'ō*, I produce). An elementary gas, without colour, taste, or smell, forming the larger portion (79 in 100) of the atmospheric air.

Nitrog'enised (*Ni'trogen*). Containing nitrogen.

Nitrog'enous (*Ni'trogen*). Containing nitrogen.

Ni'trous (*Nitre*). Pertaining to nitre ; applied to an acid containing less oxygen than nitric acid.

Nodal (Lat. *nodus*, a knot). Relating to a knot ; applied to the points and lines at which the vibrations of a body become arrested, and which assume various regular forms.

Node (Lat. *nodus*, a knot). A small oval figure made by the intersection of one branch of a curve with another ; in *astronomy*, the point at which the moon or a planet crosses the ecliptic ; in *botany*, the point in a stem from which a leaf-bud proceeds.

Nodo'se (Lat. *nodus*, a knot). Knotty.

Nod'ule (Lat. *nodus*, a knot ; *ule*, denoting smallness). A little knot ; an irregular concretion of rocky matter round a central nucleus.

Nomad'ic (Gr. νομος, *nom'os*, a pasture). Wandering ; subsisting on cattle, and wandering for the sake of pasture.

No'menclature (Lat. *nomen*, a name ; *calo*, from Gr. καλεω, *kaleō*, I call). The collection of names peculiar to science in general, or to any branch of science.

Nom'inative (Lat. *nomen*, a name). Naming ; applied to the first case of nouns, which denotes the name of the person or thing.

Non-conductor. A substance which does not conduct heat, electricity, &c.

Normal (Lat. *norma*, a rule). According to rule ; regular : a perpendicular, especially to a curve at a given point.

Nosog'raphy (Gr. νοσος, *nos'os*, disease ; γραφω, *graph'ō*, I write). A description of diseases.

Nosolog'ical (Gr. νοσος, *nos'os*, disease ; λογος, *logos*, discourse). Relating to a classification of diseases.

Nosol'ogy (Gr. νοσος, *nos'os*, disease ; λογος, *logos*, discourse). The branch of medical science which distributes diseases into classes, orders, genera, and species, and distinguishes diseases by their proper names.

Nostal'gia (Gr. νοστος, *nostos*, return ; ἀλγος, *algos*, pain). Homesickness ; a desire to return to one's country, amounting to disease.

Notal (Gr. νωτος, *nōtos*, the back). Belonging to the back.

Nota'tion (Lat. *noto*, I mark). The marking or reading anything by figures or other characters.

No'tochord (Gr. νωτος, *nōtos*, the back ; χορδα, *chorda*, a cord). The fibro-cellular gelatinous column which forms the primary condition of the spine in vertebrate animals.

Notorhi'zal (Gr. νωτος, *nōtos*, the back ; ῥιζα, *rhiza*, a root). Having the radicle in the embryonic plant on the back of the cotyledons.

Nubec'ula (Lat. a little cloud). In *astronomy*, a name given to the Magellanic clouds, or two extensive nebulous patches of stars.

Nu'chal (Lat. *nucha*, the back of the neck). Belonging to the neck.

Nu'clear (Lat. *nu'cleus*.) Formed of nuclei.

Nu'cleated (Lat. *nu'cleus*, a kernel). Having a nucleus, or central particle.

Nu'cleolus (*Nu'cleus*). A little nucleus ; a small body sometimes observed within the nucleus of an animal or vegetable cell.

Nu'cleus (Lat. a kernel). A body about which matter is collected ; a small compact body found in animal and vegetable cells ; in *astronomy*, the bright central spot sometimes seen in the nebulous or misty matter forming the head of a comet.

Nudibra'chiate (Lat. *nudus*, naked ; *bra'chium*, an arm). Having naked arms ; applied to polypi, the tentacles of which are not covered with cilia.

Nudibran'chiate (Lat. *nudus*, naked ; Gr. βραγχια, *bran'chia*, gills). Having exposed gills ; applied to an order of gasteropodous mollusca which have no shell, and have the gills exposed.

Numera'tion (Lat. *nu'merus*, a number). The art of reading or writing numbers.

Nu'merator (Lat. *nu'merus*, a number). The number in fractions which shows how many of the parts are to be taken.

Numer'ical Method. The branch of science which treats of the right manner of deriving conclusions from the collected numerical statement of the results of certain forces or causes.

Numismat'ic (Lat. *numis'ma* ; from Gr. νομισμα, *nomis'ma*, money). Relating to coins or money.

Numismatol'ogy (Lat. *numis'ma* ; Gr. λογος, *logos*, discourse). The science of describing coins and medals.

Num'mulated (Lat. *nummus*, money). Having some resemblance to a coin.

Num'mulite (Lat. *nummus*, money ; λιθος, *lith'os*, a stone). A fossil shell resembling a coin, found in the limestone in the tertiary strata.

Nuta'tion (Lat. *nuto*, I nod). In *astronomy*, the alternate approach and departure of the pole of the equator to and from the pole of the ecliptic, combined with the alternate increase and decrease of its retrogressive motion ; in *botany*, applied to a property which some flowers have of following the apparent motion of the sun.

Nu′trient (Lat. *nu′trio*, I nourish). Nourishing.

Nu′triment (Lat. *nu′trio*, I nourish). Food ; the material supplied for repairing the waste or promoting the growth of living bodies.

Nutrit′ion (Lat. *nu′trio*, I nourish). The process by which animals or vegetables appropriate to their repair or growth material taken from external organic substances.

Nyctalo′pia (Gr. *νυξ, nux*, the night ; ἀλαομαι, *ala′omai*, I grope about ; ὠψ, *ōps*, the eye). A defect of vision, in which the patient can see by day, but not by night.

O.

Ob (Lat.) A preposition in compound words, signifying against, reversed, or contrary.

Obcompress′ed (Lat. *ob ; com primo*, I press together). Flattened in front and behind.

Obcor′date (Lat. *ob*, against ; *cor*, the heart). Like a heart reversed ; applied in *botany* to leaves shaped like a heart, with the apex next the stem.

Ob′elisk (Gr. ὀβελος, *ob′elos*, a spit). A four-sided column, of one stone, rising in the form of a pyramid, and having a smaller pyramid at the top.

Obe′sity (Lat. *obe′sus*, fat), An excessive fatness.

Ob′ject (Lat. *ob*, against ; *jac′io*, I throw). That which is acted on by the senses, the mental faculties, or other agents.

Object-glass. The lens in a telescope or microscope which first receives the rays of light coming from an object and collects them to a focus or central point, where they form an image which is viewed through the eye-piece.

Objec′tive (Lat. *ob*, against ; *jac′io*, I throw). Belonging to an object ; in *medicine*, applied to symptoms observed by the physician ; in *grammar*, denoting the case which is acted on.

Obla′te (Lat. *ob*, against ; *latus*, borne or carried). Flattened at the poles ; applied to spherical bodies flattened at the poles or ends, like an orange.

Obli′que (Lat. *obli′quus*, sideways) Neither perpendicular nor parallel.

Ob′olite Grit. In *geology*, the lower Silurian sandstones of Sweden and Russia, from the abundance of shells of the *obolus*, a brachiopod mollusk.

Obo′vate (Lat. *ob ; ovate*). Reversely ovate, the broad end of the egg being uppermost.

Observa′tion (Lat. *obser′vo*, I observe). The art of observing ; one of the processes by which natural phenomena are to be investigated.

Obser′vatory (Lat. *obser′vo*, I observe). A place or building constructed for astronomical observations.

Obsid′ian (Lat. *obsidia′num vitrum*, a kind of thick glass). A glassy lava, much resembling artificial glass, but usually black and nearly opaque ; it consists of silica and alumina, with a little potash and oxide of iron.

Obsoles′cence (Lat. *obsoles′co*, I grow out of use). The state of becoming disused ; in *medicine*, applied to the stage in diseased formations at which they cease to undergo further change.

Ob′solete (Lat. *obsoles′co*, I grow out of use). In *botany*, imperfectly developed or abortive.

Obstet′ric (Lat. *obstet′rix*, a midwife). Relating to midwifery.

Obtec′ted (Lat. *ob′tego*, I cover over). Covered over ; applied to a form of metamorphosis in insects in which the wings and limbs are lodged in recesses in the integument of the pupa.

Ob'turator (Lat. *obturo*, I stop up). That which stops up ; a name applied to two muscles, which arise near an opening in the pelvis called the obturator or thyroid foramen.

Obtusan'gular (Lat. *obtu'sus*, blunt ; *an'gulus*, an angle). Having angles larger than right angles.

Obtu'se (Lat. *obtu'sus*, blunt). In *geometry*, applied to angles which are larger than right angles.

Ob'verse (Lat. *ob*, opposite ; *verto*, I turn.) The side of a coin which has the face or head on it.

Ob'volute (Lat. *ob*, against ; *volvo*, I roll). Rolled into ; in *botany*, applied to an arrangement of leaves in buds in which the margins of one leaf alternately overlap those of the leaf opposite to it.

Occiden'tal (Lat. *oc'cidens*, the west ; from *ob*, down ; *cado*, I fall, in allusion to the setting of the sun). Relating to or produced in the west.

Occip'ital (Lat. *oc'ciput*, the back of the head). Belonging to the back of the head.

Oc'ciput (Lat. *ob*, opposite ; *cap'ut*, the head). The back part of the head.

Occulta'tion (Lat. *occul'to*, I hide). A hiding ; the concealment from sight of a star or planet, by the interposition of another body.

Ocel'lus (Lat. *oc'ulus*, an eye). A little eye ; one of the small eyes of which the compound organs of vision are formed in many invertebrate animals.

Ochle'sis (Gr. ὀχλος, *ochlos*, a multitude). A crowding together.

Ochre (Gr. ὠχρος, *ōchros*, pale). A fine clay, coloured by more or less peroxide of iron.

O'chrea or O'crea (Lat. a boot). In *botany*, the tube formed in some plants by the growing together of the stipules, through which the stem passes.

Oct- or Octo- (Gr. ὀκτω, *oktō*, eight). A prefix in compound words implying eight.

Oc'tagon (Gr. ὀκτω, *oktō*, eight ; γωνια, *gōnia*, an angle). A figure having eight angles.

Octagyn'ia (Gr. ὀκτω, *oktō*, eight ; γυνη, *gunē*, a female). An order of plants in the Linnean system, having eight pistils.

Octahed'ron (Gr. ὀκτω, *oktō*, eight ; ἑδρα, *hed'ra*, a base). A solid figure bounded by eight equal sides, each of which is an equilateral triangle.

Octan'dria (Gr. ὀκτω, *oktō*, eight ; ανηρ, *anēr*, a man). A class of plants in the Linnean system having eight stamens.

Octan'gular (Lat. *octo*, eight ; *an'gulus*, an angle). Having eight angles.

Oc'tant (Lat. *octo*, eight). The eighth part of a circle ; the aspect of two planets in which they are distant from each other the eighth part of a circle, or forty-five degrees.

Oc'tastyle (Gr. ὀκτω, *oktō*, eight ; στυλος, *stulos*, a pillar). A building having eight columns in front.

Oc'tave (Lat. *octa'vus*, the eighth). In *music*, a collection of eight consecutive notes, of which the eighth (or highest) is produced by twice the number of vibrations which form the first or lowest.

Oc'topod (Gr. ὀκτω, *oktō*, eight ; πους, *pous*, a foot). An animal having eight feet or legs ; a tribe of cephalopods so called.

Oc'ular (Lat. *oc'ulus*, an eye). Relating to the eyes.

Oc'uliform (Lat. *oc'ulus*, an eye ; *forma*, form). Having the form of an eye.

Oc'ulist (Lat. *oc'ulus*, an eye). A person who treats disorders of the eyes.

-Ode or -Odes (Gr. ωδης, *ōdēs*). A termination generally denoting abundance of that substance which is implied by the previous part of the word.

Ode'um (Gr. ὠδειον, *ōdeion* ; from ωδη, *ōdē*, a song). A small theatre for the recitation of musical compositions.

Odom'eter (Gr. ὁδος, *hod'os*, a way ; μετρον, *met'ron*, a measure). An instrument for measuring the distance travelled over by the wheels of a carriage.

Odontal'gia (Gr. ὁδους, *od'ous*, a tooth ; ἀλγος, *algos*, pain). Toothache.

Odon'tograph (Gr. ὁδους, *od'ous*, a tooth ; γραφω, *graph'ō*, I write). An instrument for measuring and designing the teeth of wheels.

Odon'toid (Gr. ὁδους, *od'ous*, a tooth ; εἰδος, *eidos*, shape). Like a tooth ; applied in *anatomy* to a process of the second vertebra of the neck, also to ligaments connected with it.

Odontol'ogy (Gr. ὁδους, *od'ous*, a tooth ; λογος, *logos*, discourse). A description of the teeth.

Odorif'erous (Lat. *odor*, smell ; *fer'o*, I carry). Giving or carrying scent.

·Œcious (Gr. οἰκος, *oikos*, a house or family). A termination used in *botany*, in reference to the arrangement of the stamens and pistils in flowers.

Œde'ma (Gr. οἰδεω, *oi'deō*, I swell). A swelling ; in *medicine*, a minor form of dropsy, consisting in a puffiness of parts from a collection of fluid in the tissue beneath the skin.

Œdem atous (Gr. οἰδεω, *oi'deō*, I swell). Having œdema.

Œnan'thic (Gr. οἰνος, *oinos*, wine ; ἀνθος, *anthos*, a flower). A term applied to a liquid or ether supposed to give its aroma to wine.

Œsoph'agus (Gr. οἰω, *oiō*, I carry ; φαγω, *phag'ō*, I eat). The gullet ; the tube which conveys the food from the mouth to the stomach.

Œsophage'al (*Œsoph'agus*, the gullet). Belonging to the œsophagus.

Œsophagot'omy (*Œsoph'agus ;* Gr. τεμνω, *temnō*, I cut). The operation of cutting into the œsophagus.

Offic'inal (Lat. *offici'na*, a workshop). Kept in shops.

Ogee. In *architecture*, a form of moulding consisting of two members, the one concave and the other convex.

·Oid (Gr. εἰδος, *eidos*, form). A ter-

mination implying likeness or alliance.

Oinoma'nia (Gr. οἰνος, *oinos*, wine ; μανια, *ma'nia*, madness). An insane desire for wine or alcoholic drinks.

Old Red Sandstone. See Sandstone.

Oleag'inous (Lat. *o'leum*, oil). Having the properties of or containing oil.

O'leate (Lat. *o'leum*, oil). A compound of oleic acid with a base.

Olec'ranon (Gr. ὠλενη, *ōlenē*, the elbow ; κρανος, *kranos*, a helmet). The projecting part of the upper end of the ulna, forming the back of the elbow.

Ole'fiant (Lat. *o'leum*, oil ; *fac'io*, I make). Making oil ; applied to a gas consisting of carbon and hydrogen, from its forming an oily liquid when mixed with chlorine.

O'leic (Lat. *o'leum*, oil). Belonging to oil : applied to an acid obtained from oil.

O'lein (Lat. *o'leum*, oil). The thin oily part of oils and fats.

Olfac'tory (Lat. *olfac'io*, I smell). Relating to the sense of smelling.

Olfac'tory Nerves. The first pair of nerves proceeding directly from the brain, being the nerves of smelling.

Oligæ'mia (Gr. ὀλιγος, *ol'igos*, little ; αἱμα, *haima*, blood). That state of the system in which there is a deficiency of blood.

Oligan'drous (Gr. ὀλιγος, *ol'igos*, few ; ἀνηρ, *anēr*, a male). Having fewer than twenty stamens.

Ol'igo- (Gr. ὀλιγος, *ol'igos*, little). A prefix in compound words, signifying defect in quantity or number.

Ol'ivary (Lat. *oli'va*, an olive). Resembling an olive.

Oma'sum. In *comparative anatomy*, the third stomach, or manyplies, of ruminant animals.

Omen'tal (*Omen'tum*). Belonging to the omentum.

Omen'tum (Lat.). The caul : a fold of the peritoneal membrane covering the intestines in front.

Omniv'orous (Lat. *omnis*, all ; *voro*, I devour.) Eating both animal and vegetable food.

Omo- (Gr. *αμος, ōmos,* the shoulder.) A prefix in compound words, signifying connection with the scapula or shoulder-blade.

Omohy'oid (Gr. *ωμος, ōmos,* the shoulder; *hyoid* bone). A name given to a muscle attached to the hyoid bone and the shoulder.

Onguic'ulate and On'gulate. *See* Unguic'ulate and Un'gulate.

Onom'atopœia (Gr. *ονομα, on'oma,* a name; *ποιεω, poi'eō,* I make). A formation of words so as to produce a real or fancied resemblance to the sounds which they are intended to describe.

Ontolog'ical (Gr. *ων, ōn,* being; *λογος, logos,* discourse). Relating to the science of beings or existing things.

Ontol'ogy (Gr. *ων, ōn,* being; *λογος, logos,* discourse). The science of being; that part of metaphysics which investigates and explains the nature of beings.

Onych'ia (Gr. *ονυξ, on'ux,* a nail.) A whitlow.

O'olite (Gr. *ωον, ōon,* an egg; *λιθος, lith'os,* a stone). Limestone composed of small rounded particles like the eggs or roe of a fish : the name *in geology* of a system of stratified rocks, characterised by the presence of limestone of this description.

Oolit'ic (Gr. *ωον, ōon,* an egg; *λιθος, lith'os,* a stone). Pertaining to the oolite.

Opales'cence (*Opal*). A coloured shining lustre reflected from a single spot in a mineral.

Oper'cular (Lat. *oper'culum,* a lid). Having, or of the nature of, a lid or cover.

Oper'culated (Lat. *oper'culum,* a lid). Provided with an operculum or cover.

Oper'culum (Lat. *oper'io,* I cover). A lid or cover.

Ophid'ians (Gr. *οφις, oph'is,* a serpent). An order of reptiles, having the serpent as the type.

Ophiol'ogy (Gr. *οφις, oph'is,* a serpent; *λογος, logos,* discourse). The description of serpents.

Oph'ite (Gr. *οφις, oph'is,* a serpent). The mineral called serpentine.

Ophthal'mia (Gr. *οφθαλμος, ophthal'mos,* the eye). Inflammation of the eye.

Ophthal'mic (Gr. *οφθαλμος, ophthal'mos,* the eye). Belonging to the eye.

Ophthalmol'ogy (Gr. *οφθαλμος, ophthal'mos,* the eye; *λογος, logos,* discourse). The part of anatomical science which describes the eyes and whatever relates to them.

Ophthalmom'eter (Gr. *οφθαλμος, ophthal'mos,* the eye; *μετρον, met'ron,* a measure). An instrument for measuring and comparing the powers of vision of the two eyes.

Ophthal'moscope (Gr. *οφθαλμος, ophthal'mos,* the eye; *σκοπεω, skop'eō,* I view). An instrument for examining the interior of the eye.

O'piate (*O'pium*). A medicine containing opium.

Opisthocœ'lian (Gr. *οπισθεν, opis'then,* backwards; *κοιλος, koilos,* hollow). Having the vertebræ hollow at the back part.

Opisthot'onos (Gr. *οπισθεν, opis'then,* backwards; *τεινω, teinō,* I stretch). A form of tetanus in which the body is bent backwards.

Opposit'ion (Lat. *ob,* against; *pono,* I place.) A standing over against; in *astronomy,* the position of a heavenly body, as seen from the earth, in the quarter directly opposite the sun, so that the earth lies in a direct line between it and the sun.

Opsiom'eter (Gr. *οψις, opsis,* vision; *μετρον, met'ron,* a measure). A measurer of sight, or of the power of vision.

Opta'tive (Lat. *opto,* I wish). Wishing : applied, in *grammar,* to that mode or form of the verb by which desire is expressed.

Optic (Gr. *οπτομαι, op'tomai,* I see). Relating to sight, or to the laws of vision.

Optic Nerves. The second pair of nerves proceeding directly from the brain, being the nerves of sight.

Optics (Gr. *οπτομαι, op'tomai,* I see).

The branch of *natural philosophy* which treats of the nature and properties of light, the theory of colours, the changes produced on light by the substances with which it comes into contact, and the structure of the eye and of instruments for aiding vision.

Optom'eter. *See* Opsiom'eter.

Oral (Lat. *os*, the mouth). Belonging to or uttered by the mouth.

Orbic'ular (Lat. *orbic'ulus*, a small round ball, from *orbis*, a round thing). Circular ; in *anatomy*, applied to the muscles which surround and close the eyelids and mouth.

Orbit (Lat. *orbis*, a wheel). In *astronomy*, the curved course in which any body, as the moon or a planet, moves in its revolution round a central body ; in *anatomy*, the cavity or socket in which the eye is situated.

Or'bital (*Orbit*). Belonging to the orbit.

Or'bito-sphenoid. A term applied to the lesser wing of the sphenoid bone, which forms part of the orbit.

Order (Lat. *ordo*). A group of genera, agreeing in more general characters, but differing in special conformation.

Or'dinate (Lat. *ordo*, order). In *conic sections*, a straight line drawn from a point in the abscissa to terminate in the curve.

Organ (Gr. ὀργανον, *or'ganon*, an instrument, from ἐργω, *ergō*, I work). A natural instrument, by which some process or function is carried on.

Organ'ic (Gr. ὀργανον, *or'ganon*, an instrument). Consisting of or possessing organs ; relating to bodies which have organs ; in *geology*, applied to the accumulations or additions made to the crust of the earth in various places by the agency of animals or vegetable matter, and to the fossil remains of animals and vegetables ; in *medicine*, applied to diseases in which the structure of an organ is evidently altered.

Or'ganism (Gr. ὀργανον, *or'ganon*, an instrument). The assemblage of living forces or instruments constituting a body.

Or'ganize (Gr. ὀργανον, *or'ganon*, an instrument). To form with suitable organs, so that the whole may work together in a body.

Organog'eny (Gr ὀργανον, *or'ganon*, an instrument ; γενναω, *genna'ū*, I produce). The development of organs.

Organog'raphy (Gr. ὀργανον, *or'ganon*, an instrument ; γραφω, *graph'ū*, I write). A description of organs ; used especially with regard to plants.

Organol'ogy (Gr. ὀργανον, *or'ganon*, an instrument ; λογος, *logos*, discourse). A description of organs, especially of the animal body.

Orien'tal (Lat. *o'riens*, the east, from *o'rior*, I arise). Eastern : relating to the east.

Ornithich'nites (Gr. ὀρνις, *ornis*, a bird ; ἰχνος, *ichnos*, a footstep). Fossil footprints of birds.

Orni'tholites (Gr. ὀρνις, *ornis*, a bird ; λιθος, *lith'os*, a stone). The fossil remains of birds.

Ornithol'ogist (Gr. ὀρνις, *ornis*, a bird ; λογος, *logos*, discourse). A person who is skilled in the knowledge of birds.

Ornithol'ogy (Gr. ὀρνις, *ornis*, a bird ; λογος, *logos*, discourse). The branch of zoology which describes birds.

Orol'ogy (Gr. ὀρος, *or'os*, a mountain ; λογος, *logos*, discourse). The science which describes mountains.

Or'rery. A machine to represent the motions and aspects of the planets in their orbits.

Ortho- (Gr. ὀρθος, *orthos*, staight). A prefix in compound words, signifying straight.

Orthocer'atite (Gr. ὀρθος, *orthos*, straight ; κερας, *ker'as*, a horn). A genus of straight horn-shaped fossil shells, with several chambers.

Orthodrom'ics (Gr. ὀρθος, *orthos*, straight ; δρομος, *drom'os*, a course). The art of sailing in the arc of a great circle, being the shortest distance between two points on the surface of the globe.

Or'thoepy (Gr. ὀρθος, orthos, right ; ἐπος, ep'os, a word). The correct pronunciation of words.

Orthog'onal (Gr. ὀρθος, orthos, straight ; γωνια, gōnia, an angle). At right angles, or perpendicular.

Orthog'raphy (Gr. ὀρθος, orthos, right ; γραφω, graph'ō, I write). The art or practice of writing words with the proper letters : in architecture, the elevation of a building, showing all the parts in their due proportions.

Orthopnœ'a (Gr. ὀρθος, orthos, upright; πνεω, pneō, I breath). A diseased state in which breathing can only be performed in the erect position.

Orthop'tera (Gr. ὀρθος, orthos, straight; πτερον, pter'on, a wing). An order of insects, which have the wings disposed, when at rest, in straight longitudinal folds ; as the cricket and grasshopper.

Orthot'ropous (Gr. ὀρθος, orthos, right ; τρεπω, trep'ō, I turn). Turned the right way ; applied in botany to the ovule where its parts undergo no change of position during growth.

Oryctog'nosy (Gr. ὀρυκτος, oruk'tos, fossil, or dug out ; γνωσις, gnōsis, knowledge). The description and classification of minerals.

Oryctol'ogy (Gr. ὀρυκτος, oruk'tos, fossil ; λογος, logos, a discourse). The description of fossils.

Oscilla'tion (Lat. oscil'lum, a swing). A swinging backwards and forwards ; centre of oscillation is the point into which the whole moving force of a vibrating body is concentrated.

Os'cula (Lat. plural of os'culum, a little mouth). The larger orifices on the surface of a sponge.

Os'mazome (Gr. ὀσμη, osmē, odour; ζωμος, zōmos, juice or soup). The name given to the extractive matter of muscular fibre, which gives the smell to boiled meat.

Os'mose (Gr. ωθεω, ōtheō, I impel). The process by which fluids and gases pass through membranes.

Os'seous (Lat. os, a bone). Formed of, or resembling bone.

Os'sicle (Lat. ossic'ulum, from os, a bone ; ulum, denoting smallness). A little bone.

Ossif'erous (Lat. os, a bone ; fer'o, I bear). Producing or containing bones.

Ossif'ic (Lat. os, a bone ; fac'io, I make). Making bone.

Ossifica'tion (Lat. os, a bone ; fac'io, I make). A change into a bony substance ; the formation of bones.

Os'sify (Lat. os, a bone ; fac'io, I make). To form bone ; to become bone.

Ossiv'orous (Lat. os, a bone ; voro, I devour). Eating bones.

Os'teal (Gr. ὀστεον, os'teon, a bone). Belonging to bone.

Os'teine (Gr. ὀστεον, os'teon, a bone). The tissue of bone.

Ostei'tis (Gr. ὀστεον, os'teon, a bone ; itis, denoting inflammation). Inflammation of bone.

Osteoden'tine (Gr. ὀστεον, os'teon, a bone ; Lat. dens, a tooth). A structure formed in teeth, in part resembling bone.

Osteog'eny (Gr. ὀστεον, os'teon, a bone ; γενναω, genna'ō, I produce). The formation or growth of bone.

Osteoid (Gr. ὀστεον, os'teon, a bone ; εἰδος, eidos, form). Resembling bone.

Osteol'ogy (Gr. ὀστεον, os'teon, a bone ; λογος, logos, discourse). A description of the bones.

Osteomala'cia (Gr. ὀστεον, os'teon, a bone; μαλακος, mal'akos, soft). A diseased softening of the bones.

Os'teophyte (Gr. ὀστεον, os'teon, a bone ; φυω, phuō, I grow). A bony tumour or projection.

Os'teotrite (Gr. ὀστεον, os'teon, a bone ; Lat. tero, I rub). An instrument for removing diseased bones.

Osteozoa'ria (Gr. ὀστεον, os'teon, a bone; ζωον, zōon, an animal). A name for the vertebrate division of the animal kingdom, comprising those animals which possess bones.

Ostra'cea (Gr. ὀστρεον, os'treon, an oyster). A family of bivalve molluscous invertebrate animals, of which the oyster is an example.

Ostrap'oda (Gr. ὀστρεον, os'treon, an

oyster; πους, *pous*, a foot). An order of entomostracous crustacea, which have the body enclosed in a bivalve shell.

Otal'gia (Gr. οὖς, *ous*, the ear; ἀλγος, *algos*, pain). Pain in the ear.

Otic (Gr. οὖς, *ous*, the ear). Belonging to the ear.

Oti'tis (Gr. οὖς, *ous*, the ear; *itis*, denoting inflammation). Inflammation of the ear.

O'tocrane (Gr. οὖς, *ous*, the ear; κρανιον, *kranion*, the skull). The part of the skull which is modified for the reception of the organ of hearing.

O'toliths (Gr. οὖς, *ous*, the ear; λιθος, *lith'os*, a stone). Ear-stones; small masses of carbonate of lime contained in the membranous labyrinth of the internal ear.

Otorrhe'a (Gr. οὖς, *ous*, the ear; ῥεω, *rheō*, I flow). A flow or discharge from the ear.

O'toscope (Gr. οὖς, *ous*, the ear; σκοπεω, *skop'eō*, I view). An instrument for listening to the sound passing through the tympanum in diseased states of the ear.

Otos'teal (Gr. οὖς, *ous*, the ear; ὀστεον, *os'teon*, a bone). The ear-bone in the skeleton of fishes.

-Ous. In *chemistry*, a termination implying that the compound has a smaller quantity of oxygen than that whose name ends in -ic.

Outcrop. In *geology*, the edge of an inclined stratum when it comes to the surface of the ground.

Out'lier. In *geology*, a patch or mass of a stratum detached from the main body of the formation to which it belongs.

Ova (Lat. plural of *ovum*, an egg). Eggs.

Oval (Lat. *ovum*, an egg). Shaped like an egg.

O'vary (Lat. *ovum*, an egg). The organ in animals in which eggs are formed and contained; in plants, the case containing the young seeds, and ultimately becoming the fruit.

Ovate (Lat. *ovum*, an egg). In *any*, like an egg, with the lower d broadest.

Overshot Wheel. A wheel which is moved by water which flows at its upper part into buckets placed round its circumference.

Ovicap'sule (Lat. *ovum*, an egg; *caps'ula*, a capsule or casket). The sac which contains the egg.

O'viduct (Lat. *ovum*, an egg; *duco*, I lead). A passage which conveys eggs from the ovary.

Ovig'erous (Lat. *ovum*, an egg; *ger'o*, I carry). Carrying eggs; applied to receptacles in which, in some animals, eggs are received after being discharged from the ovary.

O'viform (Lat. *ovum*, an egg; *forma*, shape). Like an egg.

Ovine (Lat. *ovis*, sheep). Pertaining to sheep.

Ovip'arous (Lat. *ovum*, an egg; *par'io*, I produce). Producing eggs; applied to animals in which the egg is hatched after extrusion from the body.

Ovipos'it (Lat. *ovum*, an egg; *pono*, I put). To lay eggs.

Oviposit'ion (Lat. *ovum*, an egg; *pono*, I put). The laying of eggs.

Ovipos'itor (Lat. *ovum*, an egg; *pono*, I put). The organ which transmits eggs to their proper place during exclusion.

Ovis (Lat., a sheep). The generic term for the animals of which the sheep is the type.

O'visac (Lat. *ovum*, an egg; *sac*). The cavity in the ovary which contains the ovum.

O'volo. In *architecture*, a round moulding, generally the quarter of a circle.

Ovovivip'arous (Lat. *ovum*, an egg; *virus*, alive; *par'io*, I produce). Hatching young from eggs in the body of the parent, but not in an uterine cavity.

Ov'ule (Lat. *ovum*, an egg). A little egg, or seed; the small body in plants which becomes a seed.

Ox'alate (*Oxal'ic*). A salt composed of oxalic acid and a base.

Oxal'ic (Lat. *ox'alis*, sorrel). Pertaining to sorrel: applied to an acid, first obtained from the sorrel, but of very common occurrence.

Ox'idate (*Oxide*). To convert into an oxide.

Oxide (*Ox'ygen*). A body formed of oxygen with another elementary body.

Ox'idize (*Ox'ygen*). To charge or impregnate with oxygen.

Oxy-. A prefix in compound words, signifying generally that oxygen enters into the composition of the substance ; sometimes also implying acuteness.

Oxyg'enate (*Ox'ygen*, from Gr. ὀξυς, *oxus*, acid ; γενναω, *genna'ō*, I produce). To unite or cause to combine with oxygen.

Oxyg'enise. *See* Oxygenate.

Oxyg'enous (*Ox'ygen*). Relating to oxygen.

Oxyhy'drogen Blowpipe. A kind of blowpipe in which oxygen and hydrogen gases are burned together, to produce intense heat.

Oxyhy'drogen Mi'croscope. A microscope illuminated by a cylinder of limestone exposed to the flame of the oxyhydrogen blow-pipe.

Ox'ysalt (*Ox'ygen ; salt*). A salt into the composition of which oxygen enters.

Oz'one (Gr. οζω, *oz'ō*, I smell). A modification of oxygen, produced by electrical action, and emitting a peculiar odour.

P.

Pab'ulum (Lat. from *pasco*, I feed). Food.

Pacchio'nian Bodies (*Pacchio'ni*, an Italian anatomist). Small fleshy looking elevations formed on the external surface of the dura mater.

Pachyder'matous (Gr. παχυς, *pach'us*, thick ; δερμα, *derma*, skin). Thick-skinned ; applied to an order of animals having hoofs, but not chewing the cud, of which the elephant, hippopotamus, horse, pig, and a large number of fossil animals, are examples.

Pacin'ian Bodies (*Paci'ni*, an Italian anatomist). Minute oval bodies, attached to the extremities of the nerves of the hand and foot, and some other parts.

Palæ'o- (Gr. παλαιος, *palai'os*, ancient). A prefix in compound words, signifying ancient.

Palæog'raphy (Gr. παλαιος, *palai'os*, ancient ; γραφω, *graph'ō*, I write). The art of deciphering and reading ancient inscriptions.

Palæol'ogy (Gr. παλαιος, *palai'os*, ancient ; λογος, *logos*, discourse). A discourse or treatise on ancient things.

Palæontol'ogy (Gr. παλαιος, *palai'os*, ancient ; ὡν, *ōn*, being ; λογος, *logos*, discourse). The branch of science which describes the fossil animals and plants found in geological strata.

Palæophytol'ogy (Gr. παλαιος, *palai'os*, ancient ; φυτον, *phuton*, a plant ; λογος, *logos*, discourse). A term proposed for that branch of palæontology which treats of fossil vegetable remains.

Palæosau'rus (Gr. παλαιος, *palai'os*, ancient ; σαυρος, *sauros*, a lizard). Ancient lizard : a fossil reptile found in the magnesian limestone of the Permian system.

Palæothe'rium (Gr. παλαιος, *palai'os*, ancient ; θηριον, *thērion*, wild beast). A fossil pachydermatous or thick-skinned animal, found in the tertiary strata.

Palæozo'ic (Gr. παλαιος, *palai'os*, ancient ; ζωη, *zōē*, life). A term applied to the lowest division of strata which contains fossil remains of animals.

Palæozool'ogy (Gr. παλαιος, *pala'ios*, ancient ; ζωον, *zōon*, an animal ; λογος, *logos*, a discourse). A term proposed for that branch of palæontology which describes fossil animal remains.

Pala'tal (Lat. *pala'tum*, the roof of the mouth). Relating to the palate : a letter formed by the aid of the palate.

Pal'atine (Lat. *pala'tum*, the roof of the mouth). Belonging to the palate.

Pal'atine (Lat. *pala'tium*, a palace). Belonging to a palace : having royal privileges : counties palatine, in England, were Chester, Durham, and Lancaster, over which the proprietors—the Earl of Chester, Bishop of Durham, and Duke of Lancaster — formerly possessed rights equal to those of the king.

Pala'to-. In *anatomy*, a prefix in compound words, signifying connection with the palate.

Pa'lea (Lat. *chaff*). A name given to a part of the flowers of grasses ; also to the small scaly plates in the receptacle of some composite flowers.

Palea'ceous (Lat. *pa'lea*, chaff). Resembling chaff ; covered with small membraneous scales.

Palim'psest (Gr. παλιν, *pal'in*, again ; ψαω, *psao*, I rub). A sort of parchment from which anything written might be rubbed out, so that it might be again written on.

Pal'lial (Lat. *pal'lium*, a mantle). Belonging to the pallium or mantle.

Palliobranchia'ta (Lat. *pal'lium*, a mantle ; Gr. βραγχια, *bran'chia*, gills). A class of molluscous invertebrate animals, having the branchiæ arranged on the inner surface of the mantle.

Pal'lium (Lat. a mantle). In *zoology*, the fleshy covering lining the interior of the shells of bivalve mollusca, and covering the body of the animal.

Pal'macites (Lat. *palma*, a palm-tree). Fossil remains which bear an analogy or resemblance to the existing palms.

Pal'mar (Lat. *palma*, the palm of the hand). Belonging to the palm.

Pal'mate (Lat. *palma*, the palm). Resembling a hand with the fingers spread ; in *botany*, applied to leaves divided into lobes to about the middle.

Palmat'ifid (Lat. *palma*, the palm ;

findo, I cleave). Divided so as to resemble a hand.

Pal'miped (Lat. *palma*, a palm ; *pes*, a foot). Web-footed ; applied to an order of birds having the toes connected by a membrane for the purpose of swimming, as the penguin, petrel, pelican, swan, goose, duck, &c.

Palpa'tion (Lat. *palpo*, I feel). Feeling : examination by means of the sense of touch.

Pal'pebra (Lat.). An eyelid.

Pal'pebral (Lat. *pal'pebra*, an eyelid). Belonging to the eyelids.

Palpi (Lat. *palpo*, I feel). Feelers : jointed filaments attached to the heads of insects and some other animals.

Palu'dal (Lat. *palus*, a marsh). Belonging to or caused by emanations from marshes.

Pam'piniform (Lat. *pam'pinus*, a tendril ; *forma*, shape). Like a tendril.

Pan-, **Pant-**, or **Panto-** (Gr. πας, *pas*, all). A prefix in compound words, signifying all, or every thing.

Panace'a (Gr. παν, *pan*, all ; ἀκεομαι, *ak'eomai*, I cure). A medicine supposed to cure all diseases.

Pan'ary (Lat. *panis*, bread). Relating to bread ; formerly applied to the fermentative process which takes place in the making of bread.

Pan'creas (Gr. παν, *pan*, all ; κρεας, *kreas*, flesh). A narrow flat gland extending across the abdomen under the stomach, and secreting a fluid which aids in the digestion of food.

Pancreat'ic (*Pan'creas*). Belonging to or produced by the pancreas.

Pandem'ic (Gr. παν, *pan*, all ; δημος, *dēmos*, people). Attacking a whole people.

Pan'duriform (Lat. *pandura*, a fiddle ; *forma*, shape). Shaped like a fiddle ; applied, in *botany*, to leaves which are contracted in the middle and broad at each end.

Pan'icle (Lat. *panic'ula*, the down upon reeds). A form of inflorescence, consisting of spikelets on long peduncles coming off in the manner of a raceme, as in grasses.

Panic'ulate (Lat. *panic'ula*). Having flowers arranged in panicles.

Panora'ma (Gr. παν, pan, all; ὁραω, horaō, I see). An entire view; a form of picture in which all the objects that can be seen from a single point are represented on the inner surface of a round or cylindrical wall.

Pan'tograph (Gr. παν, pan, all; γραφω, graphō, I write). An instrument for copying drawings.

Pantom'eter (Gr. παν, pan, all; μετρον, met'ron, a measure). An instrument for measuring all kinds of elevations, angles, and distances.

Papavera'ceous (Lat. papa'ver, a poppy). Belonging to the order of plants of which the poppy is the type.

Papiliona'ceous (Lat. papil'io, a butterfly). Resembling a butterfly: applied to plants of the leguminous order, as the pea, from the shape of the flowers.

Papil'la (Lat. a nipple). A small conical or cylindrical projection of the skin or mucous membrane, containing blood-vessels and nerves, and serving sometimes to extend the surface, and sometimes for receiving impressions made on the extremities of the nerves.

Papil'lary (Lat. papil'la). Consisting of or provided with papillæ.

Pap'illated or **Pap'illose** (Lat. papil'la). Covered with small nipple-like prominences.

Pappose (Lat. pappus, down). Downy.

Pap'ulæ (Lat. plural of pap'ula, a kind of pimple). Pimples.

Papyra'ceous (Lat. papy'rus, paper). Papery: of the nature or consistence of paper.

Par'a- (Gr. παρα, par'a). A Greek preposition used in compound words, signifying close to, side by side, beyond, passing through, or contrary.

Parab'ola (Gr. παρα, par'a, beyond; βαλλω, ballō, I cast; probably from being the curve described in the motion of projectiles). The figure produced by cutting a cone by a plane parallel to one of its sides.

Parabol'ic (Parab'ola). Having the form of, or relating to, a parabola.

Parab'oloid (Parab'ola; Gr. εἰδος, eidos, form). The solid body produced by the revolution of a parabola about its axis.

Paracente'sis (Gr. παρα, par'a, beyond; κεντεω, ken'teō, I pierce). The operation of perforating a part of the body to allow the escape of fluid.

Paracen'tric (Gr. παρα, par'a, beyond; κεντρον, kentron, a centre). Deviating from the curve which would form a circle.

Par'adox (Gr. παρα, par'a, beyond; δοξα, doxa, opinion). Something that seems at first to be contrary to received opinion, or absurd.

Par'affin (Lat. parum, little; affi'nis, allied to). A substance obtained from tar, remarkable for its resistence to strong chemical agents, and for not being known to combine in a definite manner with any other body.

Parago'ge' (Gr. παρα, par'a, beyond; αγω, agō, I draw). The addition of a letter or syllable to the end of a word.

Parallac'tic (Gr. παρα, par'a, beyond; αλλασσω, allas'sō, I change). Belonging to the parallax. Parallactic inequality in the moon's course is the inequality dependent on the difference between the disturbing forces exercised by the sun in conjunction and opposition.

Par'allax (Gr. παρα, par'a, beyond; ἀλλασσω, allas'sō, I change). The apparent change in the position of an object, according to the point from which it is viewed. Diurnal parallax is the difference between the place of a celestial body as seen from the surface, and that in which it would appear if seen from the centre, of the earth. Horizontal parallax is the greatest amount of diurnal parallax, occurring when the object is in the horizon. Annual parallax is the apparent displacement of a celestial body arising from its being viewed from different parts of the earth's orbit.

Par'allel (Gr. παρα, *par'a*, opposite ; ἀλληλων, *allēlōn*, one another). Extending in the same direction and equally distant in every part.

Parallel'ogram (Gr. παραλληλος, *parallēlos*, parallel ; γραφω, *graph'ō*, I write). A figure with four straight sides, having the opposite sides equal and parallel.

Parallelopi'ped (Gr. παραλληλος, *parallēlos*, parallel ; ἐπιπεδος, *epip'edos*, level). A solid figure bounded by six parallelograms, parallel to each other two and two, as in a brick.

Paral'ysis (Gr. παρα, *par'a*, from ; λυω, *luō*, I loosen). Palsy ; a loss of power of voluntary motion or sensation, or both, in any part of the body.

Paralyt'ic (Gr. παρα, *par'a*, from ; λυω, *luō*, I loosen). Affected with palsy.

Par'alyse (Gr. παρα, *par'a*, from ; λυω, *luō*, I loosen). To render incapable of motion or sensation.

Paramagnet'ic (Gr. παρα, *par'a*, by ; μαγνης, *magnēs*, a magnet). A term applied to bodies which are attracted by both poles of the magnet, and which then arranges itself parallel to the straight line joining the poles.

Paraple'gia (Gr. παρα, *par'a*, across ; πλησσω, *plēssō*, I strike). Palsy of the lower half of the body, or of both lower limbs.

Parapoph'ysis (Gr. παρα, *par'a*, beyond ; *apoph'ysis*). A name given to the transverse process of an ideal typical vertebra.

Parasele'ne' (Gr. παρα, *par'a*, beyond ; σεληνη, *selēnē*, the moon). A mock moon ; a luminous ring surrounding the moon.

Par'asite (Gr. παρα, *par'a*, by ; σιτος, *sitos*, corn : applied originally to a class of public servants, who were maintained at the tables of the richer people). Any plant or animal which lives and feeds on the body of another plant or animal.

Parasit'ic (*Parasite*). Living on some other body, and deriving nutriment from it.

Paratonnerre (Gr. παρα, *par'a*, from ; Fr. *tonnerre*, thunder). A lightning conductor ; a pointed metallic rod erected over a building or other object to protect it from lightning.

Paregor'ic (Gr. παρηγορεω, *parēgoreō*, I mitigate). Mitigating pain.

Paren'chyma (Gr. παρα, *par'a*, by ; ἐγχυμα, *en'chuma*, a tissue). A term used to denote either the solid part of a gland, including all its tissues, or any substance lying between the ducts, vessels, and nerves.

Parenchy'matous (*Paren'chyma*). Consisting of parenchyma ; or affecting parts formed of parenchyma.

Paren'thesis (Gr. παρα, *par'a*, beyond ; ἐν, *en*, in : τιθημι, *tithēmi*, I place). An insertion of words in the body of a sentence, giving some explanation or comment, but not forming a part of its grammatical structure.

Parhe'lion (Gr. παρα, *par'a*, beyond ; ἡλιος, *hēlios*, the sun). A mock sun ; a meteor appearing as a bright light near the sun, sometimes tinged with colours like a rainbow.

Pari'etal (Lat. *par'ies*, a wall). Relating to or acting as a wall : in *anatomy*, applied to a large flat bone at each side of the head ; in *botany*, applied to any organ which grows from the sides or walls of another.

Par'ietes (Lat. plural of *par'ies*, a wall). The enclosing walls of any cavity.

Parisyllab'ic (Lat. *par*, equal ; Gr. συλλαβη, *sul'labē*, a syllable). Having an equal number of syllables.

Paronoma'sia (Gr. παρα, *par'a*, near ; ὀνομαζω, *onoma'zō*, I name). A figure by which words nearly alike in sound, but of different meanings, are used in relation to each other in the same sentence.

Parot'id (Gr. παρα, *par'a*, near ; οὑς, *ous*, the ear). Near the ear ; applied to one of the salivary glands from its situation.

Paroti'tis (Lat. *paro'tis*, the parotid gland; *itis*, denoting inflammation). Mumps; inflammation of the parotid gland.

Par'oxysm (Gr. παρα, *par'a*, beyond; ὀξυς, *oxus*, sharp). A fit of any disease, coming on after a period of intermission or suspension.

Paroxys'mal (*Paroxysm*). Occurring in paroxysms or fits.

Parthenogen'esis (Gr. παρθενος, *par'-thenos*, a virgin; γενναω, *genna'ō*, I produce). The successive production of animals or vegetables from a single ovum.

Par'ticle (Lat. *pars*, a part : *cle*, denoting smallness). A minute part of a body.

Par'tite (Lat. *par'tio*, I divide) In *botany*, divided to near the base.

Partu'rient (Lat. *partu'rio*, I bring forth). Bringing forth young.

Parturit'ion (Lat. *partu'rio*, I bring forth). The act of bringing forth young.

Pas'seres (Lat. *passer*, a sparrow). An order of birds, characterised by slender legs, feeble, straight or nearly straight bill, sufficiently large wings, and small or moderate size; including the sparrow, swallow, blackbird, and numerous other birds.

Pas'serine (Lat. *passer*, a sparrow). Belonging to the order *passeres*, of which the sparrow is a type.

Patel'la (Lat. a dish with a broad brim). The knee-pan.

Pathogenet'ic (Gr. παθος, *path'os*, suffering; γενναω, *genna'ō*, I produce). Producing disease : relating to the production of disease.

Pathog'eny (Gr. παθος, *path'os*, suffering; γενναω, *genna'ō*, I produce). The study of the seats, nature, general forms, and varieties of disease.

Pathognomon'ic (Gr. παθος, *path'os*, suffering; γινωσκω, *ginōskō*, I know). Peculiar to any special disease, and distinguishing it from all others.

Pathol'ogy Gr. παθος, *path'os*, suffer-

ing; λογος, *logos*, discourse). The branch of medical science which treats of the nature and constitution of disease.

Patholog'ical (Gr. παθος, *path'os*, suffering; λογος, *logos*, a discourse). Relating to the study of the nature of disease.

Pat'ulous (Lat. *pat'eo*, I am open). Spreading open.

Paucispi'ral (Lat. *paucus*, few; *spira*, a spire). Having few spiral turns.

Pavement Epithe'lium. A form of epithelium in which the particles have the form of small angular masses or thin scales.

Pe'cilopods. See Pœ'cilopods.

Pec'ora (Lat. *pec'us*, cattle). A name given by Linnæus to the ruminating mammals.

Pec'tin (Gr. πηκτος, *pēktos*, solid, congealed). The jelly of fruits.

Pec'tinate (Lat. *pecten*, a comb). Resembling the teeth of a comb.

Pectine'al (Lat. *pecten*, a comb). In *anatomy*, applied to a line forming a sharp ridge on the pubic bone of the pelvis.

Pectinibranchia'ta (Lat. *pecten*, a comb; Gr. βραγχια, *bran'chia*, gills). An order of gasteropodous molluscous animals, which have the gills in a comb-like form, usually seated in a cavity behind the head.

Pec'tiniform (Lat. *pecten*, a comb; *forma*, shape). Resembling a comb.

Pec'toral (Lat. *pectus*, the breast). Belonging to or situated on the region of the breast; the pectoral fins in fishes are the anterior fins, which represent the fore limbs of the higher vertebrate animals.

Pectoril'oquy (Lat. *pectus*, the breast; *loquor*, I speak). A direct transmission of the sound of the voice from the chest to the ear, heard on listening over the chest in certain diseased states.

Pectus (Lat.) The breast.

Pedate (Lat. *pes*, the foot). Having divisions like the toes.

Ped'icle (Lat. *pes*, the foot). A subdivision of a peduncle or stem.

Ped'iform (Lat. *pes*, a foot; *forma*, shape). Shaped like a foot.

K

Pedig'erous (Lat. *pes*, a foot ; *gero*, I bear). Carrying feet.

Pedilu'vium (Lat. *pes*, a foot ; *lavo*, I wash). A foot-bath.

Ped'iment (Lat. *pes*, a foot). In *architecture*, the triangular surface formed by the vertical termination of a roof consisting of two sloping sides, and bounded by three cornices.

Pedipal'pi (Lat. *pes*, a foot ; *palpi*, feelers). A section of arachnida, remarkable for the large size of their palpi, which are furnished with claws or pincers, as the scorpion.

Ped'uncle (Lat. *pes*, a foot ; *cle*, denoting smallness). A stem.

Pedun'culated (*Ped'uncle*). Growing or supported on a stem.

Peg'matite (Gr. πηγμα, *pēgma*, anything fastened together). A form of granite, being a fine-grained compound of feldspar and quartz, with minute scales of mica.

Pelag'ic (Gr. πελαγος, *pel'agos*, the open sea). Belonging to the deep sea.

Pollag'ra (Lat. *pellis ægra*, diseased skin). Italian leprosy ; a disease of the skin common in the north of Italy.

Pel'licle (Lat. *pellis*, a skin ; *cle*, denoting smallness). A thin skin or film ; in *botany*, the outer covering of plants.

Pellu'cid (Lat. *per*, through ; *lucidus*, light). Clear ; transparent.

Pel'tate (Lat. *pelta*, a target). Having the shape of a round shield or target ; in *botany*, applied to leaves having the stem inserted at or near the middle of the under surface.

Pelvic (*Pelvis*). Belonging to the pelvis.

Pelvis (Lat. a basin). In *anatomy*, the cavity or inclosure in the animal body made up of the innominate bones, the sacrum, and the coccyx, and supporting the lower organs of the abdomen on the inside, and the lower limbs on the outside.

Pemphi'gus (Gr. πεμφιξ, *pemphix*, a small blister). A disease of the skin, consisting in an eruption of blisters of various sizes, from the size of a sixpence to that of a half-crown.

Pencil of Rays. In *optics*, a collection of rays of light radiating from or converging to a common point, and included within the surface of a cone or other regular limit.

Pendant (Fr. hanging, from Lat. *pen'deo*, I hang). An ornament used in the vaults and ceilings of Gothic architecture.

Pen'dulous (Lat. *pen'deo*, I hang). Hanging.

Pen'dulum (Lat. *pen'deo*, I hang). A body suspended so that it may vibrate about some fixed point by the action of gravity.

Penicil'late (Lat. *penicillus*, a small brush). Having the form of a pencil or small brush.

Penin'sula (Lat. *pene*, almost ; *in'sula*, an island). A portion of land nearly or in great part surrounded by water, and joined to the mainland by a part narrower than the tract itself.

Pennate (Lat. *penna*, a feather). Winged.

Pen'nifer (Lat. *penna*, a feather ; *fer'o*, I bear). Covered with feathers.

Pen'niform (Lat. *penna*, a feather ; *forma*, shape). Having the shape of a feather ; in *anatomy*, applied to muscles of which the fibres pass out on each side from a central tendon.

Pen'ninerved (Lat. *penna*, a feather ; *nervus*, a nerve). In *botany*, applied to leaves which have the nerves or veins arranged like the parts of a feather.

Pennule (Lat. *penna*, a feather ; *ule*, denoting smallness). A small feather, or division of a feather.

Penta- (Gr. πεντε, *pente*, five). A prefix in compound words, signifying five.

Pentac'rinites (Gr. πεντε, *pente*, five ; κρινον, *krinon*, a lily). A tribe of echinoderms, mostly fossil, in which the animal consists of a jointed flexible column fixed at the base, and supporting a concave disc or body, with five jointed cylindrical arms.

Pentadac'tyle (Gr. πεντε, *pente,* five; δακτυλος, *dak'tulos,* a finger). Having five fingers or toes.

Pen'tagon (Gr. πεντε, *pente,* five; γωνια, *gōnia,* an angle). A figure having five angles.

Pen'tagraph. *See* Pantagraph.

Pentagyn'ia (Gr. πεντε, *pente,* five; γυνη, *gunē,* a female). A term applied in the Linnean system to those classes of plants which have five pistils.

Pentahed'ral (Gr. πεντε, *pente,* five; ἑδρα, *hed'ra,* a base). Having five equal sides.

Pentahed'ron (Gr. πεντε, *pente,* five; ἑδρα, *hed'ra,* a base). A solid figure, having five equal sides.

Pentam'era (Gr. πεντε, *pente,* five; μερος, *mer'os,* a part). Having five parts; in *zoology,* a section of the coleoptera or beetle tribe, having the tarsi of all the feet five-jointed.

Pentam'eter (Gr. πεντε, *pente,* five; μετρον, *met'ron,* a measure). A verse of five feet.

Pentan'dria (Gr. πεντε, *pente,* five; ἀνηρ, *anēr,* a man). A class of plants in the Linnæan system, having five distinct stamens.

Pentan'gular (Gr. πεντε, *pente,* five; Lat. *an'gulus,* an angle). Having five angles.

Pentaphyl'lous (Gr. πεντε, *pente,* five; φυλλον, *phullon,* a leaf). Having five leaves.

Pentasper'mous (Gr. πεντε, *pente,* five; σπερμα, *sperma,* a seed). Having five seeds.

Pen'tastyle (Gr. πεντε, *pente,* five; στυλος, *stulos,* a pillar). A building having five columns in front.

Penul'timate (Lat. *pene,* almost; *ul'timus,* last). Last but one.

Penum'bra (Lat. *pene,* almost; *um-bra,* a shadow). Partial shade or shadow; in *optics* and *astronomy,* a space on each side of a perfect shadow or eclipse, from which the rays of light are partially cut off by the opaque body; in *painting,*

the part where the shade and light blend with each other.

Pepsine (Gr. πεπτω, *peptō,* I digest). The active principle of the gastric juice, which effects digestion.

Pep'tic (Gr. πεπτω, *peptō,* I digest). Promoting digestion.

Per- (Lat.) A preposition used in compound words, signifying through, thoroughly, very, in excess.

Per Annum (Lat.) By the year.

Per Cap'ita (Lat.). By the head.

Percep'tion (Lat. *per,* by or through; *cap'io,* I take). The process by which the mind takes notice of external objects.

Perchlo'rate (Lat. *per,* through; *chlorine*). A salt consisting of perchloric acid and a base.

Perchlo'ric (Lat. *per,* very; *chlorine*). A term applied to an acid consisting of one equivalent of chlorine and seven of oxygen.

Per'colate (Lat. *per,* through; *colo,* I strain). To strain through.

Percola'tion (Lat. *per,* through; *colo,* I strain). The act of straining.

Percur'rent (Lat. *per,* through; *curro,* I run). Running through from top to bottom.

Percus'sion (Lat. *percut'io,* I strike). A striking.

Peren'nial (Lat. *per,* through; *an-nus,* a year). Lasting through several or many years.

Perennibran'chiate (Lat. *peren'nis,* lasting; Gr. βραγχια, *bran'chia.* gills). Having lasting gills; applied to batrachian reptiles in which the gills remain throughout life.

Perfo'liate (Lat. *per,* through; *fo'-lium,* a leaf). Applied to leaves which have the lobes at the base united, so as to surround the stem, as if the stem ran through them.

Peri- (Gr. περι, *per'i,* around). A preposition in compound words, signifying around.

Per'ianth (Gr. περι, *per'i,* about; ἀνθος, *anthos,* a flower). A term applied to the calyx and corolla of flowers; especially when they cannot be easily distinguished from each other.

Pericar'dial (*Pericar'dium*). Belonging to or produced in the pericardium.

Pericardi'tis (*Pericar'dium ; itis*, denoting inflamation). Inflammation of the pericardium or membrane covering the heart.

Pericar'dium (Gr. περι, *per'i*, around ; καρδια, *kar'dia*, the heart). The serous membrane covering the heart.

Per'icarp (Gr. περι, *per'i*, around ; καρπος, *karpos*, fruit). The seed-vessel, or shell of the fruit of plants.

Perichon'drium (Gr. περι, *per'i*, around ; χονδρος, *chondros*, cartilage). The membrane covering cartilages.

Pericra'nium (Gr. περι, *per'i*, around ; κρανιον, *kra'nion*, the skull). The membrane immediately covering the bones of the skull.

Per'iderm (Gr. περι, *per'i*, about ; δερμα, *derma*, skin). In *botany*, the outer layer of bark.

Per'igee (Gr. περι, *per'i*, about ; γη, *gē*, the earth). The point in the moon's path which is nearest to the earth, and where it therefore appears largest.

Per'igone (Gr. περι, *per'i*, about ; γονη, *gon'ē*, a pistil). A term for the floral envelopes : sometimes restricted to cases in which the flower bears pistils only.

Perig'ynous (Gr. περι, *per'i*, about ; γυνη, *gunē*, a female). Growing on some part that surrounds the ovary in a flower ; applied to the corolla and stamens when they are attached to the calyx.

Perihe'lion (Gr. περι, *per'i*, about ; ηλιος, *hēlios*, the sun). The point of its orbit in which a planet or comet is nearest to the sun.

Perim'eter (Gr. περι, *per'i*, around ; μετρον, *met'ron*, a measure). The bounds or limits of a body : in a circle, the circumference.

Pe'riod (Gr. περι, *per'i*, about ; δδος, *hodos*, a way). A circuit : a stated portion of time.

Period'ic or **Period'ical** (Gr. περι, *per'i*, about ; δδος, *hodos*, a way). Performed in a regular circuit in a given time ; occurring at regular intervals.

Period'ic (Lat. *per*, very ; *i'odine*). A term applied to an acid containing an equivalent of iodine and seven of oxygen.

Periodic'ity (*Period*). The disposition of certain things, or circumstances, to return at stated intervals.

Periodon'tal (Gr. περι, *per'i*, about ; δδους, *odous*, a tooth). Surrounding the teeth.

Periœ'ci (Gr. περι, *per'i*, round about ; οικεω, *oi'keō*, I dwell). The inhabitants of the earth who live in the same latitudes, but whose longitudes differ by 180 degrees, so that when it is noon with one it is midnight with the other.

Periosti'tis (*Perios'teum : itis*, denoting inflammation). Inflammation of the periosteum.

Perios'teum (Gr. περι, *per'i*, around ; οστεον, *os'teon*, a bone). The fibrous membrane which invests the bone.

Perios'tracum (Gr. περι, *per'i*, around ; οστρακον, *os'trakon*, a shell). The membrane which covers shells.

Peripatet'ic (Gr. περιπατεω, *peripat'eō*, I walk about). Walking about : a term applied to the philosophy of Aristotle, because taught during walking in the Lyceum at Athens.

Periph'eral (Gr. περι, *per'i*, around ; φερω, *pher'ō*, I bear). Belonging to the periphery or circumference.

Periph'ery (Gr. περι, *per'i*, around ; φερω, *pher'ō*, I bear). The circumference.

Periph'rasis (Gr. περι, *per'i*, about ; φραζω, *phrazō*, I speak). Circumlocution : the use of more words than are necessary to express an idea.

Per'iplus (Gr. περι, *per'i*, around ; πλεω, *pleō*, I sail). A sailing round a certain sea or coast.

Peripneumo'nia. *See* Pneumo'nia.

Perisc'ii (Gr. περι, *per'i*, around ; σκια, *skia*, a shadow). A name given to the inhabitants of the frigid zones whose shadows move round, and at certain times in the year describe a circle during the day.

Per'iscope (Gr. περι, per'i, about; σκοπεω, skop'eō, I look). A general view.

Per'isperm (Gr. περι, per'i, about; σπερμα, sperma, seed). The albumen or nourishing matter stored up with the embryo in a seed.

Peris'sodactyle (Gr. περισσος, peris'sos, odd, or uneven; δακτυλος, dak'tulos, a finger). Having an uneven number of toes on the hind feet.

Peristal'tic (Gr. περι, per'i, about; στελλω, stellō, I send). Sending round : applied to a motion like that of a worm, such as takes place in the intestines and other internal muscular organs, by the contraction of successive portions.

Per'istome (Gr. περι, per'i, around; στομα, stom'a, a mouth). The ring of bristles situated close round the orifice of the seed-vessel in mosses.

Per'istyle (Gr. περι, per'i, around; στυλος, stulos, a pillar). A range of columns surrounding any thing.

Perit'omous (Gr. περι, per'i, around; τεμνω, temnō, I cut). In mineralogy, cleaving in more directions than one parallel to the axis, the faces being all of one quality.

Peritone'al (Peritone'um). Belonging to the peritoneum.

Peritone'um (Gr. περι, per'i, about; τεινω, teinō, I stretch). The serous membrane which lines the cavity of the abdomen, and is reflected over the organs contained therein, so as to hold them in their place, and at the same time allow free movement where required.

Peritoni'tis (Peritone'um; itis, denoting inflammation). Inflammation of the peritoneum.

Peritre'ma (Gr. περι, per'i, around; τρημα, trēma, a hole). The raised margin which surrounds the breathing holes of scorpions.

Per'meable (Lat. per, through; meo, I pass). Capable of being passed through without rupture or apparent displacement of parts.

Per'meate (Lat. per, through; meo, I pass). To pass through without rupture or apparent displacement, as water through porous stones, or light through transparent bodies.

Permuta'tion (Lat. per, through; muto, I change). An exchange ; the different combination of any number of quantities, taking a certain number at a time, with reference to their order.

Perone'al (Gr. περονη, per'onē, the fibula, or small bone of the leg). Belonging to, or lying near the fibula.

Perox'ide (Lat. per, very; oxide). The oxide of a substance which contains most oxygen, but has not acid characters.

Perpendic'ular (Lat. perpendic'ulum, a plumb-line). Hanging in a straight line towards the centre of the earth or of gravity ; meeting another line at right angles.

Persis'tent (Lat. persis'to, I continue). In botany, applied to parts which remain attached to the axis.

Per'sonate (Lat. perso'na, a mask). In botany, applied to an irregular corolla with the petals inverted, and having the lower lip projecting so as to close the opening between the lips.

Perspec'tive (Lat. per, through; spec'to, I look). The science which teaches the representation of an object or objects on a surface, so as to affect the eye in the same manner as the objects themselves.

Perspira'tion (Lat. per, through ; spi'ro, I breathe). The exhalation of vapour or fluid through the skin.

Persul'phate (Lat. per, very ; sulphate). A combination of sulphuric acid with a peroxide.

Perturba'tion (Lat. per ; turbo, I disturb). A disturbing ; in astronomy, applied to the deviation, produced by the gravitation of a body external to the orbit, of a planet or other revolving body, from the path which it would follow if regulated solely by the attraction of a central body.

Pertus'sis (Lat. *per*, very ; *tussis*, cough). Hooping-cough.

Pestif'erous (Lat. *pestis*, plague; *fer'o*, I bring). Injurious to health ; producing disease.

Pestilen'tial (Lat. *pestis*, plague). Partaking of the nature of, or tending to produce, an infectious disease.

Pet'al (Gr. πεταλον, *pet'alon*, a leaf). A flower-leaf, or part of the corolla, generally coloured.

Pet'aloid (Gr. πεταλον, *pet'alon*, a leaf or petal ; ειδος, *eidos*, shape). Like a petal or leaf.

Pete'chia. A small red spot like a flea-bite.

Pete'chial (*Pete'chia*). Belonging to petechiæ, or characterised by their presence.

Pet'iolate (*Petiole*). Having a stalk or petiole.

Pet'iole (Lat. *pet'iolus*, the stalk of fruits ; probably diminutive of *pes*, a foot). The stem of a leaf.

Petit'io Princip'ii (Lat. a demand of the principle). A species of faulty reasoning, which consists in taking the question in dispute as settled, and drawing conclusions from it.

Petrifac'tion (Lat. *petra*, a stone or rock ; *fac'io*, I make). A changing into stone ; a process effected by the entrance of particles of stony matter in solution into the pores of an animal or vegetable body, taking the place of the organic matter.

Pet'rify (Lat. *petra*, a stone or rock ; *fac'io*, I make). To change into stone.

Petro'sal (Lat. *petra*, a stone or rock). A name given to the ossified portion in the fish, corresponding to the petrous portion of the temporal bone in the higher vertebrates.

Pet'rous (Lat. *petra*, a stone or rock). Like stone ; applied to a portion of the temporal bone, from its hardness.

Phænog'amous (Gr. φαινω, *phainō*, I show ; γαμος, *gam'os*, marriage). Having conspicuous flowers.

Phagedæna (Gr. φαγω, *phag'ō*, I eat). A rapidly spreading malignant ulcer.

Phagede'nic (Gr. φαγω, *phagō*, I eat). Of the nature of a spreading ulcer.

Phalange'al (Gr. φαλαγξ, *phalanx*, a line of battle). Belonging to the phalanges, or small bones of the fingers and toes.

Phalanx (Gr. φαλαγξ, *phalanx*, a line of battle). A name applied to the small bones forming the fingers and toes, which are arranged in three rows.

Phanerog'amous (Gr. φανερος, *phan'eros*, manifest ; γαμος, *gam'os*, marriage). Having conspicuous flowers.

Phantasmago'ria (Gr. φαντασμα, *phantas'ma*, an appearance ; αγοραομαι, *agora'omai*, I meet). An optical instrument, consisting of a magic lantern which is made to to recede from or approach a screen, so as to magnify or diminish the appearance of objects, and give them an appearance of motion.

Pharmaceu'tic (Gr. φαρμακον, *phar'makon*, a drug). Relating to the art of preparing drugs.

Pharmaceu'tist (Gr. φαρμακον, *phar'makon*, a drug). One who prepares drugs.

Pharmacopœ'ia (Gr. φαρμακον, *phar'makon*, a drug ; ποιεω, *poi'eō*, I make). A book which teaches the method of preparing drugs for use as medicines.

Phar'macy (Gr. φαρμακον, *phar'makon*, a drug). The art of collecting and preparing drugs for use as medicine.

Pharynge'al (*Pharynx*). Belonging to the pharynx.

Pharyngot'omy (Gr. φαρυγξ, *pharunx*, the pharynx ; τεμνω, *temnō*, I cut). The operation of cutting open the pharynx.

Pharynx (Gr. φαρυγξ, *pharunx*). The muscular organ or tube at the back part of the mouth, which leads into the œsophagus or gullet.

Phase (Gr. φασις, *phasis*, an appearance). An appearance ; in *astronomy*, applied to the different appearances which the moon or a planet presents, according to its position with respect to the sun and the earth.

Phenom'enon (Gr. φαινομαι, phai'no-mai, I appear). That which appears; whatever is presented to the senses by observation or experiment, or is discovered to exist.

Philol'ogy (Gr. φιλος, phil'os, a friend; λογος, logos, a word). The branch of literature which comprehends a knowledge of the etymology and structure of words; the science of language.

Philos'ophy (Gr. φιλος, phil'os, a friend; σοφια, soph'ia, wisdom). Love of wisdom; but applied generally to an investigation of the causes of all phenomena, both of mind and of matter.

Phlebi'tis (Gr. φλεψ, phleps, a vein; itis, denoting inflammation). Inflammation of a vein or of veins.

Phleb'olites (Gr. φλεψ, phleps, a vein; λιθος, lith'os, a stone). Small dense masses found in veins.

Phlebot'omy (Gr. φλεψ, phleps, a vein; τεμνω, temnō, I cut). The act or practice of opening a vein to let blood.

Phlegma'sia (Gr. φλεγω, phleg'ō, I burn). Inflammation accompanied by fever.

Phleg'mon (Gr. φλεγω, phleg'ō, I burn). An inflammatory swelling on the external surface.

Phleg'monous (Gr. φλεγω, phleg'ō, I burn). Having the nature of phlegmon.

Phlogis'tic (Phlogis'ton). Belonging or relating to phlogiston.

Phlogis'ton (Gr. φλογιζω, phlogi'zō, I inflame). A name formerly given to what was supposed to be pure fire fixed in combustible bodies.

Phlyctæ'na (Gr. φλυω, phluō, I boil up). A vesicle containing serous fluid.

Phonet'ic (Gr. φωνη, phōnē, sound). Belonging to sound; applied to written characters which represent sounds.

Phon'ic (Gr. φωνη, phōnē, sound). Belonging to sound.

Phonocamp'tic (Gr φωνη, phōnē, sound; καμπτω, kamptō, I bend). Having the power to turn sound from its direction.

Phonog'raphy (Gr. φωνη, phōnē, sound; γραφω, graph'ō, I write). A description of the sounds uttered by the organs of speech; a system of writing, in which every sound of the voice has its own character.

Phon'olite (Gr. φωνη, phōnē, sound; λιθος, lith'os, a stone). A species of basaltic greenstone, so called from its ringing sound when struck.

Phon'otypy (Gr. φωνη, phōnē, sound; τυπος, tu'pos, a type). A proposed system of printing, in which each letter represents a single sound.

-Phore (Gr. φερω, pher'ō, I bear). A termination in compound words, signifying a bearer or supporter.

Phos'gene (Gr. φως, phōs, light; γενναω, genna'ō, I produce). Producing light, or produced by light.

Phos'phate (Phos'phorus). A salt consisting of phosphoric acid combined with a base.

Phos'phene (Gr. φως, phōs, light; φαινομαι, phai'nomai, I appear). An appearance of light in the eye.

Phos'phite (Phos'phorus). A salt consisting of phosphorous acid combined with a base.

Phosphores'cence (Gr. φως, phōs, light; φερω, pher'ō, I bear). A faint luminous appearance presented in the dark by certain bodies, not accompanied by sensible heat.

Phosphores'cent. Shining with a faint light.

Phosphor'ic (Phos'phorus). Belonging to phosphorus; applied to an acid containing one equivalent of phosphorus and five of oxygen.

Phos'phorous (Phos'phorus). A term applied to an acid containing one equivalent of phosphorus and three of oxygen.

Phos'phorus (Gr. φως, phōs, light; φερω, pher'ō, I bear). An elementary non-metallic substance, having the property of burning at a low temperature, so as to produce a luminous appearance in the dark.

Phos'phuretted (Phos'phorus). Combined with phosphorus.

Photo- (Gr. φως, phōs, light). A prefix in compound words, denoting relation to or connection with light.

Photogen'ic (Gr. φως, *phōs*, light; γενναω, *genna'ō*, I produce). Producing light; produced by light.

Pho'tograph (Gr. φως, *phōs*, light; γραφω, *graph'ō*, I write). A representation of an object, produced by the action of light.

Photog'raphy (Gr. φως, *phōs*, light; γραφω, *graph'ō*, I write). The process of producing representations of objects by the action of light on a surface coated with a preparation capable of being acted on by certain rays of the sun.

Photol'ogy (Gr. φως, *phōs*, light; λογος, *log'os*, a discourse). The science which describes light.

Photomag'netism (Gr. φως, *phōs*, light; *mag'netism*). The branch of science which describes the relation of the phenomena of magnetism to those of light.

Photom'eter (Gr. φως, *phōs*, light; μετρον, *met'ron*, a measure). An instrument for measuring the intensity of light.

Photom'etry (Gr. φως, *phōs*, light; μετρον, *met'ron*, a measure). The art of measuring the intensity of light by observation.

Photopho'bia (Gr. φως, *phōs*, light; φοβος, *phob'os*, fear). Dread of light.

Phragma (Gr. φρασσω, *phrassō*, I divide). A transverse division or false dissepiment in fruits.

Phrag'mocone (Gr. φρασσω, *phrassō*, I divide; κωνος, *kōnos*, a cone). The chambered cone of the shell of the belemnite cephalopods.

Phren'ic (Gr. φρην, *phrēn*, the diaphragm). Of or belonging to the diaphragm.

Phreni'tis (Gr. φρην, *phrēn*, the mind; *itis*, denoting inflammation). Inflammation of the brain.

Phrenol'ogy (Gr. φρην, *phrēn*, the mind; λογος, *log'os*, discourse). Literally, the science of the human mind; but applied especially to a doctrine of mental philosophy, founded on a presumed knowledge of the functions of different parts of the brain, obtained by comparing their apparent relative forms and magnitudes in different individuals with the mental propensities and powers which these individuals are found to possess.

Phthi'sic or **Phthis'ical** (Gr. φθιω, *phthiō*, I consume). Belonging to or affected with phthisis or tubercular disease.

Phthi'sis (Gr. φθιω, *phthiō*, I consume). The disease commonly known as consumption, connected with a morbid deposit in the lungs, called tubercle.

Phycol'ogy (Gr. φυκος, *phu'kos*, seaweed; λογος, *log'os*, discourse). The study of algæ or sea-weeds.

Phyllo'dium (Gr. φυλλον, *phullon*, a leaf; ειδος, *eidos*, form). A leaf-stalk enlarged so as to resemble a leaf.

Phyll'ogen (Gr. φυλλον, *phullon*, a leaf; γενναω, *genna'ō*, I produce). The terminal bud from which the leaves of palms grow.

Phyll'oid (Gr. φυλλον, *phullon*, a leaf; ειδος, *eidos*, form). Like a leaf.

Phylloplas'tic (Gr. φυλλον, *phullon*, a leaf; πλασσω, *plas'sō*, I form). Forming leaves.

Phyllopto'sis (Gr. φυλλον, *phullon*, a leaf; πτωσις, *ptōsis*, a falling). The fall of the leaf.

Phyllotax'is (Gr. φυλλον, *phullon*, a leaf; τασσω, *tassō*, I arrange). The arrangement of leaves on the axis or stem.

Phys'ical (Gr φυσις, *phu'sis*, nature). Belonging to natural or material things, as opposed to moral or imaginary; applied also to those properties of bodies which are directly perceptible to the senses, in opposition to those which are known as chemical or vital.

Phys'ico-Mathematics. The branch of mathematical science which investigates the laws and actions of bodies and their combinations, by means of data drawn from observation and experiment.

Phys'ics (Gr. φυσις, *phu'sis*, nature). In its literal sense, the science of nature and natural objects, implying the study or knowledge of everything that exists. In modern acceptation, however, the word is limited to that department of science commonly known also as natural philosophy, which describes the general properties of bodies, their mutual action on each other, their causes, effects, phenomena, and laws.

Physiogn'omy (Gr. φυσις, phu'sis, nature ; γνωμων. gnōmōn, one who knows). The general appearance of an animal or vegetable being, without reference to special anatomical or botanical characters.

Physiolog'ical (Gr. φυσις, phu'sis, nature ; λογος, log'os, discourse). Relating to the science of the properties and functions of living beings.

Physiol'ogy (Gr. φυσις, phu'sis, nature ; λογος, log'os, discourse). Literally, a treatise on nature ; but now applied to the science which investigates the functions of organised beings and of their several parts, and their relations to each other and to external objects.

Physiophilos'ophy (Gr. φυσις, phu'sis, nature ; φιλοσοφια, philosoph'ia, philosophy). Natural philosophy.

Phy'sograde (Gr. φυσαω, phusa'ō, I blow ; Lat. gradus, a step). Moving in the water by air-bladders ; applied to a tribe of acalephæ or sea-nettles.

Phytiv'orous (Gr. φυτον, phu'ton, a plant ; Lat. vo'ro, I devour). Living on plants or herbage.

Phyto- (Gr. φυτον, phu'ton, a plant). A prefix in compound words, signifying plant.

Phytogen'esis (Gr. φυτον, phu'ton, a plant ; γενναω, genna'ō, I produce). The development of plants.

Phytogeograph'ical (Gr. φυτον, phu'ton, a plant ; geogra̧hy). Relating to the distribution of plants on the surface of the globe.

Phytog'raphy (Gr. φυτον, phu'ton, a plant ; γραφω, graph'ō, I write). A description of plants.

Phy'toid or **Phytoi'dal** (Gr. φυτον, phu'ton, a plant ; ειδος, eidos, form). Resembling plants.

Phytol'ogy (Gr. φυτον, phu'ton, a plant ; λογος, log'os, a discourse). A discourse or treatise on plants.

Phytoph'agous (Gr. φυτον, phu'ton, a plant ; φαγω, phag'ō, I eat) Eating or living on plants.

Phytophysiol'ogy (Gr. φυτον, phu'ton, a plant ; physiology). The physiology of plants ; the doctrine of their intimate structure and functions.

Phytot'omy (Gr. φυτον, phu'ton, a plant ; τεμνω, temnō, I cut). The dissection of plants.

Phytozo'a (Gr. φυτον, phu'ton, a plant ; ζωον, zōon, an animal). Moving filaments in the antheridia or analogues of flowers in cryptogamic plants.

Pia Mater. A name given to the membrane immediately investing the brain, and which consists chiefly of blood-vessels finely divided before entering the substance of the organ.

Pigment (Lat. pin'go, I paint). In anatomy, applied to the material, contained in minute cells, which gives colour to various parts of the body, as the interior of the eye, the skin in coloured races, &c.

Pi'leate (Lat. pi'leus, a cap). Having the form of a cap or cover for the head.

Pi'leiform (Lat. pi'leus, a cap ; for'ma, shape). Resembling a cap or hat.

Pi'lifer (Lat. pi'lus, hair ; fer'o, I bear). Covered with hair.

Pi'liform (Lat. pi'lus, hair ; for'ma, shape). Resembling hairs.

Pi'lose (Lat. pi'lus, hair). Provided with hairs.

Pinacothe'ca (Gr. πιναξ, pin'ax, a picture ; θηκη, thēkē, a repository). A picture gallery.

Pi'neal (Lat. pi'nus, a pine). Belonging to, or resembling the fruit of the pine.

Pinen'chyma (Gr. πιναξ, pinax, a tablet ; εγχυμα, en'chuma, a type). A term applied to the cellular tissue of plants when arranged in a tabular form.

Pi'nites (Lat. pi'nus, the fir-tree). A generic term for fossil remains of plants allied to the coniferous order.

Pin'na (Lat. a fin or wing). In *anatomy*, the part of the external ear which projects beyond the head ; in *botany*, a division of a pinnate leaf.

Pin'nate (Lat. *pin'na*, a feather). Like a feather ; in *botany*, applied to leaves which have a series of leaflets on each side of the petiole.

Pinnat'ifid (Lat. *pin'na*, a feather ; *fin'do*, I cleave). In *botany*, applied to leaves which are irregularly divided, to about the midrib, into segments or lobes.

Pinnatipar'tite (Lat. *pin'na*, a feather ; *par'tio*, I divide). In *botany*, applied to leaves cut into lateral segments nearly to the central rib.

Pinnat'iped (Lat. *pin'na*, a feather ; *pes*, a foot). Having the toes bordered by membranes.

Pis'ces (Lat. *pis'cis*, a fish). Fishes : a class of oviparous vertebrate animals, inhabiting the water, breathing by gills, having a heart with two cavities, and the body generally covered with scales.

Pis'cine (Lat. *pis'cis*, a fish). Relating to fish.

Pisciv'orous (Lat. *pis'cis*, a fish ; *vo'ro*, I devour). Living on fishes.

Pi'siform (Lat. *pi'sum*, a pea ; *for'ma*, shape). Resembling a pea.

Pi'solite (Lat. *pi'sum*, a pea ; Gr. λιθος, *lith'os*, a stone). A mineral called peastone, consisting of carbonate of lime with a little oxide of iron, occurring in small globular masses.

Pis'til (Lat. *pistil'lum*, a pestle). In *botany*, the central organ of a flowering plant, consisting of the ovary, style, and stigma.

Pistil'lary (Lat. *pistil'lum*, a pistil). Belonging to a pistil.

Pistil'late (Lat. *pistil'lum*, a pistil). Bearing pistils.

Pistillid'ium (Lat. *pistil'lum*, a pistil). An organ in cryptogamic or flowerless plants, supposed to be the analogue of the pistil.

Pistillif'erous (Lat. *pistil'lum*, a pistil ; *fer'o*, I bear). Producing pistils.

Pis'ton (Lat. *pin'so*, I pound). A short cylinder fitting exactly into a tube, and used for the purpose of forcing air or fluid into or out of the latter.

Pitch'stone. A rocky compound of silica and alumina, having a compact texture and a pitchy glassy lustre.

Pitu'itary (Lat. *pitui'ta*, phlegm). Secreting phlegm or mucus ; applied especially to the membrane lining the nose : also to a small oval body at the base of the brain, formerly supposed to secrete the mucus of the nostrils.

Pitu'itous (Lat. *pitui'ta*, phlegm or mucus). Consisting of, or resembling mucus.

Pityri'asis (Gr. πιτυρον, *pit'uron*, bran). A disease of the skin, characterised by the appearance of patches of bran-like scales.

Placen'ta (Gr. πλακους, *plakous*, a flat cake). In *anatomy*, the mass or cake, consisting principally of blood-vessels, by which a connection is maintained between the mother and the foetus ; in *botany*, that part of a seed-vessel or fruit to which the ovules or seeds are attached.

Placen'tal (*Placenta*). Belonging to the placenta.

Placenta'tion (*Placenta*). The function and arrangement of the placenta.

Placentif'erous (Lat. *placenta; fer'o*, I bear). Bearing a placenta.

Placogan'oid (Gr. πλαξ, *plax*, a flat thing ; γανος, *gan'os*, splendour ; ειδος, *eidos*, form). A suborder of fossil fishes, covered with large ganoid plates.

Pla'coid (Gr. πλαξ, *plax*, a flat thing ; ειδος, *eidos*, form). A term applied to an order of fishes, having the body covered with irregular plates of enamel.

Plag'iostome (Gr. πλαγιος, *plag'ios*, oblique ; στομα, *stom'a*, a mouth). Oblique-mouthed ; applied to certain fossil obliquely compressed oval bivalve mollusca ; also to an order of fishes.

Plane (Lat. *planus*, flat). A level surface, such that a straight line, drawn between any two points on it, will altogether lie on the surface ; applied also to an imaginary flat surface supposed to pass through a body.

Plane Geometry. The geometry of plane or flat surfaces, in opposition to that of solids.

Plan'et (Gr. πλαναομαι, plana'omai, I wander). A globe revolving round the sun in an elliptic orbit ; the name having been given by the ancients to such bodies on account of the apparent irregularity of their motions.

Plan'etary (Gr. πλανητης, planētēs, a planet). Consisting of, or relating to planets.

Plan'etoid (Gr. πλανητης, planētēs, a planet ; ειδος, eidos, shape). A name given to the bodies found by astronomers in the space between Mars and Jupiter, where, on mathematical reasoning, a planet would be expected.

Planim'etry (Lat. pla'nus, flat ; μετρον, met'ron, a measure). The measuring of plane surfaces.

Pla'no-con'cave (Lat. pla'nus, flat ; con'cavus, hollowed out). Flat on one side and concave on the other.

Pla'no-con'ical (Lat. pla'nus, flat ; co'nus, a cone). Flat on one side and conical on the other.

Pla'no-convex' (Lat. pla'nus, flat ; convex'us, convex). Flat on one side and convex on the other.

Plantar (Lat. plan'ta, the sole of the foot). Belonging to the sole.

Plan'tigrade (Lat. plan'ta, the sole of the foot ; grad'ior, I step). Walking on the sole of the foot, as the bear.

Plas'ma (Gr. πλασσω, plassō, I form). The colourless part of the blood, being the material from which the tissues are nourished.

Plas'tic (Gr. πλασσω, plassō, I form). Capable of being moulded into a form ; giving a definite form.

Plas'tron. The floor, in tortoises and turtles, of the bony encasement of which the carapace forms the upper part.

Plat'y- (Gr. πλατυς, plat'us, flat).

A prefix in compound words, signifying flat.

Platycœ'lion (Gr. πλατυς, plat'us, flat ; κοιλος, koi'los, hollow). A term applied to some fossil crocodilian reptiles, in which one end of the body of a vertebra was flat and the other concave.

Platys'ma (Gr. πλατυνω, platu'nō, I widen). An expansion ; in anatomy, a broad thin muscular expansion lying under the skin at each side of the neck.

Plectogna'thous (Gr. πλεκω, plek'ō, I connect ; γναθος, gnath'os, the jaw). Applied to an order of fishes which have the upper jaw firmly attached to the skull.

Plei'ades (Gr. πλεω, pleō, I sail). A cluster of seven stars in the neck of the constellation Taurus ; the rising of which, to the Greeks, indicated the time of safe navigation.

Plei'ocene (Gr. πλειων, plei'ōn, more ; καινος, kai'nos, new). A term in geology for the upper tertiary group, containing more of recent than of extinct species.

Pleis'tocene (Gr. πλειστος, pleis'tos, most ; καινος, kai'nos, new). A term applied in geology to the upper or post-tertiary group, implying that the organic remains almost entirely represent existing species.

Ple'onasm (Gr. πλεοναζω, pleona'zō, I am more than enough). The use of more words than are necessary to express an idea.

Pleonas'tic (Gr. πλεοναζω, pleona'zō, I am more than enough). Belonging to pleonasm ; redundant.

Plesiomor'phism (Gr. πλησιος, plēsios, near ; μορφη, morphē, form). Close but not identical resemblance in form ; applied to certain crystals.

Plesiomor'phous (Gr. πλησιος, plēsios, near ; μορφη, morphē, form). Nearly of the same form.

Pleth'ora (Gr. πληθω, plēthō, I become full). Fulness ; in medicine, fulness of blood ; a full habit of body.

Pletho'ric (Gr. πληθω, plēthō, I become full). Having a full habit of body.

Pleu'ra (Gr. πλευρα, *pleu'ra*, a rib). The serous membrane which lines the interior of the chest and covers the lungs.

Pleural'gia (Gr. πλευρα, *pleu'ra*, a rib; άλγος, *alg'os*, pain). Pain in the side.

Pleurapoph'ysis (Gr. πλευρα, *pleu'ra*, a rib; *apoph'ysis*). A name given to the bone projecting from the typical vertebra, which forms the first part of the hæmal arch on each side; a rib.

Pleuren'chyma (Gr. πλευρα, *pleu'ra*, a rib; εγχυμα, *en'chuma*, a tissue). Woody tissue in plants.

Pleu'risy (*Pleura*). Inflammation of the pleura or serous lining of the chest.

Pleurit'ic (*Pleuri'tis*). Belonging to or having pleurisy.

Pleuri'tis (*Pleura; itis*, denoting inflammation). Pleurisy.

Pleu'rodont (Gr. πλευρα, *pleu'ra*, a rib or the side; όδους, *odous*, a tooth). A term applied to saurian reptiles which have the teeth anchylosed to the bottom of an alveolar groove, and supported by its side.

Pleurorhi'zal (Gr. πλευρα, *pleu'ra*, a rib; ριζα, *rhiza*, a root). Having the radicle applied to the edges of the cotyledons.

Plex'iform (Lat. *plex'us*, a network; *for'ma*, shape). Having the form of a network.

Plexus (Lat., a network). An interweaving or network; in *anatomy*, a term applied to an arrangement of blood-vessels, absorbent vessels, or nerves in the form of a network.

Pli'cate (Lat. *pli'ca*, a fold). Folded.

Plinth (Gr. πλινθος, *plinth'os*, a brick or tile). In *architecture*, the flat square table under the moulding of the base and pedestal of a column, serving as the foundation.

Pli'ocene. See Plei'ocene.

Plu'mose (Lat. *plu'ma*, a small feather, or down). Feathery; resembling feathers.

Plu'mule (Lat. *plu'mula*, a little

feather). In *botany*, the growing point of the embryo in the seed, representing the future stem of the plant.

Plural (Lat. *plus*, more). Relating to more than one; but, in the grammars of the Greek and some other languages, expressing more than two.

Pluri- (Lat. *plus*, more). A prefix in compound words, signifying several.

Plurilit'eral (Lat. *plus*, more; *lit'era*, a letter). Containing more than three letters.

Pluton'ic (Lat. *Pluto*, the god of the lower regions). In *geology*, applied to rocks formed by the agency of fire at some depth below the surface of the land or sea.

Plu'vial (Lat. *plu'via*, rain). Rainy; relating to rain.

Pluviam'eter (Lat. *plu'via*, rain; Gr. μετρον, *met'ron*, a measure). A rain-gauge; an instrument for measuring the amount of rain which falls.

Pneumat'ic (Gr. πνευμα, *pneu'ma*, air). Consisting of, or pertaining to air; moved by means of air.

Pneumat'ic Trough. A trough filled with water or mercury, and provided with a perforated shelf for holding inverted jars or receivers, used in chemistry for collecting gases.

Pneumat'ics (Gr. πνευμα, *pneu'ma*, air.) The branch of natural philosophy which describes the mechanical properties of air and gases, as well as those machines which act by application of these properties.

Pneu'mato- (Gr. πνευμα, *pneu'ma*, air). A prefix in compound words, implying relation to, or connection with air or breath.

Pneumatochem'ical (Gr. πνευμα, *pneu'ma*, air; *chem'ical*). Relating to the chemistry of air or gases.

Pneumatol'ogy (Gr. πνευμα, *pneu'ma*, air; λογος, *log'os*, discourse). A description of air or breath.

Pneumatotho'rax or **Pneumotho'rax** (Gr. πνευμα, *pneu'ma*, air; θωραξ, *thōrax*, the chest). Air in the chest, between the walls of the cavity and its contents.

Pneumatol'ogy (Gr. πνευμα, *pneu'ma*, air ; λογος, *log'os*, a discourse). The doctrine of the properties of airs or gases.

Pneu'mo- (Gr. πνευμων, *pneumōn*, a lung). A prefix in compound words, implying connection with, or relation to lungs.

Pneumogas'tric (Gr. πνευμων, *pneumōn*, the lungs ; γαστηρ, *gustēr*, the stomach). Belonging to the lungs and stomach ; applied to a nerve which supplies these organs.

Pneumon'ic (Gr. πνευμων, *pneumōn*, a lung). Belonging to the lungs.

Pneumo'nia (Gr. πνευμων, *pneumūn*, a lung). Inflammation of the lungs.

Po'acites (Gr. ποα, *poa*, grass). In *geology*, the generic term for all fossil monocotyledonous leaves, having the veins parallel, simple, and equal, and not connected transversely.

Podag'ra (Gr. πους, *pous*, a foot; αγρα, *agra*, a seizing). The gout.

Pod'ocarp (Gr. πους, *pous*, a foot ; καρπος, *kar'pos*, fruit). The stem supporting the fruit.

Podophthalma'ria (Gr. πους, *pous*, a foot ; οφθαλμος, *ophthal'mos*, an eye). A group of crustacea, having the eyes placed on moveable peduncles or stalks.

Pod'osperm (Gr. πους, *pous*, a foot ; σπερμα, *sper'ma*, a seed). In *botany*, the little bud connecting an ovule with its placenta.

Pœ'cilopods (Gr. ποικιλος, *poi'kilos*, varied ; πους, *pous*, a foot). Crustaceous animals having the fore-feet adapted either for swimming or seizing.

Polar (Lat. *pol'us*, a pole). Belonging to one of the poles of the earth ; or to the magnetic pole.

Polar Circles. Two small circles of the earth, parallel to the equator, and surrounding the poles, north and south.

Polar'iscope (Lat. *pola'ris*, belonging to a pole ; Gr. σκοπεω, *skop'eō*, I view). An optical instrument for observing the phenomena of the polarisation of light.

Polar'ity (Lat. *pol'us*, a pole). The property by which the particles of many bodies arrange themselves in fixed directions to given poles.

Polariza'tion (Lat. *pol'us*, a pole). The act of giving polarity to a body.

Polarization of Light. The process by which a ray of light acquires new properties when submitted, under peculiar conditions, to reflection or refraction.

Pole (Gr. πολος, *pol'os*, an axis or pole). The extremity of the axis of a spherical body, or of a straight line passing through the centre of such a body. Each pole is 90 degrees distant from any part of the equatorial circumference. Magnetic poles are two poles in a loadstone corresponding to the poles of the earth. The poles of a Voltaic battery are the ends of the wires that connect its opposite ends.

Polem'ic (Gr. πολεμος, *pol'emos*, war). Controversial : disputative.

Pollen (Lat. fine flour or dust). The fine dust on the anther of flowers.

Pol'y- (Gr. πολυς, *pol'us*, much). A prefix in compound words, signifying much or many.

Polyadel'phia (Gr. πολυς, *pol'us*, many ; αδελφος, *adel'phos*, a brother). A name given to a class of plants in the Linnæan system, in which the stamens are collected into several parcels.

Polyan'dria (Gr. πολυς, *pol'us*, many ; ανηρ, *anēr*, a male). A name given to a class of plants in the Linnæan system, having twenty or more stamens on the receptacle.

Pol'ybasic (Gr. πολυς, *pol'us*, many ; βασις, *bas'is*, a base). A term applied to acids which require two or more equivalents of a base for neutralisation.

Polycar'pous (Gr. πολυς, *pol'us*, many ; καρπος, *kar'pos*, fruit). Having many fruit.

Polychromat'ic (Gr. πολυς, *pol'us*, many ; χρωμα, *chrōma*, colour). Having many colours ; showing a play of colours.

Polycotyle'donous (Gr. πολυς, *pol us*, many ; κοτυληδων, *kotulēdōn*, a seed-lobe). Having more than two lobes to the seed.

Polydac'tylous (Gr. πολυς, *pol'us*, many ; δακτυλος, *dak'tulos*, a finger). Having many fingers.

Polyem'bryony (Gr. πολυς, *pol'us*, many ; εμβρυον, *em'bruon*, an embryo). In *botany*, the presence of several embryos in the same ovule.

Polygam'ia (Gr. πολυς, *pol'us*, many ; γαμος, *gam'os*, marriage). A name applied to a class of plants in the Linnæan system, which have neutral flowers, with male or female flowers or both, not collected in the same calyx, but scattered on the same, or on two or three distinct individuals.

Polygas'tric (Gr. πολυς, *pol'us*, many ; γαστηρ, *gastēr*, a stomach). Having, or appearing to have, many stomachs.

Pol'yglot (Gr. πολυς, *pol'us*, many ; γλωττα, *glōtta*, a tongue). Containing or written in many languages.

Pol'ygon (Gr. πολυς, *pol'us*, many ; γωνια, *gōnia*, an angle). A figure of more than four sides and angles.

Polyg'onal (Gr. πολυς, *pol'us*, many ; γωνια, *gōnia*, an angle). Having, or capable of being arranged in, the form of a polygon.

Polygyn'ia (Gr. πολυς, *pol'us*, many ; γυνη, *gunē*, a female). A name given to an order of plants in the Linnæan system, which have more than twelve pistils or styles.

Polyhed'ron (Gr. πολυς, *pol'us*, many ; εδρα, *hed'ra*, a base). A solid figure having many angles and sides.

Polymer'ic (Gr. πολυς, *pol'us*, many ; μερος, *mer'os*, a part). Having many parts.

Polymor'phous (Gr. πολυς, *pol'us*, many ; μορφη, *morphē*, shape). Having many shapes.

Polyne'sia (Gr. πολυς, *pol'us*, many ; νησος, *nēsos*, an island). A large collection of islands.

Polyno'mial (Gr. πολυς, *pol'us*, many ; ονομα, *on'oma*, a name). In *algebra*, a quantity or expression which consists of several terms.

Polynom'ic (Gr. πολυς, *pol'us*, many ; νομος, *nom os*, a region). In *botany*, applied to plants which are distributed over several regions of the globe.

Pol'ypary (*Pol'ypus*). The organ of support, or coral, of a polype.

Polypet'alous (Gr. πολυς, *pol'us*, many ; πεταλον, *pet'alon*, a petal). Having many petals.

Polyphyl'lous (Gr. πολυς, *pol'us*, many ; φυλλον, *phullon*, a leaf). Having many leaves or leaflets

Polyp'idom (*Pol'ypus*; Lat. *do'mus*, a house). The stony or coralline structure inhabited by polypes.

Polypif'erous (*Pol'ypus*; Lat. *fer'o*, I bear). Producing polypes.

Pol'ypus (Gr. πολυς, *pol'us*, many ; πους, *pous*, a foot). A small soft-bodied water animal, generally having a cylindrical, oval, or oblong body, with an aperture at one end surrounded by radiating filaments or tentacles ; in *surgery*, a kind of tumour.

Polysep'alous (Gr. πολυς, *pol'us*, many ; *sep'al*). Having the sepals distinct from each other.

Polysper'mal or **Polysper'mous** (Gr. πολυς, *pol'us*, many ; σπερμα, *sper'ma*, a seed). Containing many seeds.

Polyste'monous (Gr. πολυς, *pol'us*, many ; στημων, *stēmōn*, a stamen). Having many stamens.

Polysyllab'ic (Gr. πολυς, *pol'us*, many ; συλλαβη, *sul'labē*, a syllable). Having many syllables.

Polytech'nic (Gr. πολυς, *pol'us*, many ; τεχνη, *technē*, art). Comprehending many arts.

Polythal'amous (Gr. πολυς, *pol'us*, many ; θαλαμος, *thal'amos*, a chamber). Having many cells or chambers.

Polyzo'nal (Gr. πολυς, *pol'us*, many ; ζωνη, *zōnē*, a belt). Composed of many zones or belts.

Pomol'ogy (Lat. *po'mum*, a fruit; Gr. λογος, *log'os*, a discourse). The branch of gardening which teaches the cultivation of fruit-trees.

Pom'pholyx (Gr. πομφος, *pomph'os*, a bubble). A disease of the skin.

Poplite'al (Lat. *po'ples*, the ham) belonging to the ham.

Pore (Gr. πορος, *por'os*, a means of passing). In *natural philosophy*, an interstice or minute space between the molecules of matter.

Po'rism (Gr. ποριζω, *pori'zō*, I bring about). In *geometry*, a proposition affirming the possibility of finding such conditions as will render a certain problem indeterminate or capable of innumerable solutions.

Poros'ity (Gr. πορος, *por'os*, a pore). The state of having pores : in *natural philosophy*, the quality of bodies in virtue of which their constituent atoms are separated by vacant spaces or pores.

Porous (Gr. πορος, *por'os*, a pore). Having pores or interstices.

Por'phyry (Gr. πορφυρα, *por'phura*, purple dye). Originally, a reddish-igneous rock : now used in *geology* to denote any rock containing imbedded crystals distinct from the main mass.

Por'tal (Lat. *por'ta*, a gate). In *anatomy*, belonging to the transverse fissure of the liver, called by old anatomists the *porta* or gate of the organ.

Posses'sive (Lat. *possid'eo*, I possess). In *grammar*, the case of nouns which denotes possession, or some relation of one thing to another.

Post- (Lat.) A Latin preposition used in the composition of many words, and signifying after or since.

Postdilu'vian (Lat. *post*, after; *dilu'vium*, a deluge). Living after the deluge.

Poste'rior (Lat. later). Later : *à posterio'ri*, a phrase signifying "from what follows," applied to an argument used to infer a cause or antecedent from an effect or consequent.

Postfron'tal (Lat. *post*, after; *frons*, the forehead). Behind the frontal bone.

Postmerid'ian (Lat. *post*, after; *meri'dies*, midday). Belonging to the afternoon.

Post Mortem. (Lat.) After death.

Postpos'itive (Lat. *post*, after; *pono*, I put). Placed after.

Pos'tulate (Lat. *pos'tulo*, I demand). A position or supposition considered too plain to require illustration ; it differs from an axiom only in being put as a request instead of an assertion.

Poten'tial (Lat. *po'tens*, able). Having the power to impress the ideas of certain qualities, though the ideas are not inherent in the thing ; existing in possibility ; in *grammar*, applied to the mood of verbs which denotes capability or power.

Præ- or **Pre-** (Lat. *præ*, before). A preposition used in compound words, signifying before or in front of.

Præcor'dia (Lat. *præ*, before ; *cor*, the heart). The region of the body in front of the heart.

Præflora'tion (Lat. *præ*, before ; *flos*, a flower). The arrangement of the parts of the flower in the flower-bud ; the same as æstivation.

Præfolia'tion (Lat. *præ*, before ; *fo'lium*, a leaf). The arrangement of the leaves in a leaf-bud ; the same as vernation.

Præno'men (Lat. *præ*, before ; *nomen*, a name). Among the Romans, a name prefixed to the family name, answering to our Christian name.

Pre- (Lat. *præ*, before). *See* Præ.

Preces'sion (Lat. *præ*, before ; *ce'do*, I go). A going before. In *astronomy*, the precession of the equinoxes is a slow retrograde motion which they undergo in a direction contrary to the order of the signs, and which makes them succeed each other sooner than they otherwise would have done.

Precip'itant (Lat. *præ'ceps*, headlong). In *chemistry*, a substance which, added to a solution of another, causes the latter to be thrown down to the bottom of the fluid.

Precip'itate (Lat. *præ'ceps*, headlong). To throw down a substance from its solution ; the substance thus thrown down.

Precor'dial (Lat. *præ*, before ; *cor*,

the heart). Belonging to the præcordia, or parts before the heart.

Preda'ceous (Lat. *præ'da*, prey). Living on prey.

Predic'ament (Lat. *præ'dico*, I affirm). In *logic*, a series or order of all the predicates or attributes contained under one genus.

Pre'dicate (Lat. *præ'dico*, I affirm). In *logic*, that which is affirmed or denied of a subject.

Predisposi'tion (Lat. *præ*, before; *dispo'no*, I put in order). An inclination or propensity.

Prefron'tal (Lat. *præ*, before; *frons*, the forehead). In front of the frontal bone : applied to the middle part of the ethmoid bone.

Prehen'sile (Lat. *prehen'do*, I take hold). Seizing or taking hold.

Prehen'sion (Lat. *prehen'do*, I take hold). A taking hold of anything.

Premon'itory (Lat. *præ*, before; *mon'eo*, I advise). Giving previous warning.

Premor'se (Lat. *præ*, before; *mor'deo*, I bite). In *botany*, applied to a root terminating abruptly, as if bitten off.

Preposit'ion (Lat. *præ*, before; *pono*, I put). A word put before another to express some relation to it.

Prepos'itive (Lat. *præ*, before; *pono*, I put). Placed before.

Presbyo'pia (Gr. *πρεσβυs*, *presb'us*, old; *ὠψ*, *ὀps*, the eye). A defect of vision common in old persons, in which, from a flattening of the cornea, near objects are seen less distinctly than those at a distance.

Preter (Lat. *præ'ter*, beyond). A Latin preposition used in compound words, signifying beyond.

Pre'terite (Lat. *præ'ter*, beyond; *eo*, I go). Past.

Prever'tebral (Lat. *præ*, before; *ver'tebra*, a bone of the spine). In front of the vertebræ or spinal bones.

Pri'mæ Viæ (Lat. The first ways). A term applied to the stomach and intestines.

Pri'mary (Lat. *pri'mus*, first). First; original ; in *astronomy*, applied to those planets which revolve round the sun ; in *ornithology*, applied to the feathers which arise from the ulnar side of the hand part of the wing of birds ; in *natural philosophy*, to those properties of matter which are inseparable from it ; in *optics*, to colours into which a ray of light may be decomposed ; in *geology*, to crystalline rocks supposed to owe their structure to the agency of fire.

Prima'tes (Lat. *pri'mus*, first). The name given by Linnæus to his first order of mammalia, including man, the apes, the lemurs and the bats.

Pri'mine (Lat. *pri'mus*, first). In *botany*, the outer covering of the ovule.

Prim'itive (Lat. *pri'mus*, first). *See* Primary.

Primor'dial (Lat. *pri'mus*, first; *or'do*, order). First in order ; appearing first.

Prism (Gr. *πρισμα*, *pris'ma*, a prism). A solid figure, the ends of which are similar, equal, and parallel plane figures, and the sides of which are parallelograms; they are triangular, square, pentagonal, &c., according to the number of sides.

Prismat'ic (*Prism*). Resembling, or formed like a prism.

Prismen'chyma (Gr. *πρισμα*, *pris'ma*, a prism; *ἐγχυμα*, *en'chuma*, tissue). In *botany*, tissue formed of prismatic cells.

Pro'blem (Gr. *προ*, *pro*, before; *βαλλω*, *ballō*, I cast). A question proposed ; a proposition in which some operation is required.

Proboscid'ian (Gr. *προβοσκιs*, *probos'kis*, a trunk or snout). A family of pachydermatous or thick skinned animals, which have the nose elongated into a flexible trunk, as the elephant.

Proboscid'iform (Gr. *προβοσκιs*, *probos'kis*, a trunk or snout ; Lat. *for'ma*, shape). Resembling a trunk or snout.

Probos'cis (Gr. *προ*, *pro*, before ; *βοσκω*, *boskō*, I feed). The snout or trunk of an elephant and analogous animals ; the flexible apparatus which some insects use in sucking; the long tongue of certain gasteropods, capable of being protruded to some distance.

Proc'ess (Lat. *proce'do*, I move forward). A proceeding or operation; in *anatomy* and *botany*, a prominence or projecting part; applied also to the parts of a vertebra which grow out from previously ossified parts.

Procliv'ity (Lat. *proclivus*, inclined). An inclination or disposition.

Procne'mial (Gr. προ, *pro*, before; κνημη, *kncme*, the knee). In front of the knee.

Procœlian (Gr. προ, *pro*, before; κοιλος, *koi'los*, hollow). Having the vertebræ concave in front.

Procum'bent (Lat. *procum'bo*, I lie down). Lying on the ground.

Progno'sis (Gr. προ, *pro*, before; γιγνωσκω, *gignōs'kō*, I know). The art of judging of the course and event of a disease by the symptoms.

Prognos'tic (Gr. προ, *pro*, before; γιγνωσκω, *gignōs'kō*, I know). Relating to foreknowledge; applied to the symptoms from which the result of a disease is predicted.

Progres'sion (Lat. *pro*, forward; *grad'ior*, I step). A moving forward or advancing; in *arithmetic*, a regular or proportional advance of numbers in a series, increasing or decreasing; in *astronomy*, the change which occurs every month in the position of the moon's apogee and perigee, in which these points appear to have moved forward, or from west to east.

Projec'tile (Lat. *pro*, forward; *jac'io*, I cast). A body impelled by force, especially through the air.

Projec'tion (Lat. *pro*, forward; *jac'io*, I cast). A throwing forward; applied also in *architecture* to a plan or delineation.

Pro'late (Lat. *pro*, forward; *la'tus*, borne). Extended beyond the line of an exact sphere.

Prolegom'ena (Gr. προ, *pro*, before; λεγω, *leg'ō*, I speak). Literally, things said first; introduc-

tory remarks prefixed to a book or treatise.

Prolegs (Lat. *pro*, for; *legs*). The tubercles representing legs on the hinder part of caterpillars.

Prolif'erous (Lat. *pro'les*, offspring; *fer'o*, I bear). Fruitful; productive; in *botany*, bearing abnormal buds.

Prolif'ic (Lat. *pro'les*, offspring; *fac'io*, I make). Fruitful; productive.

Prolig'erous (Lat. *pro'les*, offspring; *ger'o*, I bear). Bearing the rudiments of the embryo or offspring.

Prona'tion (Lat. *pro'nus*, having the face downward). The position of the arm and hand in which the palm is turned downwards.

Prona'tor (Lat. *pro'nus*, with the face downwards). A muscle which turns the arm so that the palm of the hand looks downwards.

Prone (Lat. *pro'nus*). Bending forward; having the face or anterior surface downwards.

Prono'tum (Gr. προ, *pro*, before; νωτος, *nōtos*, the back). The upper half of the anterior division of the thorax in insects.

Prop'erty (Lat. *pro'prius*, proper). A peculiar quality of anything; that which is inherent in, or naturally essential to, a substance.

Prophylac'tic (Gr. προ, before; φυλασσω, *phulas'sō*, I guard). In *medicine*, preserving from disease.

Prophylax'is (Gr. προ, *pro*, before; φυλασσω, *phulas'sō*, I guard). The art of preventing or defending against diseases.

Prop'olis (Gr. προ, *pro*, before; πολις, *pol'is*, a city). A thick substance formed by bees, and used as a kind of mortar or cement to their hives.

Propor'tion (Lat. *pro*, for; *por'tio*, a share). The comparative relation of one thing to another; in *arithmetic*, the identity or similitude of two or more ratios.

Proposit'ion (Lat. *pro*, forward; *po'no*, I put). A thing proposed or put forward; in *logic*, a sentence or statement in which something is affirmed or denied of a subject; in

L

mathematics, a statement of a truth to be proved—theorem, or of an operation to be performed—problem.

Pros- (Gr. πρos, *pros*, towards). A preposition in compound words, signifying towards or near.

Prosec'tor (Lat. *pro'seco*, I cut off). An anatomist; one who dissects the body for a lecturer on anatomy.

Prosencephal'ic (Gr. πρos, *pros*, near; ἐγκεφαλον, *enkeph'alon*, the brain). Seated before the brain.

Prosen'chyma (Gr. πρos, *pros*, towards; ἐγχυμα, *en'chuma*, a tissue). Vegetable tissue formed of spindle-shaped cells, generally applied closely together.

Pros'ody (Gr. πρos, *pros*, to; ᾠδη, *ōdē*, an ode or singing). The part of grammar which treats of the quantity of syllables, and of the laws of versification.

Proster'num (Gr. πρo, *pro*, before; στερνον, *ster'non*, the breast). The lower half of the anterior division of the thorax in insects.

Pros'thesis (Gr. πρos, *pros*, to; τιθημι, *tithēmi*, I place). In *grammar*, the adding of one or more letters to the beginning of a word.

Pro'tein (Gr. πρωτος, *prōtos*, first). A substance consisting of oxygen, hydrogen, carbon, and nitrogen, produced by the action of alkali or acetic acid on albumen, fibrin, and casein.

Proth'esis (Gr. πρo, *pro*, before; τιθημι, *tithēmi*, I place). *See* Prosthesis.

Protho'rax (Gr. πρo, *pro*, before; θωραξ, *thōrax*, a breast-plate). The anterior segment of the thorax in insects, bearing the anterior pair of legs.

Protich'nites (Gr. πρωτος, *prōtos*, first; ἰχνος, *ichnos*, a footstep). Imprints of the feet of early fossil animals.

Proto- (Gr. πρωτος, *prōtos*, first). A prefix used in compound words, signifying first; frequently employed in chemical nomenclature.

Pro'toplasm (Gr. πρωτος, *prōtos*, first; πλασσω, *plas'sō*, I form).

The material which appears to be concerned in the early formation of simply organised bodies.

Protox'ide (Gr. πρωτος, *prōtos*, first; *oxide*). The degree of oxidation which possesses the most strongly marked basic properties.

Protozo'a (Gr. πρωτος, *prōtos*, first; ζωον, *zōon*, an animal). The lowest division of the animal kingdom, consisting of creatures of very low organisation, apparently occupying a neutral ground between animals and vegetables.

Protozo'ic (Gr. πρωτος, *prōtos*, first; ζωον, *zōon*, an animal). In *geology*, applied to the strata containing the earliest traces of animal life.

Protrac'tile (Lat. *pro*, forward; *tra'ho*, I draw). Having the power of lengthening or drawing out.

Protu'berance (Lat. *pro*. before; *tu'ber*, a bunch or knob). A prominence.

Prox'imate (Lat. *prox'imus*, nearest). Nearest; proximate principles are those compounds which exist ready formed in animals and vegetables, as albumen, casein, sugar, gum, starch, &c.

Pruri'go (Lat. *pru'rio*, I itch). An eruptive disease of the skin, accompanied by much itching.

Prus'siate (*Prussic* acid). A term formerly given to supposed compounds of prussic acid with bases, but now known as cyanides of metals.

Prus'sic. A name sometimes given to hydrocyanic acid.

Pseud- or **Pseudo-** (Gr. ψευδος, *pseu'dos*, a falsehood). A prefix in some compound words, signifying false or counterfeit.

Pseudomor'phous (Gr. ψευδος, *pseu'dos*, a falsehood; μορφη, *morphē*, form). Not having the true form; applied to minerals, the form of which has not been derived from true crystallisation.

Pseudosper'mous (Gr. ψευδος, *pseu'dos*, falsehood; σπερμα, *sper'ma*, seed). Having single-seeded fruits resembling seeds.

Psoas (Gr. ψοα, *psoa*, the loin). A

name given to certain muscles in the region of the loins.

Psori'asis. A disease of the skin consisting of irregular patches covered with white scales.

Psy'chical (Gr. ψυχη, *psuché*, the soul). Relating to the doctrine of the nature and properties of the soul.

Psycholog'ical (Gr. ψυχη, *psuché*, the soul ; λογος, *log'os*, discourse). Relating to the doctrine of the mind or soul.

Psychol'ogy (Gr. ψυχη, *psuché*, the soul ; λογος, *log'os*, discourse). The doctrine of the nature and properties of the soul ; generally applied with regard to the faculties of the mind.

Psychop'athy (Gr. ψυχη, *psuché*, the soul ; παθος, *path'os*, suffering). Mental disease.

Psychrom'eter (Gr. ψυχρος, *psu'chros*, cold or cool ; μετρον, *met'ron*, a measure). A hygrometer, the indications of which depend on the depression of temperature procured by evaporation in an atmosphere not perfectly saturated with moisture.

Pter-, -pter'a, or pter'o- (Gr. πτερον, *pter'on*, a wing). A prefix, or a termination, in compound words, signifying relation or likeness to a wing.

Pterocar'pous (Gr. πτερον, *pter'on*, a wing ; καρπος, *kar'pos*, fruit). Having winged fruits.

Pterodac'tyle (Gr. πτερον, *pter'on*, a wing ; δακτυλος, *dak'tulos*, a finger). A fossil flying reptile, with an elongated wing-finger.

Pter'opods (Gr. πτερον, *pter'on*, a wing ; πους, *pous*, a foot). A class of molluscous animals, having a distinct head formed for floating and swimming by means of two fins, one being placed on each side of the neck.

Pterosau'ria (Gr. πτερον, *pter'on*, a wing ; σαυρος, *sau'ros*, a lizard). An order of fossil reptiles, having the anterior limbs adapted for flying.

Pter'ygoid (Gr. πτερυξ, *pter'ux*, a wing ; ειδος, *ei'dos*, shape). Like

a wing ; applied to a part of the sphenoid bone, having some resemblance to a wing ; also to muscles, vessels, nerves, &c., having connection with, or relation to, this part.

Ptolema'ic (Gr. Πτολεμαιος, *Ptolemai'os*, a Greek geographer and astronomer). According to Ptolemy; the Ptolemaic system *in astronomy* was that which supposed the earth to be fixed in the centre of the universe, and the other bodies to revolve round it.

Pto'sis (Gr. πτωσις, *pto'sis*, a falling). A paralysis of the upper eyelid, so that it falls over the eye, and cannot be raised.

Pty'alism (Gr. πτυαλιζω, *ptuali'zo*, I spit often). An excessive flow of saliva.

Pu'berty (Lat. *puber*, ripe of age). The period at which childhood ends and adolescence begins.

Pubes'cence (Lat. *pu'bes*, the down of plants). The downy substance, or short and soft hairs, on plants.

Pubes'cent (Lat. *pu'bes*, down). In *botany*, applied to plants covered with soft, short, downy hairs.

Pud'dling. In iron manufacture, the process by which the oxygen and carbon of cast iron are expelled ; the metal being reduced by heat to a pasty condition, and stirred so as to expose every part to the action of the air.

Pug-mill. A machine for mixing and tempering clay, consisting of an iron cylinder, in which the clay is cut and kneaded by a series of knives revolving on an axis within the cylinder.

Pul'mograde (Lat. *pul'mo*, a lung ; *gra'dior*, I step). Moving by lungs ; applied to a tribe of invertebrate animals which swim by means of the disc on which the respiratory apparatus is placed.

Pul'monary (Lat. *pul'mo*, a lung). Relating to the lungs.

Pulmon'ic (Lat. *pul'mo*, a lung). Relating to the lungs.

Pulmonif'erous (Lat. *pul'mo*, a lung ; *fer'o*, I bear). Provided with lungs.

Pul'sate (Lat. *pul'so*, I beat). To beat or throb.

Pulsa'tion (Lat. *pul'so*, I beat). A beating; the act of beating or throbbing of the heart or an artery, in the process of the circulation of the blood.

Pulse (Lat. *pul'so*, I beat). The phenomenon produced in an artery by its extension with each beat of the heart, and the resistance of the flow of blood to pressure.

Pulta'ceous (Lat. *puls*, a kind of gruel). Softened; nearly fluid.

Pul'verize (Lat. *pul'vis*, powder). To reduce to powder.

Pul'vinate (Lat. *pulvi'nar*, a pillow). Like a cushion or pillow.

Pul'vinated (Lat. *pulvi'nar*, a pillow). In *architecture*, a term used to denote a swelling in any portion of an order.

Pulvis (Lat.). A powder.

Punc'tated (Lat. *punc'tum*, a point). Dotted.

Punctua'tion (Lat. *punc'tum*, a point). In *grammar*, the art of marking with points the divisions of a writing into sentences and members of sentences.

Pu'pa (Lat. a puppet or baby). A term applied to the third or chrysalis state of an insect.

Pupil (Lat. *pupil'la*). The round opening in the centre of the iris of the eye.

Pupip'arous (Lat. *pu'pa*; *par'io*, I bring forth). Producing young in the pupa state.

Purg'ative (Lat. *pur'go*, I cleanse). Having the power of cleansing; especially applied to medicines which act on the intestines.

Pur'pura (Lat. purple). A diseased state of the blood, allied to scurvy.

Purpu'ric (Lat. *pur'pura*, purple). A name applied to an acid which forms deep red or purple compounds with most bases.

Pu'rulent (Lat. *pus*). Of the nature of or containing pus.

Pus (Lat.). A peculiar fluid, yielded from the blood in consequence of inflammation, containing minute cells.

Puta'men (Lat. the shell of a nut). The hard covering of some fruits.

Putrefac'tion (Lat. *pu'tris*, putrid; *fac'io*, I make). A spontaneous change, to which complicated organic bodies are subject, consisting in changes occurring in the presence of moisture; the effect being a transposition of the elements of the body so as to form new compounds.

Putrefac'tive (Lat. *pu'tris*, putrid; *fac'io*, I make). Belonging to, or promoting putrefaction.

Pu'trefy (Lat. *pu'tris*, putrid; *fio*, I become). To dissolve and return to the original distinct elements, or to less complex compounds, as in animal and vegetable substances.

Putres'cent (Lat. *putres'co*, I become putrid). Passing from an organised state, having complex chemical combinations, to mere constituent elements, or comparatively simple combinations of these.

Puzzola'na (*Puzzuoli*, in Italy). A volcanic ash, used in the manufacture of Roman cement.

Pyæ'mia (Gr. πυος, *pu'os*, pus; αιμα, *hai'ma*, blood). A dangerous disease occurring after injuries and wounds, consisting of a peculiar alteration of the blood, and attended by great depression of the powers of life and the formation of more or less numerous abscesses in various parts of the body.

Pyc'nodonts (Gr. πυκνος, *puk'nos*, thick; ὀδους, *odous*, a tooth). A family of fossil fishes, occurring mostly in the oolite formation, and characterised by blunt rounded teeth.

Pyeli'tis (Gr. πυελος, *pu'elos*, a basin; *itis*, denoting inflammation). Inflammation of the pelvis, or expanded open space of the kidney.

Pylor'ic (*Pylo'rus*). Belonging to, or connected with the pylorus.

Pylo'rus (Gr. πυλωρος, *pulōros*, a gate-keeper). The part of the stomach through which the food passes into the intestines.

Pyogen'ic (Gr. πυος, *pu'os*, pus; γενναω, *genna'ō*, I produce). Forming or yielding pus.

Pyogen esis (Gr. πυος, pu os, pus; γενεσις, gene'sis, a production). The formation of pus.

Pyohæ'mia. See Pyæmia.

Pyr'amid (Generally said to be from Gr. πυρ, pur, fire; but uncertain). A solid body, having a plane base, with any number of sides and angles, the sides consisting of planes meeting in a vertex or point.

Pyretol'ogy (Gr. πυρετος, pu'retos, a fever; λογος, log'os, a discourse). A treatise on fevers, or the doctrine of fevers.

Pyrex'ia (Gr. πυρ, pur, fire; ἑξις, hexis, a holding). A state of fever.

Py'riform (Lat. py'rus, a pear; for'-ma, shape). Shaped like a pear.

Pyri'tes (Gr. πυρ, pur, fire). Fire-stone; a name given to the native sulphurets of copper and iron.

Pyro- (Gr. πυρ, pur, fire). A prefix in compound words, signifying fire; in *chemistry*, signifying that the substance named has been formed at a high temperature.

Pyrog'enous (Gr. πυρ, pur, fire; γεννaω, genna'ō, I produce). Produced by fire.

Pyrolig'neous or **Pyrolig'nous** (Gr. πυρ, pur, fire; Lat. lig'num, wood). Procured by the distillation of wood; applied to the acid liquor which passes over with the tar when wood is subjected to destructive distillation.

Pyrolig'nite. A salt formed by the combination of pyroligneous acid with a base.

Pyrol'ogy (Gr. πυρ, pur, fire; λογος, log'os, a discourse). A treatise on heat.

Pyroma'nia (Gr. πυρ, pur, fire; μανια, ma'nia, madness). An insane desire for burning houses, &c.

Pyrom'eter (Gr. πυρ, pur, fire; μετρον, met'ron, a measure). An instrument for measuring the expansion of bodies by heat; or for measuring degrees of heat above those indicated by the mercurial thermometer.

Pyromor'phous (Gr. πυρ, pur, fire; μορφη, morphē, form). Having the property of being crystallised by fire.

Pyroph'orous (Gr. πυρ, pur, fire; φερω, pher'ō, I bear). A substance which takes fire on exposure to the air, or which maintains or retains light.

Pyrophos'phate. A compound of pyrophosphoric acid with a base.

Pyrophosphor'ic (Gr. πυρ, pur, fire; phosphor'ic acid). An acid procured by exposing phosphoric acid to heat, and differing from it in uniting with two equivalents of base.

Py'roscope (Gr. πυρ, pur, fire; σκοπεω, skop'eō, I view). An instrument for measuring the intensity of heat radiating from a fire.

Pyro'sis (Gr. πυρωσις, purō'sis, a burning). A diseased state of the stomach attended with severe pain and the ejection of a large quantity of watery fluid; water-brash.

Pyrotech'nic (Gr. πυρ, pur, fire; τεχνη, technē, art). Relating to the art of making fireworks.

Pyroxyl'ic (Gr. πυρ, pur, fire; ξυλον, xulon, wood). A term applied to a spirit produced by the destructive distillation of wood.

Pyrox'ylin (Gr. πυρ, pur, fire; ξυλον, xulon, wood). Gun-cotton.

Pyr'rhonism (Pyrrho, the founder of a sect). Scepticism: universal doubt.

Pyxid'ium (Lat. pyx'is, a small box). In *botany*, a fruit, consisting of a capsule with a lid.

Q.

Quad'ra (Lat., a square). In *architecture*, a square frame or border.

Quad'rangle (Lat. quat'uor, four; an'gulus, an angle). A figure having four sides and four angles.

Quad'rant (Lat. quad'ro, I make square). A fourth part; the fourth part of the circumference of a circle, or 90 degrees; also the space included between the arc and two radii drawn from its extremities to the centre of the circle; an instrument consisting of a graduated quarter circle, used for taking the altitude of the sun or stars.

Quad'rate (Lat. *quad'ra*, a square). A square; square.

Quadrat'ic (Lat. *quad'ra*, a square). Denoting, or pertaining to a square; quadratic equations are those which contain the square of the quantity, the value of which is to be found.

Quad'rature (Lat. *quad'ra*, a square). The reduction of a figure to a square; in *astronomy*, the position of a planet when the lines from the earth to the sun and it form an angle of 90 degrees.

Quadra'tus (Lat. *quad'ra*, a square). Square; a name applied to several muscles of the body, from their shape.

Quadren'nial (Lat. *quat'uor*, four; *an'nus*, a year). Comprising four years; occurring every four years.

Quadri-(Lat. *quat'uor*, four). A prefix in compound words, signifying four.

Quadrifa'rious (Lat. *quadrifa'riam*, in four ways). In four rows.

Quad'rifid (Lat. *quad'ra*, four; *findo*, I cleave). Four-cleft.

Quadriju'gate (Lat. *quat'uor*, four; *ju'gum*, a yoke). Having four pairs of leaflets.

Quadrifur'cate (Lat. *quat'uor*, four; *fur'ca*, a fork). Doubly forked.

Quadrigem'inal (Lat. *quat'uor*, four; *gem'ini*, twins). Fourfold; having four similar parts.

Quadrilat'eral (Lat. *quat'uor*, four; *la'tus*, a side). Having four sides.

Quadrilit'eral (Lat. *quat'uor*, four; *lit'era*, a letter). Consisting of four letters.

Quadrilo'bate (Lat. *quat'uor*, four; *lo'bus*, a lobe). Having four lobes.

Quadriloc'ular (Lat. *quat'uor*, four; *loc'ulus*, a little space). Having four cells or chambers.

Quadripar'tite (Lat. *quat'uor*, four; *par'tio*, I divide). Divided deeply into four parts.

Quadrip'licate (Lat. *quat'uor*, four; *plic'a*, a fold). Having four plaits or folds.

Quadru'mana (Lat. *quat'uor*, four; *man'us*, a hand). An order of mammals, characterised by the presence of thumbs on all the four limbs, as the monkeys.

Quad'ruped (Lat. *quat'uor*, four; *pes*, a foot). Having four legs and feet.

Quadru'plicate (Lat. *quat'uor*, four; *plic'o*, I fold). Fourfold; four times repeated.

Quaquaver'sal (Lat. *quaqua*, on every side; *versus*, turned). Dipping on all sides; applied in *geology* to strata that dip on all sides from a common centre.

Quar'antine (Italian *quaranti'na*, forty). Properly, a space of forty days; but now applied to any term, during which a ship on arriving at port, if suspected of being infected with contagious disease, is obliged to forbear all intercourse with the place.

Quar'tan (Lat. *quar'tus*, fourth). Occurring every fourth day; applied especially to a form of ague.

Quarta'tion (Lat. *quar'tus*, fourth). A process in chemistry by which the quantity of one thing is made equal to the fourth part of another.

Quar'tite (Lat. *quar'tus*, fourth). In *astronomy*, an aspect of the planets when they are distant from each other a quarter of a circle.

Quar'tine (Lat. *quar'tus*, fourth). In *botany*, the fourth coat of the ovule.

Quartz. Crystallised silica; silica in its purest rock-form.

Quasi (Lat. as if). A word used to express resemblance.

Quater'nary (Lat. *quat'uor*, four). Consisting of fours; in *geology*, applied to the accumulations above the true tertiary strata.

Queen-post. In *architecture*, the suspending posts in the framed principal of a roof, where there are two such posts.

Quies'cent (Lat. *qui'es*, rest). Being at rest; having no sound.

Qui'nary (Lat. *qui'ni*, five by five). Composed of five parts; arranged in fives.

Quin'cunx (Lat. *quin'que*, five). An arrangement of five objects in a square, one at each corner, and one in the middle.

Quindec'agon(Lat.*quin'decim*, fifteen; Gr. γωνια, *gōnia*, an angle). A plane figure with fifteen sides and fifteen angles.

Quinquan'gular (Lat. *quinq'ue*, five; *an'gulus*, an angle). Having five angles.

Quin'que (Lat. five). A prefix in compound words, signifying five.

Quin'quefid (Lat. *quin'que*, five; *fin'do*, I cleave). Five-cleft.

Quinquelo'bate (Lat. *quin'que*, five; *lo'bus*, a lobe). Having five lobes.

Quinqueloc'ular (Lat. *quin'que*, five; *loc'ulus*, a little space). Having five cells or chambers.

Quinquepar'tite (Lat. *quin'que*, five; *par'tio*, I divide). Divided deeply into five parts.

Quin'sy (Corrupted from *Cynanche*; Gr. κυων, *kuōn*, a dog; ἀγχω, *anchō*, I strangle). Acute inflammation of the tonsils; inflammatory sore throat.

Quin'tile (Lat. *quin'tus*, fifth). The position of the planets when they are distant 72 degrees, or the fifth part of a circle from each other.

Quin'tine (Lat *quin'tus*, fifth). In *botany*, the fifth coat of the ovule.

Quin'tuple (Lat. *quin'tus*, the fifth; *pli'co*, I fold). Five-fold.

Quotid'ian (Lat. *quo'tus*, how many; *dies*, a day). Occurring every day; applied especially to a form of ague.

Quo'tient (Lat. *quo'ties*, how often). The number showing how often one number is contained in another.

R.

Rab'ies (Lat. fury). The disease known as hydrophobia.

Rac'eme (Lat. *race'mus*, a cluster of grapes). In *botany*, a form of inflorescence, consisting of a common peduncle or stem, with short equal lateral pedicels, as in the hyacinth.

Race'mose (*Race'me*). Bearing flowers in racemes.

Rachis (Gr. ῥαχις, *rha'chis*, the spine). In *botany*, a term applied to the stems of ferns, and the axis or stem of an inflorescence.

Rachit'ic Gr. ῥαχις, *rha'chis*, the spine). Pertaining to the back; rickety.

Rachi'tis (Gr. ῥαχις, *rha'chis*, the spine; *itis*, denoting inflammation). Literally, inflammation of the spine; but applied to the diseased state of the bones, called rickets.

Ra'dial (Lat. *ra'dius*, a ray; or one of the bones of the arm). Having the quality or appearance of a ray; in *anatomy*, belonging or attached to the radius, or outer bone of the forearm; in *astronomy*, applied, in the theory of variable orbits, to that component part of the disturbing force which acts in the direction of the radius vector.

Ra'diant (Lat. *ra'dius*, a ray). Sending out rays, as from a centre.

Radia'ta (Lat. *ra'dius*, a ray). A subdivision of invertebrate animals, characterised by having the parts of the body regularly disposed round a common centre; as the star-fish.

Ra'diated (Lat. *ra'dius*, a ray). Having rays or lines proceeding from a centre.

Ra'diation (Lat. *ra'dius*, a ray). The shooting of anything, as light, from a centre; the emission of light and heat, or sound, in all directions, like rays, from a body.

Rad'ical (Lat. *ra'dix*, a root). Belonging to or arising from the root; in *philology*, a primitive or original word; in *chemistry*, a compound body which enters into combination after the manner of a simple body;

in *botany*, applied to hair-like projections on young roots, and to leaves arising from the root ; radical sign in *algebra*, the sign $\sqrt{}$ with a number prefixed thus, $\sqrt[3]{}$, placed before any quantity to show what root is to be extracted.

Rad'icle (Lat. *radic'ula*, a little root). The part of the embryo in plants which becomes the root ; the end of roots, absorbing nutriment.

Ra'diolites (Lat. *ra'dius*, a ray ; Gr. λιθος, *lith'os*, a stone). In *geology*, a genus of bivalves in the chalk-formation, having a radiated structure of the outer layer of the upper valve.

Radiom'eter (Lat. *ra'dius*, a ray : Gr. μετρον, *met'ron*, a measure). An instrument formerly used for taking the altitude of celestial bodies.

Ra'dius (Lat. a ray). In *geometry*, a straight line drawn from the centre to the circumference of a circle ; in *anatomy*, the outer bone of the forearm, reaching from the elbow to the wrist above the thumb.

Ra'dius Vector (Lat. a carrying radius). A straight line drawn to any body moving in a curvilinear path, from a fixed point considered as the centre of the motion.

Radix (Lat. a root). In *etymology*, a primitive word from which other words spring ; in *arithmetic*, a number which is arbitrarily made the base of any system of computation.

Rain-gauge. An instrument for measuring the quantity of rain which falls at any place.

Ra'mal (Lat. *ra'mus*, a branch). Belonging to branches.

Ramen'ta (Lat. *ramen'tum*, a little scraping). Scrapings ; in *botany*, applied to thin brown leafy scales found on young shoots and other parts.

Ram'ification (Lat. *ra'mus*, a branch ; *fac'io*, I make). A branching : the manner in which a tree produces its branches.

Ram'ify (Lat. *ra'mus*, a branch ; *fac'io*, I make). To make branches, or shoot into branches.

Ramollis'sement (French, from the Latin *mollis*, soft). Softening ; a diseased condition occurring in various parts of the body, in which they become softer than is natural.

Ra'mous (Lat. *ra'mus*, a branch). Having or belonging to branches.

Ra'mus (Lat. a branch). In *anatomy*, applied to branches of arteries or other organs.

Rani'dæ (Lat. *ra'na*, a frog). The family of batrachian reptiles, having as its type the frog.

Ra'nine (Lat. *ra'na*, a frog, or a swelling of the tongue). Belonging to a frog ; in *anatomy*, applied to an artery of the tongue.

Ra'nula (Lat. a little frog). A kind of swelling under the tongue.

Ra'phe' (Gr. ραφη, *rhaphē*, a seam). A term applied to parts which look as if they had been sewn together.

Raph'ides (Gr. ραφις, *rhaph'is*, a needle). Minute crystals, like needles, lying in the tissues of plants.

Rapto'res (Lat. *rap'io*, I snatch). An order of birds characterised by the strength of their claws and bill, and the general strength of their bodies : the birds of prey ; as the eagle, vulture, hawk, &c.

Rarefac'tion (Lat. *rarus*, rare or thin ; *fac'io*, I make). A making thin ; an increase of the intervals between the particles of matter, so that the same amount is made to occupy a larger space ; applied especially to airs and gases ; also the state of the lessened density.

Ra'refy (Lat. *rarus*, thin ; *fac'io*, I make). To make or become thin.

Raso'res (Lat. *rado*, I scratch). The order of birds, including pigeons and gallinaceous birds, which seek their food by scratching the ground.

Ratchet. A piece of mechanism, one end of which abuts against a tooth of a wheel called a ratchet-wheel.

Ratchet-wheel. A wheel with pointed teeth, on which a ratchet abuts.

Ra'tio (Lat. *reor*, I think or suppose). The relation of two quantities of the same kind to one another ; the rate in which one quantity exceeds or is less than another.

Rat'ional(Lat. *ra'tio*, reason). Having the faculty of reason; in *algebra* and *arithmetic*, applied to definite quantities, or to those of which an exact root can be found; in *chemistry*, applied to formulæ which aim at describing the exact composition of one equivalent or combining portion of a substance, by stating the absolute number of equivalents of each of its elements necessary to its formation.

Re- or **Red-**. (Lat. back). A preposition used in compound words, signifying return or repetition.

React (Lat. *re; ag'o*, I act). To return an impulse or impression.

Reac'tion (Lat. *re; ag'o*, I act). The resistance made by a body to the action or impulse of another body.

Rea'gent (Lat. *re; ag'o*, I act). In *chemistry*, a substance used to detect the presence of other bodies.

Recep'tacle (Lat. *recip'io*, I receive). That which receives or contains; in *botany*, the shortened axis of a flower-stem, bearing numerous flowers.

Recip'ient (Lat. *recip'io*, I receive). That which receives or takes.

Recip'rocal (Lat. *recip'rocus*, moving backwards and forwards). Acting alternately; interchangeable; in *arithmetic*, applied to the quotient of one or unity divided by any quantity, thus the reciprocal of 4 is $\frac{1}{4}$; and to quantities which when multiplied together produce unity; applied also to a form of proportion in which the first term has to the second the same ratio as the fourth to the third, or as the reciprocal of the third has to the reciprocal of the fourth.

Recip'rocally (Lat. *recip'rocus*, moving backwards and forwards). Interchangeably; applied to quantities which are so related, that when one increases the other diminishes.

Recip'rocating Motion. A form of action illustrated in the suspension of a rigid bar on an axis, so that the parts on each side of the axis take alternately the position of those on the other.

Rec'linate (Lat. *re*, back; *clino*, I lean). In *botany*, applied to leaves which are folded longitudinally from apex to base in the bud.

Reclina'tion (Lat. *re : clino*, I lean). A leaning; in *surgery*, an operation for the cure of cataract, in which the crystalline lens is moved downwards from its place, and laid horizontally.

Rec'ondite (Lat. *recon'do*, I hide). Hidden.

Rec'tangle (Lat. *rectus*, right; *an'gulus*, an angle). A four-sided figure, having all its angles right angles.

Rectan'gular (Lat. *rectus*, right; *an'gulus*, an angle). Having right angles.

Rectifica'tion (Lat. *rectus*, right; *fac'io*, I make). A correcting or making right; in *chemistry*, the purification of any substance by repeated distillation; in *geometry*, the determination of a straight line, the length of which is equal to a portion of a curve.

Rec'tify (Lat. *rectus*, right; *fac'io*, I make). To make right; in *chemistry*, to purify a substance by repeated distillation; in *astronomy*, to rectify the globe is to bring the sun's place in the ecliptic to the brass meridian, or to adjust it for the solution of a problem.

Rectilin'ear (Lat. *rectus*, straight; *lin'ea*, a line). Contained in or consisting of straight lines.

Rectiros'tral (Lat. *rectus*, straight; *rostrum*, a beak). Having a straight beak.

Rectise'rial (Lat. *rectus*, straight, *se'ries*, a row). Disposed in a rectilinear or straight series.

Rectum (Lat. straight). The last part of the large intestines.

Rectus (Lat. straight). A name given to several muscles of the body, on account of their direction.

Recum'bent (Lat. *re*, back; *cumbo*, I lie down). Leaning or lying on anything.

Recur'rent (Lat. *re*, back ; *curro*, I run). Returning ; in *anatomy*, applied to a branch of the pueumogastric nerve, which is given off in the upper part of the chest and runs up along the trachea and larynx.

Recur'ring (Lat. *re*, back ; *curro*, I run). Returning ; in *arithmetic*, applied to decimals in which the figures are continually repeated in the same order.

Recur'vate (Lat. *re*, back ; *curvus*, crooked). Bent backwards.

Reduc'tion (Lat. *re*, back ; *duco*, I bring). In *chemistry*, the bringing back a metal to its simple state from a compound ; in *surgery*, the restoration to its place of a dislocated bone or other part.

Redu'plicate (Lat. *re*, back ; *duplex*, double). In *botany*, applied to a form of æstivation in which the edges of the sepals or petals are turned downwards.

Reflecting Goniom'eter. An instrument for measuring the angles of crystals by means of rays of light reflected from their surface.

Reflec'tion (Lat. *re*, back ; *flecto*, I bend). The act of throwing back; in *natural philosophy*, applied to the motion of light, heat, or sound, by which either of them rebounds from a body against which it has struck, making an equal angle with that at which it has fallen on the body.

Reflec'tor (Lat. *re*, back ; *flecto*, I bend). That which reflects or bends back ; a surface of polished metal or other suitable material for the purpose of throwing back rays of light, heat, or sound, in any required direction.

Re'flex (Lat. *re*, back; *flecto*, I bend). Bent back ; in *physiology*, applied to a class of actions in which an impression is carried by a nerve to the nervous centre, whence a nerve of motion conveys the impulse of motion to certain muscles, which thus act without the will of the individual.

Reflex' (Lat. *re*, back; *flecto*, I bend).

In *painting*, the illumination of one body by light reflected from another body in the same piece.

Re'flux (Lat. *re*, back ; *flu'o*, I flow). A flowing back.

Refrac'tion (Lat. *re*, back ; *fran'go*, I break). The change in direction which a moving body, especially light, undergoes in passing from any medium into one of different density.

Refrac'tive (Lat. *re*, back ; *fran'go*, I break). Allowing or favouring refraction.

Refrac'tory (Lat. *re*, against ; *fran'go*, I break). In *chemistry*, applied to substances which resist the action of heat or other agencies.

Refrangibil'ity (Lat. *re*, back; *fran'go*, I break). The disposition of rays of light to be turned from their direct course in passing from one medium to another ; especially the degree of that disposition possessed by the coloured rays.

Refrig'erant (Lat. *re*, back ; *fri'gus*, cold). Abating heat ; cooling.

Refrigera'tion (Lat. *re*, back ; *fri'gus*, cold). Cooling ; the removal of heat.

Regenera'tion (Lat. *re*, again ; *gen'ero*, I produce). In *physiology*, the renewal of a portion of lost or removed tissue by the formation of a new portion of tissue of the same kind.

Reg'imen (Lat. *reg'o*, I rule or govern). In *medicine*, regulation of diet and habit; in *grammar*, the regulation of the dependence of words on each other.

Reg'ister Pyrom'eter. An instrument for measuring high temperatures by the linear expansion of bars of metal.

Reg'ister Thermom'eter. A thermometer which records its own indications.

Regres'sion (Lat. *re*, back ; *grad'ior*, I step). A moving backwards.

Reg'ular (Lat. *reg'ula*, a rule). According to rule ; in *geometry*, applied to bodies the sides and angles of which are equal.

Relaxa'tion (Lat. *re*, back ; *lax'o*, I loosen). A loosening, or letting loose.

Relief Valve. A valve in an air-pump, to prevent the momentary condensation of air in the receiver when the piston descends.

Re'miges (Lat. *re'mex*, a rower). The large quills of the wings of birds.

Remit'tent (Lat. *re ; mitto*, I send). Ceasing for a time ; applied to diseases of which the symptoms alternately diminish and return, but without ever leaving the patient quite free.

Renais'sance (French, from *renaître*, to be born again). The revival of anything which has long been in decay, or obsolete.

Re'niform (Lat. *ren*, a kidney ; *form'a*, shape). Resembling a kidney.

Reo-. For words with this beginning, *see* Rhe'o-.

Repeat'er (Lat. *rep'eto*, I seek again, or repeat). That which repeats ; in *arithmetic*, a decimal in which the same figure continually recurs.

Re'pent (Lat. *re'po*, I creep). In *natural history*, creeping.

Rep'etend (Lat. *rep'eto*, I repeat). That part of a repeating decimal which recurs continually.

Reproduc'tion (Lat. *re ; produ'co*, I produce). The art or process of producing again.

Rep'tiles or Reptilia (Lat. *re'po*, I creep). Cold-blooded vertebrate animals, breathing air incompletely from birth, and having the circulation so arranged that a portion of the venous blood mixes unchanged with the arterial ; as the serpent, crocodile, and tortoise.

Repul'sion (Lat. *re ; pel'lo*, I drive). A driving back ; the power or principle by which bodies, or the particles of bodies, under certain circumstances recede from each other.

Resid'ual (Lat. *resid'uus*, that which is left). Remaining after a part is taken.

Resid'uum (Lat.). A remainder.

Res'inous Electricity. A name given to negative electricity, from its being developed by the friction of resinous substances.

Resolu'tion (Lat. *re ; solvo*, I loosen). The process of separating the parts which form a complex substance or idea ; in *mathematics*, the enumeration of things to be done in order to obtain what is required in a problem ; in *dynamics*, the revolution of forces is the dividing of any single force or motion into two or more others which, acting in different directions, shall produce the same effect as the given motion or force.

Respira'tion (Lat. *re ; spiro*, I breathe). The act of breathing, or the process by which the blood is brought under the action of air for the purpose of purification.

Res'tiform (Lat. *restis*, a cord ; *forma*, shape). Like a cord.

Resul'tant (Lat. *resul'to*, I leap back). In *dynamics*, the force which results, or arises from, the composition or putting together of two or more forces acting from different directions on the same point.

Resuscita'tion (Lat. *re ; sus'cito*, I raise). The act of raising from apparent death.

Retarda'tion (Lat. *re ; tardus*, slow). A making slow.

Rete Mirab'ile (Lat. a wonderful net). An arrangement of blood-vessels, in which an artery suddenly divides into small anastomosing branches which, in many cases, unite again to form a trunk.

Re'te Muco'sum (Lat. *rete*, a net; *muco'sus*, mucous). The mucous network : a name sometimes given to the soft under layer of the epidermis or scarf-skin.

Retic'ular (Lat. *retic'ulum*, a small net). Having the form of a network.

Reticula'ted (Lat. *retic'ulum*, a small net). Arranged like a network.

Retic'ulum (Lat. a little net). The second, or honeycombed cavity in the compound stomach of ruminant animals.

Re'tiform (Lat. *re'te*, a net ; *forma*, shape). Having the form of a net.

Ret'ina (Lat. *re'te*, a net). One of the coats of the eye, consisting of the expansion of the optic nerve in the form of a fine network; it is the part of the nervous system which receives the first perception of the rays of light.

Retinac'ulum (Lat. a band). In *botany*, the viscid matter by which the pollen-masses in orchids adhere to a prolongation of the anther.

Retini'tis (Lat. *ret'ina; i'tis*, denoting inflammation). Inflammation of the retina.

Retor't (Lat. *re; tor'queo*, I twist or bend). In *chemistry*, a globular vessel with a long neck employed in distillations.

Re'tro- (Lat. backwards). A preposition used in compound words, signifying backward or back.

Retroce'dent (Lat. *re'tro*, backwards; *ce'do*, I go). In *medicine*, applied to diseases which move from one part of the body to another, as gout.

Retroces'sion (Lat. *re'tro; ce'do*, I go). A moving backwards.

Re'troflex (Lat. *re'tro*, backwards; *flecto*, I bend). Bent backwards; in *botany*, bent this way and that.

Re'trofract (Lat. *retro*, backwards; *fran'go*, I break). Bent backwards as if broken.

Re'trograde (Lat. *re'tro*, backwards; *grad'ior*, I step). Moving backwards; in *astronomy*, apparently moving in the contrary direction to the order of the signs of the zodiac, in which the sun appears to move.

Retrogres'sion (Lat. *re'tro*, backwards; *grad'ior*, I step). A moving backwards; in *astronomy*, the change of position undergone by the moon's nodes, in a direction contrary to the motion of the sun.

Retropul'sive (Lat. *re'tro*, backwards; *pel'lo*, I drive). Driving back.

Re'trorse (Lat. *re'tro*, backwards; *versus*, turned). Turned backwards.

Retrover'sion (Lat. *re'tro*, backwards; *ver'to*, I turn). A turning backwards.

Re'trovert (Lat. *re'tro*, backwards; *ver'to*, I turn). To turn back.

Re'tuse (Lat. *re; tundo*, I bruise).

Having a broad, blunt, and slightly depressed apex.

Rever'berate (Lat. *re; ver'bero*, I beat). To beat back or return.

Reverbera'tion (Lat. *re; ver'bero*, I beat). A beating back.

Rever'beratory (Lat. *re; ver'bero*, I beat). Applied to a furnace or oven, in which a crucible or other object is heated by flame or hot air reverberated or beaten back from the roof.

Reviv'ification (Lat. *re; vi'vus*, alive; *fac'io*, I make). Restoration of life.

Re'volute Lat. *re; vol'vo*, I roll). Rolled backwards.

Revolu'tion (Lat. *re; vol'vo*, I roll). Rotation; the circular movement of a body round a centre.

Rhachi'tis (Gr. ραχις, *rhach'is*, the spine). *See* Rachi'tis.

Rheom'eter (Gr. ρεος, *rhe'os*, a current; μετρον, *met'ron*, a measure). An apparatus for measuring the intensity of a galvanic current.

Rheom'etry (Gr. ρεος, *rhe'os*, a current; μετρον, *met'ron*, a measure). The differential and integral calculus; the method of determining the force of galvanic currents.

Rheomo'tor (Gr. ρεος, *rhe'os*, a current; Lat. *mov'eo*, I move). Any apparatus by which an electrical or galvanic current is originated.

Rhe'oscope (Gr. ρεος, *rhe'os*, a current; σκοπεω, *skop'eō*, I view). An apparatus for ascertaining the pressure of a galvanic current.

Rhe'ostat (Gr. ρεος, *rhe'os*, a current; ιστημι, *his'tēmi*, I make to stand). An apparatus for enabling a galvanic needle to be kept at the same point during an experiment.

Rhe'otome (Gr. ρεος, *rhe'os*, a current; τεμνω, *tem'nō*, I cut). An instrument for periodically interrupting an electric current.

Rhe'otrope (Gr. ρεος, *rhe'os*, a current; τρεπω, *trep'ō*, I turn). An instrument for reversing the direction of a voltaic current.

Rhet'oric (Gr. ρεω, *rhe'ō*, I flow). The art of speaking with propriety, elegance, and force.

Rheumat'ic (Gr. ρευμα, *rheuma*, watery fluid). Belonging to or having rheumatism.

Rheu'matism (Gr. ρευμα, *rheu'ma*, watery fluid). A painful disease affecting the muscles and joints.

Rhipip'tera (Gr. ριψ, *rhips*, a matwork or fan ; πτερον, *pter'on*, a wing). An order of insects having only two wings, folded longitudinally like a fan.

Rhinenceph'alic (Gr. ριν, *rhin*, the nose ; εγκεφαλον, *enkeph'alon*, the brain). Belonging to the nose and brain : applied to the prolongation of brain-substance which forms the so-called olfactory nerves.

Rhi'zanths (Gr. ριζα, *rhi'za*, a root ; ανθος, *anthos*, a flower). A class of plants occupying a position between the flowering and the non-flowering species.

Rhi'zogen (Gr. ριζα, *rhi'za*, a root ; γενναω, *genna'ō*, I produce). Producing roots.

Rhizocar'pous (Gr. ριζα, *rhi'za*, a root ; καρπος, *kar'pos*, fruit). In *botany*, applied to plants whose root lasts many years, but whose stem perishes annually.

Rhi'zome (Gr. ριζωμα, *rhi'zōma*, a root). In *botany*, a thick stem running along and partly under the ground, sending forth shoots above and roots below.

Rhi'zopods (Gr. ριζα, *rhi'za*, a root ; πους, *pous*, a foot). A class of simple organic beings, consisting of minute gelatinous masses, generally covered by a shell, and often provided with long, slender, contractile filaments.

Rhizotax'is (Gr. ριζα, *rhi'za*, a root ; τασσω, *tassō*, I arrange). The arrangement of roots.

Rhomb (Gr. ρομβω, *rhom'bō*, I whirl round). A four-sided figure, with the sides equal, and the opposite sides parallel, but with unequal angles.

Rhombigan'oid (Gr. ρομβος, *rhom'bos*, a rhomb ; γανος, *gan'os*, splendour ; ειδος, *ei'dos*, shape). Having ganoid or shining scales of a lozenge shape.

Rhombohcd'ral (Gr. ρομβος, *rhom'bos*, a rhomb ; εδρα, *hed'ra*, a base). Of the nature of a rhombohedron.

Rhombohed'ron (Gr. ρομβος, *rhom'bos*, a rhomb ; εδρα, *hed'ra*, a base). A solid figure, bounded by six planes in the form of rhombs.

Rhom'boid (Gr. ρομβος, *rhom'bos*, a rhomb ; ειδος, *ei'dos*, form). A four-sided figure, having neither equal sides nor equal angles.

Rhon'chus (Gr. ρογχος, *rhon'chos*). A rattling or wheezing sound ; in *medicine*, applied to any unnatural sound produced in the air-passages, by obstructions to the passage of the breath.

Rhyn'cholites (Gr. ρυγχος, *rhun'chos*, a beak ; λιθος, *lith'os*, a stone). Fossil remains of the beaks of certain cephalopods.

Rhythm (Gr. ρυθμος, *rhuth'mos*, measured motion, proportion). The agreement of measure and time in poetry, prose, music, and motion.

Rhyth'mical (Gr. ρυθμος, *rhuth'mos*, measured motion, proportion). Having one sound proportioned to another ; regulated by cadences, accents, and quantities.

Rhythmom'eter (Gr. ρυθμος, *rhuth'mos*, measured motion ; μετρον, *met'ron*, a measure). An instrument for marking time to movements in music.

Rickets (Gr. ραχις, *rhach'is*, the spine). A diseased state of the bones in infancy and childhood, consisting in a deficiency of earthy and other essential matters, and leading to distortion.

Rig'id (Lat. *rig'idus*, stiff). Stiff ; applied to bodies which have become so from a naturally flexible state.

Rigid'ity (Lat. *rigidus*, stiff). Stiffness arising in bodies that are naturally flexible.

Rin'gent (Lat. *rin'go*, I grin). In *botany*, applied to forms of labiate corolla, where the upper lip is much arched, and the lips are separated by a distinct gap.

Ring-Mountains. In *astronomy*, circular formations on the surface of the moon, of the same nature as bulwark plains, but smaller and more regular in outline.

Ri'sus Sardon'icus (Lat. Sardon'ic laugh). A kind of convulsive grin observed in some diseases: so called because supposed to be produced by a species of ranunculus growing in Sardinia.

Ro'dent (Lat. ro'do, I gnaw). Gnawing; applied to an order of mammals which nibble and gnaw their food, as the squirrel, rat, hare, &c.

Root. In *arithmetic*, the root of any quantity is that which, if multiplied into itself a certain given number of times, will exactly produce the quantity.

Rosa'ceous (Lat. *rosa*, a rose). Belonging to the rose tribe of plants; like a rose.

Rostel'lum (Lat. a little beak, from *ros'trum*, a beak). A beak-shaped process.

Ros'tral (Lat. *ros'trum*, a beak). Belonging to a beak.

Ros'trate (Lat. *ros'trum*, a beak). Having a beak, or process resembling a beak.

Ros'trum (Lat. a beak). A beak; anything projecting or shaped like a beak.

Rota'tion (Lat. *ro'ta*, a wheel). The movement of a body on its axis; in *agriculture*, the mode in which different kinds of crops are made to succeed each other in the same ground.

Rota'tor (Lat. *ro'ta*, a wheel). That which gives a circular or rolling motion; applied to certain muscles of the body.

Ro'tatory (Lat. *ro'ta*, a wheel). Turning on an axis; moving in succession.

Rötheln (Germ.). A form of eruptive febrile disease, partaking of the characters of both measles and scarlet fever.

Rotif'era (Lat. *ro'ta*, a wheel; *fer'o*, I bear). Wheel-bearers; a class of animalcules, which have circles of cilia, appearing under the microscope like wheels in motion.

Rotund (Lat. *rotun'dus*, round).

Round; bounded by a curve without angles.

Rouleaux (Fr.). Rolls.

Rubefa'cient (Lat. *ruber*, red; *fac'io*, I make). Making red; an application which produces redness of the skin, not followed by a blister.

Rube'ola (Lat. *ruber*, red). A term often used for measles, but now applied to the eruptive disease called rötheln, which presents the characters of both measles and scarlet fever.

Rubes'cent (Lat. *rubes'co*, I become red). Becoming red; tending to a red colour.

Ru'diment (Lat. *rudimen'tum*). A first principle or element; the original of anything in its first or most simple form.

Rudimen'tary (Lat. *rudimen'tum*, a first principle). Belonging to or consisting in first principles; in an original or simple state; arrested in development.

Rugæ (Lat. plaits or folds). The folds into which the mucous membrane of some organs is thrown, when they are not distended, by contraction of the external coats.

Ru'gate (Lat. *ruga*, a wrinkle). Wrinkled.

Ru'gose (Lat. *ruga*, a wrinkle). Full of wrinkles.

Ru'minant (Lat. *rumen*, the cud). Chewing the cud; applied to an order of herb-eating animals, of which the camel, cow, and sheep, are examples.

Ru'minate. In *botany*, applied to the albumen of the seed when it presents a mottled appearance, as in the nutmeg.

Run'cinate (Lat. *runci'na*, a large saw). In *botany*, applied to pinnatifid leaves with more or less triangular divisions, pointed downwards towards the base, as the dandelion.

Ru'nic (Icelandic *runa*, a furrow or line). A term applied to the alphabet of the ancient Scandinavians, consisting of letters of peculiar shape, principally formed of straight lines cut on wood or stone.

S.

Sab'ulous (Lat. *sab'ulum*, sand). Sandy.

Sac (Lat. *saccus*, a bag). A bag.

Sac'cate (Lat. *saccus*, a bag). Having a bag, or formed into a bag.

Sac'charic (Lat. *sac'charum*, sugar). Belonging to sugar; applied to an acid formed from sugar.

Sacchariferous (Lat. *sac'charum*, sugar; *fer'o*, I bear). Producing sugar.

Sac'charine (Lat. *sac'charum*, sugar). Belonging to, or having the properties of sugar.

Sac'charoid (Lat. *sac'charum*, sugar; Gr. ειδος, *eidos*, shape). Resembling loaf-sugar in texture.

Saccharom'eter (Lat. *sac'charum*, sugar; Gr. μετρον, *met'ron*, a measure). An instrument for measuring the specific gravity of brewers' and distillers' worts, and thus determining the amount of sugar contained in them.

Saccholac'tic (Lat. *sac'charum*, sugar; *lac*, milk). A term applied to an acid obtained from the sugar of milk.

Sac'ciform (Lat. *saccus*, a bag; *forma*, shape). Resembling a sac or bag.

Sac'cular (Lat. *sac'culus*, a little bag). Belonging to, or formed of little sacs or bags.

Sa'cral (*Sa'crum*). Belonging to the os sacrum.

Sa'crum (Lat. *sacer*, sacred; because originally offered in sacrifices). The largest piece of the vertebral column, placed at the upper and back part of the pelvis.

Safety Lamp. A lamp surrounded by fine wire-gauze, invented by Sir H. Davy, to indicate danger in mines from explosion of firedamp.

Safety Valve. A contrivance for preventing or diminishing the risk of explosion in steam-boilers, formed on the principle of applying such a force as will yield to the pressure from within before the latter reaches the point of danger.

Saga. An heroic tale, among the northern nations.

Sagit'tal (Lat. *sagit'ta*, an arrow). Like an arrow; in *anatomy*, applied to the suture which unites the parietal bones of the head, its direction being on the centre of the skull from before backwards.

Sagit'tate (Lat. *sagit'ta*, an arrow). Shaped like the head of an arrow; in *botany*, applied to leaves having two long sharp lobes projecting backwards from the insertion of the petiole into the leaf.

Sa'lient (Lat. *sal'io*, I leap). Leaping; beating; springing up or out; in *geometry*, applied to projecting angles.

Saliferous (Lat. *sal*, salt; *fer'o*, I bear). Producing salt.

Salifi'able (Lat. *sal*, salt; *fi'o*, I become). Capable of forming a salt by combining with an acid.

Saline (Lat. *sal*, salt). Containing or having the properties of salt.

Salinom'eter (Lat. *sali'nus*, saline; Gr. μετρον, *met'ron*, a measure). An apparatus for indicating the density of brine in the boilers of marine steam-engines, so as to show when they should be cleaned.

Sal'ivary (Lat. *sali'va*). Belonging to or conveying saliva.

Sal'ivary Glands. The glands which secrete the saliva; being the parotid, sublingual, and submaxillary.

Sal'ivate (Lat. *sali'va*). To produce an excessive flow of saliva.

Saliva'tion (Lat. *sali'va*). The process of producing an excessive flow of saliva.

Salpingo- (Gr. σαλπιγξ, *salpinx*, a tube). In *anatomy*, a prefix in some compound words, denoting connection with a tube, generally the Eustachian tube.

Salt (Lat. *sal*, common salt). In popular language, chloride of sodium; in *chemistry*, any substance resulting from the combination of two oxides or analogous bodies, of which one is highly basic and the other highly acid.

Salt-rad'ical. In *chemistry*, an element, such as chlorine or iodine, which forms a salt by combination with a metal.

Sal'tant (Lat. *salto*, I leap). Leaping.

Salta'tion (Lat. *salto*, I leap). The act of leaping or jumping.

Saltato'rious (Lat. *salto*, I leap). Having the power of, or formed for, leaping.

Sal'tigrade (Lat. *salto*, I leap; *grad'-us*, a step). Formed for leaping; advancing by leaping.

Sal Volat'ile (Lat. volatile salt). The popular name for carbonate of ammonia.

San'atory (Lat. *sano*, I heal). Healing.

Sand. In *geology*, an aggregation of water-worn particles derived from pre-existing rocks and other mineral substances.

Sandstone. In *geology*, sand of which the particles have been consolidated together by pressure.

Sanguif'erous (Lat. *san'guis*, blood ; *fer'o*, I carry). Conveying blood.

San'guification (Lat. *san'guis*, blood ; *fac'io*, I make). The making of blood ; the process by which blood is formed from chyle.

Sanguig'enous (Lat. *san'guis*, blood ; *gig'no*, I produce). Forming blood.

Sanguin'eous (Lat. *san'guis*, blood). Belonging to, or abounding in, blood ; constituting blood.

Sanguiniv'orous (Lat. *san'guis*, blood ; *voro*, I devour). Eating blood.

Sanguin'olent (Lat. *san'guis*, blood). Bloody.

Sa'nies (Lat.). A thin reddish discharge from wounds or sores.

Sa'nious (*Sa'nies*). Having the properties of, or pouring out, sanies.

San'itary (Lat. *san'itas*, health). Relating or conducing to the preservation of health.

Saphe'nous (Gr *σαφηνης*, *saphēnēs*, open, manifest). A name given to the superficial vessels and nerves of the thigh and leg.

Sap'id (Lat. *sap'io*, I taste). Capable of exciting the sense of taste.

Sapona'ceous (Lat. *sa'po*, soap). Soapy ; resembling soap.

Saponifi'able (Lat. *sa'po*, soap ; *fi'o*, I become). Capable of being converted into soap.

Sapon'ification (Lat. *sa po*, soap ; *fac'io*, I make). The change which fats undergo in contact with alkaline solutions at high temperatures ; the formation of soap.

Sapon'ify (Lat. *sa'po*, soap ; *fac'io*, I make). To convert into soap.

Saporif'ic (Lat. *sap'or*, taste ; *fac'io*, I make). Producing taste.

Sarco- (Gr. *σαρξ*, *sarx*, flesh). A prefix in compound words, denoting relation or similarity to flesh.

Sar'cocarp (Gr. *σαρξ*, *sarx*, flesh ; *καρπος*, *kar'pos*, fruit). The fleshy part of fruits, lying between the epicarp and the endocarp ; a fleshy succulent mesocarp.

Sar'code (Gr. *σαρξ*, *sarx*, flesh). The simple gelatinous structure of which some of the lowest organic beings are formed.

Sar'coderm (Gr. *σαρξ*, *sarx*, flesh ; *δερμα*, *der'ma*, skin). The middle covering of a seed when it becomes succulent or juicy.

Sarcolem'ma (Gr. *σαρξ*, *sarx*, flesh ; *λεμμα*, *lem'ma*, a husk or peel). The proper tubular sheath of muscular fibre.

Sarcol'ogy (Gr. *σαρξ*, *sarx*, flesh ; *λογος*, *log'os*, a discourse). The part of anatomy which describes the soft parts of the body.

Sarco'ma (Gr. *σαρξ*, *sarx*, flesh). A fleshy tumour.

Sarcoph'agous (Gr. *σαρξ*, *sarx*, flesh ; *φαγω*, *phag'ō*, I eat.) Eating flesh.

Sarco'sis (Gr. *σαρξ*, *sarx*, flesh). The production of flesh.

Sar'cosperm (Gr. *σαρξ*, *sarx*, flesh : *σπερμα*, *sper'ma*, a seed). The mesosperm or middle covering of a seed, when it becomes fleshy.

Sarcot'ic (Gr. *σαρξ*, *sarx*, flesh). Inducing the growth of flesh.

Sarmen'tous (Lat. *sarmen'tum*, a twig). In *botany*, applied to a stem which is long and almost destitute of leaves and buds.

Sarmen'tum (Lat. a twig). A running stem giving off leaves or roots at intervals, as the strawberry; sometimes also a twining stem supporting itself by means of others.

Sarto'rius (Lat. *sar'tor*, a tailor). In *anatomy*, a name applied to a muscle of the thigh, which turns the leg obliquely inwards and over the other.

Sat'ellite (Lat. *satel'les*, an attendant). A secondary planet or moon revolving round a primary planet: in *anatomy*, applied to the veins which accompany the arteries in the limbs.

Sat'urate (Lat. *sa'tur*, full). To supply until no more can be received: to neutralise; thus an acid is saturated by an alkali, or *vice versâ*, when no portion of either is left uncombined.

Satura'tion (Lat. *sa'tur*, full). A supplying to fulness; in *chemistry*, the solution of one body in another until no more can be contained in union by the receiving body.

Satur'nian System. In *astronomy*, the system composed of the planet Saturn, together with its rings and satellites.

Sau'rian (Gr. σαυρος, *sau'ros*, a lizard). The term designating the family of lizards.

Sau'roid (Gr. σαυρος, *sau'ros*, a lizard; ειδος, *ei'dos*, form). Like a lizard: applied to fishes which approach in structure to lizards, as the sturgeon.

Sauroidich'nites (Gr. σαυρος, *sau'ros*, a lizard; ειδος, *ei'dos*, form; ιχνος, *ich'nos*, a footstep). Fossil footprints of reptiles.

Sca'brous (Lat. *sca'ber*, rough). Rough; having small elevations.

Scagl'iola (Italian *scagl'ia*, a scale or chip). In *architecture*, a composition in imitation of marble, laid on bricks in the manner of stucco.

Scala'riform (Lat. *sca'la*, a ladder; *form'a*, shape). Having bars like a ladder.

Scale'ne (Gr. σκαληνος, *skalēnos*, uneven). Unequal: applied to triangles, of which the three sides are unequal; in *anatomy*, applied to certain muscles, from their shape.

Scan'dent (Lat. *scan'do*, I climb). Climbing.

Scanso'res (Lat. *scan'do*, I climb). Climbers; an order of birds having the power of turning one of the front toes backwards, so as to be able to lay hold of and climbing trees: as the parrot, woodpecker, and cuckoo.

Scanso'rial (Lat. *scan'do*, I climb). Climbing, or fitted for climbing.

Scape (Lat. *sca'pus*, an upright stalk or stem). In *botany*, a naked flower-stalk bearing one or more flowers arising from a short axis, as the primrose.

Scaph'ite (Gr. σκαφη, *skaph'ē*, a skiff or boat). In *geology*, a chambered fossil shell, so called from its boat-like appearance.

Scaph'oid (Gr. σκαφη, *skaph'ē*, a skiff or boat; ειδος, *ei'dos*, shape). Resembling a boat.

Scap'ula (Probably allied to Gr. σκαπανη, *skap'anē*, a spade, from its shape). The shoulder-blade.

Scap'ular (Lat. *scap'ula*, the shoulder-blade). Belonging to the scapula or shoulder-blade.

Scapula'riæ (Lat. belonging to the shoulder-blade; scil. *pennœ*, feathers). The feathers which lie over the humerus in the wings of birds.

Scar'ification (Lat. *scarif'ico*, I make an incision). The operation of making several incisions or punctures in any part of the body, to let out blood or fluid.

Scar'ificator (Lat. *scarif'ico*, I make incision). An instrument for making several incisions in any part of the body.

Sca'rious (*Scar*). Like a dry scale; membranous, dry, and shrivelled.

Scarlati'na or Scarlet Fever. An infectious or contagious febrile disease, characterised by a scarlet eruption.

Schindyle'sis (Gr., a slit or fissure). In *anatomy*, a form of articulation in which a ridge in one bone is received into a groove in another.

Schist (Gr. σχιζω, *schi'zō*, I split). In *geology*, properly applied to rocks

M

which have a leafy structure and split up in thin irregular plates.

Schist'ose (*Schist*).   Fissile ; having a slaty texture.

Schneide'rian Membrane. The mucous membrane lining the nose.

Scho'liast (Gr. σχολιον, *schol'ion*, an interpretation).   A commentator ; one who writes notes upon the works of another.

Scho'lium (Gr. σχολιον, *schol'ion*).   An explanatory observation or remark.

Sciat'ic (Gr. ισχιον, *is'chion*, the hip).   Belonging to the hip.

Sciat'ica (Gr. ισχιον, *is'chion*, the hip).   A painful rheumatic affection of the hip.

Sci'ence (Lat. *sci'o*, I know).   Knowledge ; in *philosophy*, a collection of the general principles or leading truths relating to any object ; any branch of knowledge which is made the subject of investigation with a view to discover and apply first principles.

Scin'tillate (Lat. *scintil'la*, a spark).   To emit sparks ; to sparkle.

Scin'tillation (Lat. *scintil'la*, a spark).   A sparkling ; the twinkling or tremulous motion of the light of the larger fixed stars.

Sciog'raphy (Gr. σκια, *skia*, a shadow ; γραφω, *graph'ō*, I write).   The art of casting and delineating shadows correctly.

Sciop'tic (Gr. σκια, *ski'a*, a shadow ; οπτομαι, *op'tomai*, I see).   Relating to the camera obscura, or to the art of viewing images through a hole in a darkened room.

Scirrhos'ity (Gr. σκιρρος, *skir'rhos*, gypsum).   A hardness.

Scir'rhous (Gr. σκιρρος, *skir'rhos*, gypsum).   Hard ; of the nature of scirrhus.

Scir'rhus (Gr. σκιρρος, *skir'rhos*, gypsum).   A hard tumour ; a kind of cancer.

Scis'sile (Lat. *scin'do*, I cleave).   Capable of being divided by a sharp instrument.

Scle'ro- (Gr. σκληρος, *sklē'ros*, hard).   A prefix in compound words, implying hardness.

Scle'roderm (Gr. σκληρος, *sklē'ros*, hard ; δερμα, *derma*, a skin).   A name given to a family of fishes having the skin covered with hard scales.

Scle'rogen (Gr. σκληρος, *sklē'ros*, hard ; γενναω, *genna'ō*, I produce).   The thickening or hardening matter of the cells of vegetables.

Sclero'sis (Gr. σκληρος, *sklē'ros*, hard).   A hardening, or hard tumour.

Scleroskel'eton (Gr. σκληρος, *sklē'ros*, hard ; σκελετον, *skel'eton*).   The portion of the skeleton which consists of bones developed in tendons, ligaments, and membranous expansions.

Sclero'tal (*Sclerot'ic*).   An ossified portion of the capsule of the eye in fishes.

Sclerot'ic (Gr. σκληρος, *sklē'ros*, hard).   Hard ; a name given to the thick white outer coat of the eye.

Scleroti'tis (*Sclerot'ic ; itis*, denoting inflammation).   Inflammation of the sclerotic coat of the eye.

Sco'biform (Lat. *scobs*, filings or sawdust ; *for'ma*, shape).   Like filings or fine sawdust.

Scolio'sis (Gr. σκολιος. *skol'ios*, crooked).   A term for distortion of the spine.

Scorbu'tic (Lat. *scorbu'tus*, scurvy).   Having or liable to scurvy ; pertaining to scurvy.

Sco'ria (Gr. σκωρ, *skōr*, refuse matter).   The dross thrown off by metals in fusion ; in plural, *scor'iæ*, the cinders of volcanic eruptions.

Scoria'ceous (*Scoria*).   Like dross or scoria.

Scorpioi'dal (Gr. σκορπιος, *skor'pios*, a scorpion ; ειδος, *ei'dos*, shape).   Like the tail of a scorpion ; applied to a peculiar twisted form of inflorescence.

Sco'riform (*Sco'ria ; forma*, shape).   Resembling scoria or dross.

Scrobic'ulate (Lat. *scrobic'ulus*, a little ditch).   Furrowed ; pitted : having small depressions.

Scrobic'ulus Cordis. (Lat. the little ditch or furrow of the heart). A name sometimes given to the epigastric region ; the pit of the stomach.

Scrof'ula (Lat.).   A peculiar diseased

state, characterised by the deposition of tubercle in the organs of the body, and a tendency to swellings of the lymphatic glands and unhealthy ulceration.

Sculp'ture (Lat. *scul'po*, I carve). The art of carving or cutting wood or stone into images of men, animals, &c.

Scurvy (Lat. *scorbu'tus*). A diseased state, characterised by an altered state of the blood, and its effusion either in livid patches under the skin or in the form of hæmorrhages from the mucous membranes; which, especially in the mouth, become spongy.

Scu'tellated (Lat. *scutel'la*, a dish). Formed like a pan : divided into small surfaces.

Scutel'lum (Lat. *scu'tum*, a shield). A little shield.

Scutibran'chiate (Lat. *scu'tum*, a buckler ; Gr. βραγχια, *bran'chia*, gills). Having the gills covered with a shell in the form of a shield ; applied to an order of gasteropods.

Scu'tiform (Lat. *scu'tum*, a buckler; *forma*, shape). Shaped like a buckler.

Scu'tiped (Lat. *scu'tum*, a buckler ; *pes*, a foot). Having the anterior part of the legs covered with segments of horny rings.

Sebac'eous (Lat. *se'bum*, tallow or suet). Made of tallow ; resembling suet ; secreting a suet-like matter.

Sebac'ic (Lat. *se'bum*, tallow). Belonging to or obtained from fat.

Se'cant (Lat. *sec'o*, I cut). Cutting ; in *geometry*, a line which divides another into two parts ; in *trigonometry*, a right line drawn from the centre of a circle, and produced until it meets a tangent to the same circle.

Secer'nent (Lat. *secer'no*, I separate). Producing secretion.

Secre'te (Lat. *secer'no*, I separate). In *physiology*, to separate some peculiar fluid or substance from the blood or nutritive fluid.

Secre'tion (Lat. *secer'no*, I separate). In *physiology*, the separation of some peculiar fluid or substance

from the blood or nutritive fluid ; the substance so separated.

Secre'tory (Lat. *secer'no*, I separate). Having the function of secreting or separating some peculiar fluid or substance.

Sec'tile (Lat. *se'co*, I cut). Capable of being cut.

Sec'tion (Lat. *se'co*, I cut). A cutting ; in *geology*, the plane which cuts through any portion of the earth's crust so as to show its internal structure.

Sec'tor (Lat. *se'co*, I cut). A part of a circle lying between two radii and an arc of the circle : a mathematical instrument, formed of two graduated rulers as radii, turning in a joint which forms the centre of a circle ; in *astronomy*, an instrument for measuring the zenith distances of stars.

Sector of a Sphere. The solid generated by the revolution of the sector of a circle round one of the radii, which remains fixed.

Sec'ular Inequalities. In *astronomy*, the inequalities in the motions of planets produced by the continual accumulation of the residual phenomena other than the variation in their relative positions ; remaining uncompensated after the disturbed and disturbing bodies have passed through all their stages of configuration.

Secunda'riæ (Lat. secondary — i.e. *pennæ*, feathers). The feathers attached to the forearm in birds.

Sec'undine (Lat. *secun'dus*, second). In *botany*, the outer but one of the coats of the ovule.

Sed'iment (Lat. *sed'eo*, I settle down). Matter settled down from suspension in water.

Seed-leaf. A primary leaf ; applied to the expanded cotyledons or seed-lobes.

Seed-lobe. A cotyledon ; one of the parts into which a seed, as the common pea, splits.

Seg'ment (Lat. *sec'o*, I cut). A part cut off : in *geometry*, generally applied to a part cut off from a circle or sphere.

Seg'mentation (Lat. *segmen'tum*, a piece cut off). A dividing or splitting into segments.

Se'gregate (Lat. *se*, denoting separation ; *grex*, a flock). To set apart ; select : in *botany*, separated from each other.

Sele'niate. A compound of selenic acid with a base.

Selen'ic (*Sele'nium*). Belonging to selenium ; applied to an acid composed of one equivalent of selenium with three of oxygen.

Sele'nious. A term applied to an acid consisting of one equivalent of selenium and two of oxygen.

Sele'niuret (*Sele'nium*). A compound of selenium with a metal or other elementary substance.

Selenog'raphy (Gr. σεληνη, *selēnē*, the moon ; γραφω, *graph'ō*, I write). A description of the moon.

Sella Tur'cica (Lat. a Turkish saddle). A portion of the sphenoid bone in the skull, so named from its shape.

Sem'aphore (Gr. σημα, *sē'ma*, a sign ; φερω, *pher'ō*, I bear). A telegraph ; a means of communicating by signals.

Semeiolog'ical (Gr. σημειον, *sēmei'on*, a sign ; λογος, *log'os*, a discourse). Relating to the doctrine of the signs or symptoms of disease.

Semeiol'ogy (Gr. σημειον, *sēmei'on*, a sign ; λογος, *log'os*, a discourse). The part of medicine which describes the signs and symptoms of disease.

Semeiot'ic (Gr. σημειον, *sēmei'on*, a sign). Relating to the signs or symptoms of disease.

Sem'i- (Lat. *sem'i*, half). A prefix in compound words signifying half.

Semicir'cular (Lat. *sem'i*, half ; *cir'culus*, a circle). Having the form of a half circle.

Semicylin'drical (Lat. *sem'i*, half ; *cyl'inder*). Like a cylinder divided evenly in two from end to end.

Sem'iformed (Lat. *sem'i*, half ; *form'a*, form). Half formed ; imperfectly formed.

Semilig'neous (Lat. *sem'i*, half ; *lig'num*, wood). Woody below and herbaceous at the top.

Semilu'nar (Lat. *sem'i*, half ; *lu'na*, a moon). Resembling a half-moon.

Semimem'branous (Lat. *sem'i*, half ; *membra'na*, membrane). Half membranous ; applied to one of the muscles of the thigh.

Sem'inal (Lat. *se'men*, a seed). Belonging to seed ; in *botany*, applied to the cotyledons or seed-leaves.

Sem'ination (Lat. *se'men*, seed). The act of sowing : in *botany*, the natural dispersion of seeds.

Sem'inude (Lat. *sem'i*, half ; *nu'dus*, naked). In *botany*, applied to seeds of which the seed-vessel opens early, as in the mignonette.

Semipal'mate (Lat. *sem'i*, half ; *pal'ma*, a palm). Having the toes connected by a web, extending along the half nearest to the foot.

Semipen'niform (Lat. *sem'i*, half ; *pen'na*, a feather ; *for'ma*, shape). Penniform on one side only ; applied, in *anatomy*, to some muscles.

Semiten'dinous (Lat. *sem'i*, half ; *ten'do*, a tendon). Half tendinous ; a name given to a muscle of the thigh, which bends the leg.

Semit'ic (*Shem*, the son of Noah). A name given to one of the great families of languages, comprehending the Assyrian, Babylonian, Syriac, Phœnician, Hebrew, and Arabic languages, with their dialects.

Sensa'tion (Lat. *sen'sus*, sense). The faculty by which an animal becomes conscious of impressions made on the extremities of the nerves either by some external body, or by some change or operation within the system.

Sense (Lat. *sen'tio*, I perceive). The faculty by which a living being receives the impression of external objects, so that they may be conveyed to the sensorium or brain.

Sensibil'ity (Lat. *sen'tio*, I perceive). The faculty by which an impression made by an external body on the parts or textures of the body is felt.

Senso'rium (Lat. *sen'tio*, I perceive). The seat of sensation ; the organ which receives the impressions made on the senses.

Sen'tient (Lat. *sen'tio*, I perceive). Capable of receiving impressions so as to be perceived.

Se'pal (Lat. *sepes*, an inclosure). A division of a calyx.

Sep'aloid (*Sepal;* Gr. εἶδος, *ei'dos*, form). Like a sepal.

Sep'arate (Lat. *se'paro*, I divide). In *botany*, applied when the stamens and pistils are in the same plant, but in different flowers.

Sep'tate (Lat. *septum*, a partition). Divided by septa or partitions.

Sep'tangular (Lat. *septem*, seven; *angulus*, an angle). Having seven angles.

Sep'temfid (Lat. *septem*, seven; *findo*, I cleave). In *botany*, applied to leaves which are divided part way through into seven lobes.

Septe'nary (Lat. *septe'ni*, series of seven). Consisting of sevens.

Septe'nate (Lat. *septe'ni*, series of seven). Arranged in sevens: applied to compound leaves with seven leaflets coming off from a point.

Septen'nial (Lat. *septem*, seven; *annus*, a year). Containing seven years: happening every seven years.

Septentrio'nal (Lat. *septen'trio*, the northern constellation called the Great Bear). Belonging to the north.

Sep'tic (Gr. σηπω, *sēpō*, I putrefy). Promoting putrefaction.

Septici'dal (Lat. *septum*, a partition; *cædo*, I cut). In *botany*, applied to fruits or seed vessels which open by dividing through the partitions of the ovary; *i.e.*, through the septa or edges of the carpels.

Septif'erous (Lat. *septum*, a partition; *fer'o*, I bear). Having partitions.

Sep'tiform (Lat. *septum*, a partition; *forma*, shape). Resembling a septum or partition.

Septif'ragal (Lat. *septum*, a partition; *frango*, I break). A form of division of a fruit in which the partitions adhere to the axis, and the valves covering the fruit are separated; the dehiscence taking place through the backs of the cells.

Septilat'eral (Lat. *septem*, seven; *latus*, a side). Having seven sides.

Septil'lion (Lat. *septem*, seven; *million*). A million multiplied seven times into itself.

Sep'tuagint (Lat. *septuagin'ta*, seventy). A Greek translation of the Old Testament, supposed to have been the work of seventy or seventy-two interpreters.

Sep'tulate (Lat. *septum*, a partition). In *botany*, applied to fruits having spurious transverse dissepiments or partitions.

Sep'tum (Lat. *se'pio*, I inclose or hedge in). A partition; in *botany*, a division in an ovary or seed vessel formed by the sides of the carpels, applied in *anatomy* to the partitions between organs in various parts.

Seque'la (Lat. *seq'uor*, I follow). That which follows; in *medicine*, applied to a diseased state following on an attack of some other disease.

Seques'trum (Lat). In *surgery*, a dead portion of bone.

Se'rial (Lat. *se'ries*, an order). Following in a determinate order or in distinct rows.

Seric'eous (Lat. *se'ricum*, silk). Silky; covered with fine closely pressed hairs.

Se'ries (Lat. an order). A continued succession or order; in *arithmetic* and *algebra*, a number of quantities succeeding each other in regular increasing or diminishing order, either by a common difference or a common multiplier.

Seros'ity (Lat. *serum*, whey). The serum of the blood, or the whey of milk.

Se'rous (Lat. *serum*, whey). Like serum or whey; secreting serum.

Se'rous Membrane. A closed membraneous bag, having its internal surface moistened with serum, and lining some cavity of the body which has no outlet.

Ser'pentine (Lat. *ser'pens*, a serpent). Like a serpent; coiled or twisted: in *geology*, a rock of flint and magnesia, of mottled colour, like the skin of a serpent.

Ser'rate (Lat. *serra*, a saw). Notched

like a saw; having sharp processes like the teeth of a saw.

Ser'ratures (Lat. *serra*, a saw). Pointed projections at the edge like the teeth of a saw.

Ser'rulate (Lat. *ser'rula*, a little saw). Having very fine notches.

Se'rum (Lat. whey). The yellowish fluid which is left in coagulation of the blood, consisting of the liquor sanguinis, or blood-fluid, deprived of fibrin.

Ses'amoid (Gr. σησαμον, *sēsamon*, a kind of small grain; ειδος, *ei'dos*, shape). Like a sesame; applied to small bones at the joints of the great toes and thumbs, and to small bodies in the valves of the aorta and pulmonary artery.

Ses'qui- (Lat. one and a half). A prefix in compound words signifying one and a half, or in the proportion of three to two.

Sesquial'teral (Lat. *sesqui*, one and a half; *alter*, the other). In *arithmetic* and *geometry*, applied to a quantity which contains one and a half of another.

Sesquiba'sic (Lat. *sesqui*, one and a half; *basis*, a base). Applied to salts containing one and a half times as much base in proportion to the acid as the neutral salt.

Sesquicar'bonate (Lat. *sesqui*, one and a half; *carbonate*). A salt consisting of three equivalents of carbonic acid with two of base.

Sesquichlo'ride (Lat. *sesqui*, one and a half; *chloride*). A compound of three equivalents of chlorine with two of another element.

Sesquidu'plicate (Lat. *sesqui*, one and a half; *duplex*, double). Having the ratio of two and a half to one.

Sesqui'odide (Lat. *sesqui*, one and a half; *iodide*). A compound of three equivalents of iodine with two of another element.

Sesqui'oxide (Lat. *sesqui*, one and a half; *oxide*). A compound of three equivalents of oxygen with two of another element.

Sesquip'licate (Lat. *sesqui*, one and a half; *plic'o*, I fold). In the ratio of one and a half to one.

Sesquisul'phate (Lat. *sesqui*, one and a half; *sulphate*). A sulphate containing three equivalents of sulphuric acid and two of base.

Sesquisul'phide (Lat. *sesqui*, one and a half; *sulphide*). A compound of three equivalents of sulphur with two of another element.

Sesquiter'tian (Lat. *sesqui*, one and a half; *tertia'nus*, tertian). Having the ratio of one and one-third.

Ses'sile (Lat. *sed'eo*, I sit). Sitting; having no stem or stalk.

Seta'ceous (Lat. *seta*, a bristle). Bristly, or resembling bristles.

Se'tiform (Lat. *seta*, a bristle; *forma*, form). Resembling a bristle.

Setig'erous (Lat. *seta*, a bristle; *ger'o*, I bear). Bearing setæ or sharp hairs.

Se'tose or **Se'tous** (Lat. *seta*, a bristle). Bristly; covered with setæ or sharp hairs.

Sex- (Lat. six). A prefix in compound words signifying six.

Sex'angular (Lat. *sex*, six; *an'gulus*, an angle). Having six angles.

Sexen'nial (Lat. *sex*, six; *annus*, a year). Lasting six years; happening once in six years.

Sexfid (Lat. *sex*, six; *findo*, I cleave). Cleft into six.

Sexloc'ular (Lat. *sex*, six; *loc'ulus*, a cell). Having six cells.

Sex'tant (Lat. *sex'tans*, a sixth). The sixth part of a circle; an instrument for measuring the angular distances of objects, having a limb of sixty degrees, or the sixth part of a circle.

Sextil'lion (Lat. *sex*, six; *million*). The sixth power of a million.

Sex'tuple (Lat. *sex*, six; *plic'o*, I fold). Six-fold.

Sex'ual (Lat. *sexus*, sex). Denoting the sexes; in *botany*, applied to a system of classification founded on the number and arrangement of the stamens and pistils.

Sex'ual System. In *botany*, the classification founded by Linnæus on the number, position, &c., of the stamens and pistils.

Shaft. In *architecture*, the body of a column between the trunk and the capital; in *mechanics*, an axle of large size.

Shale (Germ. *scha'len*, to peel off). In *geology*, applied to all argillaceous or clayey strata which split up or peel off in thin laminæ.

Shemit'ic. *See* Semitic.

Shingle. In *geology*, loose imperfectly rounded stones and pebbles.

Sial'agogue (Gr. σιαλον, *si'alon*, saliva; αγω, *ag'ō*, I lead). Promoting a flow of saliva.

Sib'ilant (Lat. *sib'ilo*, I hiss). Making a hissing sound.

Sidera'tion (Lat. *sidus*, a star). A blasting or blast in plants; a sudden deprivation of sense.

Side'real (Lat. *sidus*, a star). Relating to, or containing stars; a sidereal day is the period between the moment at which a star is in the meridian of a place, and that at which it arrives at the meridian again; a sidereal year is the period in which the fixed stars apparently complete a revolution; sidereal period is the time which a planet takes to make a complete revolution round the sun.

Siderog'raphy (Gr. σιδηρον, *sidēron*, iron; γραφω, *graph'ō*, I write). The art of engraving on steel.

Sigilla'ria (Lat. *sigil'lum*, a seal). In *geology*, a large genus of fluted tree-stems having seal-like punctures on the ridges.

Sig'moid (C, the old form of the Greek letter, σιγμα, *sigma*; ειδος, *ei'dos*, form). Like the Greek letter C, or sigma; applied in *anatomy* to several structures in the body.

Sign (Lat. *signum*, a mark). In *astronomy*, the twelfth part of the ecliptic; in *algebra*, a character indicating the relation between quantities; in *medicine*, anything by which the presence of disease is made known; physical signs are phenomena taking place in the body in accordance with physical laws, and capable of being perceived by the senses of the observer.

Sil'ica (Lat. *silex*, flint). The com-pound of silicon with oxygen, forming pure flint or rock-crystal.

Sil'icate (Lat. *silex*, flint). A compound of silicic acid with a base.

Silic'eous (Lat. *silex*, flint). Belonging to or containing silex or flint; having a flinty texture.

Silic'ic (Lat. *silex*, flint). Belonging to flint; silicic acid, a name applied to silica, or a compound of silicon and oxygen having certain of the properties of an acid.

Silicif'erous (Lat. *silex*, flint; *fer'o*, I bear). Producing silex or flint.

Silicifica'tion (Lat. *silex*, flint; *fac'io*, I make). Petrifaction; the conversion of any substance into a flinty mass.

Sili'cified (Lat. *silex*, flint; *fac'io*, I make). Converted into flinty matter.

Silic'ula (Lat. a little pod). A fruit resembling a siliqua, but broader and shorter.

Silic'ulose (Lat. *silic'ula*, a little pod). Bearing siliculæ or silicles.

Sil'iqua (Lat. a pod). A form of fruit consisting of two long cells, divided by a partition, having seeds attached on each side, as in the cabbage and turnip.

Sil'iquose (Lat. *sil'iqua*, a pod). bearing a siliqua.

Silt. In *geology*, properly the fine mud which collects in lakes and estuaries, but generally used to designate all calm and gradual deposits of mud, clay, or sand.

Silu'rian (Lat. *Silu'res*, the ancient inhabitants of South Wales). Applied in *geology* to a system of slaty, gritty, and calcareous beds, containing occasional fossils, and largely developed in South Wales.

Sin'apism (Gr. σιναπι, *sina'pi*, mustard). A mustard poultice.

Sin'ciput (Lat.) The fore part of the head.

Sine (Lat. *sinus*). In *trigonometry*, the straight line drawn from one extremity of the arc of a circle, perpendicular to the diameter passing through the other extremity.

Sin'ical (Lat. *sinus*, a sine). Belonging to a sine.

Sinis'tral (Lat. *sinis'ter*, left). Having spiral turns towards the left.

Sinis'trorse (Lat. *sinis'ter*, left; *versus*, towards). Turned towards the left.

Sin'uate (Lat. *sinus*, a bay or indentation). Having large curved breaks in the margin.

Sinuos'ity (Lat. *sinus*, an indentation). A winding in and out.

Sin'uous (Lat. *sinus*, an indentation). Winding; crooked; having a wavy or flexuous margin.

Sinus (Lat. a bay or indentation). In *anatomy*, a cavity in a bone, widest at the bottom ; a dilated form of vein, mostly found in the head ; in *surgery*, an elongated cavity containing pus.

Si'phon (Gr. σιφων, *siphōn*, a reed). A bent tube with legs of unequal length, used for drawing liquid from a vessel.

Siphon Barometer. A barometer in which the lower end of the tube is bent upwards in the form of a siphon.

Siphon Gauge. A glass siphon partly filled with mercury, used for indicating the degree of rarefaction, which has been produced in the receiver of an air-pump.

Sipho'nal (Gr. σιφων, *siphōn*, a siphon or reed). Of the nature of a siphon.

Siphuncle (Gr. σιφων, *siphōn*, a reed; *cle*, denoting smallness). A small siphon.

Siphonibran'chiate (Gr. σιφων, *si'phōn*, a tube ; βραγχια, *bran'chia*, gills). Having a siphon or tube, by which water is carried to the gills.

Siphonos'tomous (Gr. σιφων, *siphōn*, a reed ; στομα, *stoma*, a mouth). Having a mouth in the shape of a siphon or tube.

Siren. In *acoustics*, an instrument for determining the number of vibrations produced by musical sounds of different pitch.

Siroc'co (Italian). An oppressive relaxing wind coming from North Africa over the Mediterranean to Sicily, Italy, &c.

Skel'eton (Gr. σκελλω, *skel'lō*, I dry). The bones of an animal, dried, and retained in their natural positions.

Slate. In *geology*, properly applied to argillaceous or clayey rocks, the lamination or arrangement in plates of which is not due to stratification but to cleavage.

Snow-line. The elevation at which mountains are covered with perpetual snow.

Soap (Lat. *sapo*). In *chemistry*, a compound of a fatty substance or an oil-acid with a base.

Soapstone. A soft variety of magnesian rock having a soapy feel.

Sob'oles (Lat. a shoot or young branch). A creeping underground stem.

Solana'ceous (Lat. *sola'num*, the nightshade). Belonging to the order of plants which includes the nightshade and potato.

Solar (Lat. *sol*, the sun). Belonging to the sun ; measured by the progress of the sun.

Solar System. In *astronomy*, the sun, with the assemblage of globes or primary planets revolving round it, and secondary planets or satellites revolving round the primary.

Sol'ecism. Impropriety in language, consisting in the use of words or expressions which do not agree with the existing rules of grammatical construction.

Solen- (Gr. σωληην, *sōlēn*, a channel or canal). A prefix in some compound words, implying the presence of a canal or pipe.

Sol'id (Lat. *sol'idus*). Having the component parts so firmly adherent that the figure is maintained unless submitted to more or less violent external action.

Solidun'gulous (Lat. *sol'idus*, solid ; *un'gula*, a hoof). Having the hoof entire or not cloven.

Sol'iped (Lat. *solus*, alone ; *pes*, a foot). Having only one apparent toe and a single hoof to each foot, as the horse.

Sol'stices (Lat. *sol*, the sun ; *sto*, I stand). In *astronomy*, the periods in winter and summer at which the centre of the disc of the sun passes through the solstitial points, or the points in the ecliptic, midway between the equatorial points, and most distant from the celestial equator.

Solstit′ial (Lat. *sol*, the sun; *sto*, I stand). Belonging to the solstice.

Solubil′ity (Lat. *solvo*, I melt). The property of being dissolved or melted in fluid.

Sol′uble (Lat. *solvo*, I melt). Capable of being dissolved or melted in a fluid.

Solu′tion (Lat. *solvo*, I melt). The act of separating the parts of any body; in *chemistry*, the melting of one substance in another in such way that the latter is not rendered opaque thereby; in *mathematics*, the finding an answer to any question, or the answer found.

Sol′vent (Lat. *solvo*, I melt). Any fluid or substance which renders other bodies liquid.

Somat′ic (Gr. σωμα, *sōma*, the body). Belonging to the body.

Somatol′ogy (Gr. σωμα, *sōma*, a body; λογος, *log′os*, description). The doctrine of bodies or material substance.

Somnam′bulism (Lat. *som′nus*, sleep; *am′bulo*, I walk). A walking in sleep.

Somnif′erous (Lat. *som′nus*, sleep; *fer′o*, I bring). Producing sleep.

Somnif′ic (Lat. *som′nus*, sleep; *fac′io*, I make). Causing sleep.

Som′nolence (Lat. *som′nus*, sleep). Drowsiness.

Som′nolent (Lat. *som′nus*, sleep). Drowsy.

Sonif′erous (Lat. *sonus*, sound; *fer′o*, I bear). Conveying sound.

Sonom′eter (Lat. *sonus*, sound; Gr. μετρον, *met′ron*, measure). An instrument for measuring sounds or the intervals of sounds; an apparatus for illustrating the phenomena exhibited by sonorous bodies.

Sonorif′ic (Lat. *sonor*, a loud sound; *fac′io*, I make). Producing sound.

Sono′rous (Lat. *sonus*, sound). Giving sound: sonorous figures, the figures which are formed by nodal lines, as when a disc of glass or metal covered with fine sand is thrown into musical vibrations.

Soph′ism (Gr. σοφισμα, *sophis′ma*, a cunning contrivance). An argument in which the conclusion is not justly deduced from the premises.

Soporif′erous (Lat. *so′por*, sound sleep; *fer′o*, I produce). Producing sleep.

Soporif′ic (Lat. *so′por*, sleep; *fac′io*, I make). Causing sleep.

Sorbefac′ient (Lat. *sor′beo*, I sup up; *fac′io*, I make). Producing absorption.

Sori′tes (Gr. σωρος, *sūros*, a heap). In *logic*, an abridged form of a series of syllogisms; or a series of propositions linked, so that the predicate of each one becomes the next subject, the conclusion being formed by joining the first subject and the last predicate.

Soro′sis (Gr. σωρος, *sō′ros*, a heap). A kind of fleshy fruit formed by the consolidation together of many flowers, seed-vessels, and receptacles; as the pine-apple.

Spa′dix (Lat.). In *botany*, a form of inflorescence in which the flowers are closely arranged round a thick fleshy axis, and the whole wrapped in a large leaf called a spathe; as in the arum or wake-robin.

Spar. In *geology*, a term applied to crystals or minerals which break up into regularly shaped forms with smooth cleavage-faces.

Spasm (Gr. σπαω, *spa′ō*, I draw). An abnormal involuntary contraction of one or more muscles or muscular fibres.

Spasmod′ic (Gr. σπασμος, *spas′mos*, spasm; ειδος, *ei′dos*, form). Resembling spasm; consisting in spasm.

Spas′tic (Gr. σπαω, *spa′ō*, I draw). Having the power of drawing to or from; applied to muscular contractions in disease.

Spatha′ceous (*Spathe*). Having the appearance and consistence of a spathe.

Spathe (Gr. σπαθη, *spathē*, a broad blade). A large membranous bract or kind of leaf, attached at the base of a spadix and enveloping it in a sheath.

Spath'ic (Gr. σπαθη, *spathē*, a broad blade). In leaves or plates.

Spath'iform (Germ. *spath*, spar; Lat. *forma*, shape). Resembling spar in form.

Spa'those (Gr. σπαθη, *spathē*, a broad blade). In *botany*, relating to or like a spathe ; in *mineralogy*, of the nature of spar.

Spat'ulate (Lat. *spat'ula*, a broad slice). Like a spatula or battle-door ; in *botany*, applied to leaves narrow at the base, and gradually widening towards a broad-crowned or straight top.

Spe'cies. In *zoology* and *botany*, a collection of individuals resembling each other so closely that they are considered to have originated from a common parent, and having the power of uniform and permanent continuance by propagation.

Specif'ic (Lat. *species*, form or figure; *fac'io*, I make). Denoting a species ; designating the peculiar property or properties which distinguish one species from another ; in *medicine*, supposed to possess a peculiar efficacy in a disease.

Specif'ic Grav'ity. The weight of a body, as compared with the weight of an equal bulk or volume of some other body (as water) taken as the standard.

Specif'ic Volume. In *chemistry*, the atomic volume, or the number representing the volume in which a body combines.

Specifica'tion (Lat. *spe'cies*, form ; *fac'io*, I make). The act of determining by a mark ; a statement of particulars, describing a work to be undertaken or an invention.

Spec'trum (Lat. *spec'to*, I behold). In *optics*, the coloured image formed on a white surface by rays of light passing through a hole, and separated by a glass prism.

Spec'ulum (Lat. *spec'to*, I behold). In *medicine*, an instrument for examining internal parts by means of light.

Spel'ter. Native impure zinc, con-
taining lead, copper, iron, arsenic, manganese, and plumbago.

Sper'moderm (Gr. σπερμα, *sper'ma*, seed ; δερμα, *der'ma*, skin). The covering of a seed.

Sphac'elate (Gr. σφακελος, *sphak'elos*, mortification). To mortify.

Sphac'elus (Gr. σφακελος, *sphak'elos*, gangrene). Death of a part of a living animal.

Sphæren'chyma (Gr. σφαιρα, *sphai'ra*, a sphere ; ἐγχυμα, *en'chuma*, tissue). Vegetable tissue composed of spherical cells.

Sphe'no- (Gr. σφην, *sphēn*, a wedge). In *anatomy*, a prefix in compound words, implying connection with, or relation to the sphenoid bone.

Sphe'noid (Gr. σφην, *sphēn*, a wedge; ειδος, *eidos*, shape). Like a wedge ; applied to a bone of the skull, which is wedged in among the other bones.

Sphe'no-maxil'lary. Belonging to the sphenoid and jaw-bones.

Sphe'no-pari'etal. Belonging to the sphenoid and parietal bones.

Sphe'no-tem'poral. Belonging to the sphenoid and temporal bones.

Sphere (Gr. σφαιρα, *sphaira*, a ball). A round body like a ball ; in *geometry*, the solid figure formed by the rotation of a semicircle about its diameter, and having a single surface, every part of which is equally distant from the centre ; in *astronomy*, the concave expanse of the heavens, having the appearance of the interior of a hollow sphere ; a right sphere being that aspect in which the circles of motion of the heavenly bodies appear at right angles with the horizon, as at the equator ; a parallel sphere, that in which the same motions appear parallel with the horizon, as at the poles ; and an oblique sphere, that in which these motions appear oblique to the horizon, as at any point between the equator and each pole.

Spher'ical (Gr. σφαιρα, *sphaira*, a sphere). Like a sphere ; globular ; relating to a sphere.

Spheric'ity (Gr. σφαιρα, *sphaira*, a sphere). Roundness.

Spher'oid (Gr. σφαιρα, *sphaira*, a ball ; ειδος, *eidos*, form). Resembling a sphere ; a body approaching a sphere in form, but not perfectly globular ; the result of the revolution of an ellipse about one of its axes.

Spherom'eter (Gr. σφαιρα, *sphaira*, a sphere; μετρον, *met'ron*, a measure). An instrument for measuring the dimensions of a sphere.

Spher'ule (Gr. σφαιρα, *sphaira*, a ball ; *ule*, denoting smallness). A little sphere or globular body.

Sphinc'ter (Gr. σφιγγω, *sphingō*, I bind). A name given to circular muscles surrounding the orifices of organs or parts of the body.

Sphygmom'eter (Gr. σφυγμος, *sphugmos*, the pulse ; μετρον, *met'ron*, a measure). An instrument for counting the pulsations of an artery by rendering the action of the pulse visible, and measuring its strength.

Spica (Lat. an ear of corn). In *surgery*, a kind of bandage, so called from its turns being thought to resemble the arrangement of the ears of corn on the stem.

Spic'ular (Lat. *spic'ulum*, a dart). Resembling a dart ; having sharp points.

Spic'ula (Lat. *spic'ulum*, a dart). In *botany*, a spikelet.

Spic'ulum (Lat. a dart). In *surgery*, a small pointed piece of bone or other hard matter.

Spike (Lat. *spica*, an ear of corn). In *botany*, a form of inflorescence in which sessile flowers are placed on a simple peduncle or stem, as in the wheat and lavender.

Spikelet. In *botany*, a small spike, or cluster of flowers, as in grasses.

Spina Bif'ida (Lat. cleft spine). A diseased state in which part of the bones of the spine are deficient, so that the membranes of the chord project in the form of a tumour.

Spinal (Lat. *spina*, the spine). Belonging to the spine or back-bone.

Spinal Chord or Marrow. The part of the nervous system contained in the canal of the vertebral column.

Spinal System of Nerves. The nerves which convey impressions to and from the spinal cord especially.

Spine (Lat. *spina*, a thorn). A thorn ; an abortive branch with a hard sharp point ; in *anatomy*, the vertebral column or back-bone ; in *zoology*, a thin pointed spike.

Spines'cent (Lat. *spina*, a thorn). Becoming thorny ; bearing spines.

Spinif'erous (Lat. *spina*, a thorn ; *fer'o*, I bear). Producing spines or thorns.

Spi'niform (Lat. *spina*, a thorn ; *forma*, shape). Like a spine or thorn.

Spin'neret (Sax. *spinnan*, to make yarn). The pointed tubes with which spiders weave their webs.

Spi'nous (Lat. *spina*, a spine or thorn). Having spines ; in *anatomy*, projecting like a spine.

Spi'racle (Lat. *spiro*, I breathe). A breathing hole ; applied to the external openings of the air-tubes of insects.

Spiral (Gr. σπειρα, *speira*, anything wound round). Winding round a fixed point, and at the same time constantly receding, as the mainspring of a watch ; winding round a cylinder, and at the same time advancing; in *architecture*, a curve winding round a cone or spire.

Spiral Vessels. In *botany*, fine transparent membranous tubes, with one or more spiral fibres coiled up in their interior.

Spirit Level. An instrument for determining a plane parallel to the horizon, consisting of a tube of glass nearly filled with spirits of wine or distilled water, and hermetically sealed, so that, when it is placed in a horizontal position, the bubble of air in the liquid stands exactly in the centre of the tube.

Spirom'eter (Lat. *spiro*, I breathe ; Gr. μετρον, *met'ron*, a measure). An instrument for measuring the quantity of air exhaled from the lungs, and thereby determining the capacity of the chest.

Spiroi'dal (Gr. σπειρα, *speira*, anything wound round; ειδος, *eidos*, shape). Like a spiral or corkscrew.

Spis'situde (Lat. *spissus*, thick). Thickness; applied to substances, &c., neither perfectly liquid nor perfectly solid.

Splanchnic (Gr. σπλαγχνον, *splanchnon*, bowels). Belonging to the viscera or intestines.

Splanchno- (Gr. σπλαγχνον, *splanchnon*, bowels). In *anatomy* and *medicine*, a prefix in compound words, implying relation to viscera.

Splanchnog'raphy (Gr. σπλαγχνον, *splanchnon*, bowels; γραφω, *graph'ō*, I write). An anatomical description of the viscera.

Splanchnol'ogy (Gr. σπλαγχνον, *splanchnon*, bowels; λογος, *log'os*, discourse). A description of the viscera.

Splan'chno-Skel'eton (Gr. σπλαγχνον, *splanchnon*, bowels; σκελετον, *skel'eton*). The bony or cartilaginous pieces which support the viscera and organs of sense in animals.

Sple'nial (Lat. *sple'nium*, a splint). Applied to a bone in the head of fishes, because applied in the manner of a splint.

Spleniza'tion (Gr. σπλην, *splēn*, the spleen). A change produced in the lungs by inflammation, so that they resemble the substance of the spleen.

Spondee (Gr. σπονδη, *spondē*, a drink offering; because solemn melodies were used on such occasions). A foot in Greek and Latin verse consisting of two long syllables.

Spongelets, *See* Spongioles.

Spongia'ria (Gr. σπογγος, *spongos*, sponge). The class of beings including sponges.

Spon'giform (Gr. σπογγος, *spongos*, a sponge; Lat. *forma*, shape). Like a sponge.

Spon'gioles (Gr. σπογγος, *spongos*, a sponge; *ole*, denoting smallness). In *botany*, the ultimate extremities of roots, composed of loose spongy cellular tissue, through which nourishment is absorbed.

Sponta'neous (Lat. *sponte*, of one's own accord). Occurring or arising apparently of itself, without any obvious cause.

Sporad'ic (Gr. σπορας, *spor'as*, scattered). Separate; scattered: applied to diseases which occur in single and scattered cases.

Sporan'gium (Gr. σπορα, *spor'a*, a seed; αγγειον, *angei'on*, a vessel). The case which contains the sporules or reproductive germs of some cryptogamic plants.

Spore (Gr. σπορα, *spor'a*, a seed). *See* Sporules.

Spor'ophore (*Spore;* Gr. φερω, *pher'ū*, I bear). A stalk supporting a spore.

Sporozo'id (*Spore;* Gr. ζωον, *zōon*, an animal; ειδος, *eidos*, shape). A spore furnished with ciliary or vibratile processes.

Spor'ules (Gr. σπορα, *spor'a*, a seed; *ule*, denoting smallness). The minute organs in flowerless plants which are the analogues of seeds in flowering plants.

Spu'riæ (Lat. spurious; sc. *pennæ*, feathers). The feathers attached to the short outer digit in the wings of birds.

Sputum (Lat. *spuo*, I spit). Spittle; in *medicine*, that. which is discharged from the mouth in disorders of the breathing organs.

Squama (Lat. a scale). A scale; a part arranged like a scale.

Squa'mifer (Lat. *squama*, a scale; *fer'o*, I bear). Covered with scales.

Squa'miform (Lat. *squama*, a scale; *forma*, shape). Like a scale.

Squamig'erous (Lat. *squama*, a scale; *ger'o*, I bear). Bearing or having scales.

Squa'mous (Lat. *squama*, a scale). Scaly; arranged in scales or like scales; squamous suture, in *anatomy*, the suture between the parietal and temporal bone, the former overlapping the latter like a scale.

Square (Lat. *quadra*). Having four equal sides and four equal angles: in *arithmetic*, applied to the product of a number multiplied into itself, the number thus multiplied being the square root of the product.

Stalac′tite (Gr. σταλαζω, *stala′zō*, I drop). A concretion of carbonate of lime hanging from the roof of a cave, produced by the filtration of water containing limy particles and its subsequent evaporation.

Stalag′mite (Gr. σταλαζω,, *stala′zō*, I drop). A concretion of carbonate of lime found on the floors of caverns, produced from the dropping and evaporation of water containing lime.

Stamen (Lat. *sto*, I stand). In a general sense, that which gives support to a body : in *botany*, the male organ in flowering plants.

Stam′inal (Lat. *stamen*). In *botany*, having stamens only.

Stamin′eous (*Stamen*). Consisting of, or having stamens.

Staminif′erous (Lat. *stamen; fer′o*, I bear). Having stamens without a pistil.

Stan′nary (Lat. *stannum*, tin). Relating to tin-works.

Stannic (Lat. *stannum*, tin). Procured from tin.

Stannif′erous (Lat. *stannum*, tin ; *fer′o*, I bear). Containing tin.

Staphylo′ma (Gr. σταφυλη, *staph′ulē*, a grape). A disease of the eye in which the cornea loses its transparency and forms a pearl-coloured projection, sometimes smooth and sometimes uneven.

Staphylor′aphy (Gr. σταφυλη, *staph′ulē*, a bunch of grapes, or the tonsils ; ῥαπτω, *rhaptō*, I sow). A surgical operation for uniting the edges of a divided palate.

Sta′sis (Gr ιστημι, *histēmi*, I make to stand). A standing or settling in one place : as of the blood.

Stat′ic (Gr. ιστημι, *histēmi*, I make to stand). Having the power of keeping in a stationary condition.

Stat′ics (Gr. ιστημι, *histēmi*, I cause to stand). The branch of mechanics which considers the action on bodies of forces at equilibrium, or producing equilibrium.

Statis′tics. The science of collecting and arranging all the numerical facts relating to any subject.

Steam Hammer. A form of forge hammer consisting of a steam cylinder and piston placed vertically over the anvil.

Ste′arate (Gr. στεαρ, *ste′ar*, suet). A salt consisting of stearic acid and a base.

Stear′ic (Gr. στεαρ, *ste′ar*, suet). An acid which is derived from certain fats.

Ste′arin (Gr. στεαρ. *ste′ar*, suet). The chief ingredient of suet and tallow.

Ste′atite (Gr. στεαρ, *ste′ar*, suet). Soap-stone : a soft unctuous mineral, consisting of a silicate of magnesia and alumina.

Steato′ma (Gr. στεαρ, *ste′ar*, suet). A tumour containing a substance resembling fat.

Steato′matous (*Steato′ma*). Of the nature of a steatoma or fatty tumour.

Stel′lar (Lat. *stella*, a star). Belonging to or containing stars.

Stel′late or **Stel′liform** (Lat. *stella*, a star). Resembling a star ; radiated.

Stem′mata (Gr. στεμμα, *stem′ma*, a chaplet). The simple minute eyes of worms, and those which are added to the large compound eyes.

Stenog′raphy (Gr. στενος, *sten′os*, narrow ; γραφω, *graph′ō*, I write). The art of writing in short hand by using abbreviations or characters for whole words.

Sterelmin′tha (Gr. στερεος, *ster′eos*, solid ; ἑλμινς, *hel′mins*, a worm). Parasitic worms, having no true abdominal cavity.

Stereograph′ic (Gr. στερεος, *ster′eos*, solid ; γραφω, *graph′ō*, I write). Delineated on a plane ; stereographic projection is the projection of a sphere delineated on the plane of one of its great circles, the eye being at the pole of the circle.

Stereog′raphy (Gr. στερεος, *ster′eos*, solid ; γραφω, *graph′ō*, I write). The art of delineating the forms of solid bodies on a plane.

Stereom′eter (Gr. στερεος, *ster′eos*, solid ; μετρον, *met′ron*, a measure). An instrument for measuring the specific gravities of various substances, solid as well as liquid.

Stereom'etry (Gr. στερεος, *ster'eos*, solid; μετρον, *met'ron*, a measure). The art of measuring solid bodies and finding their solid contents.

Ster'eoscope (Gr. στερεος, *ster'eos*, solid; σκοπεω, *skop'eō*, I view). An optical instrument by which we look on two pictures taken under a small difference of angular view, each eye looking on one picture only; so that, as in ordinary vision, two images are conveyed to the brain and unite in one impression.

Stereot'omy (Gr. στερεος, *ster'eos*, solid; τεμνω, *tem'nō*, I cut). The art of cutting solids into certain figures or sections.

Ster'eotype (Gr. στερεος, *ster'eos*, solid; τυπος, *tu'pos*, type). A fixed metal type; a plate of the size of a page, cast from a mould in which an exact representation of the types set in order by a printer has been produced.

Ster'nal (Lat. *ster'num*, the breast-bone). Belonging to the breast-bone.

Ster'no- (Lat. *ster'num*, the breast-bone). A prefix in compound words, signifying relation to the sternum or breast-bone.

Ster'num (Lat.) The breast-bone to which the ribs are jointed in front.

Stern'utatory (Lat. *stern'uo*, I sneeze). Producing sneezing.

Stethom'eter (Gr. στηθος, *stē'thos*, the chest; μετρον, *met'ron*, a measure). An instrument for measuring the movements of the chest on the outside.

Steth'oscope (Gr. στηθος, *stē'thos*, the chest; σκοπεω, *skop'eō*, I view). A cylindrical instrument of light wood or gutta percha, generally hollow, for listening to the sounds produced in the chest or other part of the body.

-Stich'ous (Gr. στιχος, *stich'os*, a row). A termination in compound words implying rows.

Sthen'ic (Gr. σθενος, *sthen'os*, strength). Attended with a morbid increase of vital action.

Stig'ma (Gr. στιζω, *sti'zō*, I prick or stick). In *botany*, the upper extremity of the pistil, or that part which receives the pollen; in the plural, *stig'mata*, it denotes the apertures in the body of insects communicating with the tracheæ or air-vessels.

Stigmat'ic (*Stigma*). Belonging to the stigma.

Stim'ulant (Lat. *stim'ulus*, a goad). In *medicine*, an article which produces a rapid and transient increase of vital energy.

Stim'ulus (Lat. a goad). In *medicine*, that which produces a rapid and transient increase of vital energy; in *botany*, a stinging hair.

Stipe (Lat. *sti'pes*, a stalk). In *botany*, applied to the stem of palms and ferns, and the stalk of agarics.

Stip'itate (Lat. *sti'pes*, a stalk). Supported on a stalk.

Stip'ular (*Stip'ule*). Resembling or consisting of stipules.

Stip'ulate (*Stip'ule*). Having stipules.

Stip'ule (Lat. *stip'ula*, a stem). In *botany*, a small leaf-like appendage to the leaf, commonly at the base of its stem.

Sto'lon (Lat. *sto'lo*, a sucker). In *botany*, a sucker, at first growing on the surface of the ground, then turning downwards and rooting.

Stolonif'erous (Lat. *sto'lo*, a sucker; *fer'o*, I produce). Producing suckers.

Stomap'oda (Gr. στομα, *stom'a*, a mouth; πους, *pous*, a foot). An order of crustacea, deriving its name from the manner in which the feet approach the mouth.

Stom'ata or **Stom'ates** (Gr. στομα, *stom'a*, a mouth). Opening between the cells of the epidermis of plants in parts exposed to the air.

Strabis'mus (Lat. *strab'o*, one who squints). Squinting; a want of coincidence in the axes of the eyes.

Strangulated (Lat. *stran'gulo*, I choke). Choked; in *surgery*, having the circulation stopped in any part.

Stratifica'tion (Lat. *stra'tum*, a layer; *fac'io*, I make). The process by which substances are formed into strata or layers; an arrangement in layers.

Stra'tiform (Lat. *stra'tum*, a layer; *for'ma*, shape). In the form of strata or layers.

Strat'ify (Lat. *stra'tum*, a layer; *fac'io*, I make). To arrange in layers.

Stra'tum (Lat. *ster'no*, I spread). A layer; in *geology*, applied to the layers in which rocks lie one above another.

Strepsip'tera (Gr. στρεφω, *streph'ō*, I turn; πτερον, *pter'on*, a wing). An order of insects in which the first pair of wings is represented by twisted rudiments.

Stri'æ (Plural of Lat. *stri'a*, a streak). Fine thread-like lines or streaks.

Stri'ated (Lat. *stri'a*, a streak). Marked with striæ or streaks, running parallel to one another.

Stri'dor (Lat.). A harsh creaking noise; a grinding.

Strigo'se (Lat. *strigo'sus*, lank, thin). Covered with rough, strong hairs, pressed together.

Strob'ile (Lat. *strob'ilus*, an artichoke). In *botany*, a large catkin, with scaly carpels bearing naked seeds, as the cone or fruit of the pine.

Strob'ilites (*Strob'ile;* Gr. λιθος, *lith'os*, a stone). Fossil remains of cone-like fruit.

Stro'phioles (Lat. *stroph'iolum*, a little garland). Small tumours or cellular bodies produced at various points on the coverings of seeds.

Stroph'ulus (Lat.). A papular eruption of various species and forms, occurring in infants.

Stru'ma. A diseased state, characterised by a tendency to the deposition of tubercle or of swelling of glands in various parts of the body; in *botany*, a cellular swelling where the leaf joins the midrib.

Stu'pose (Lat. *stu'pa*, tow). Having a tuft of hairs.

Style (Gr. στυλος, *stu'los*, a column). In *botany*, the part of the pistil consisting of the column proceeding upwards from the ovary and supporting the stigma.

Sty'liform (Lat. *sty'lus*, a pen or bodkin; *forma*, form). Resembling a style or pen; pointed.

Stylo- (Gr. στυλος, *stu'los*, a style or pen). In *anatomy*, a prefix in some compound words, denoting attachment to the styloid process of the temporal bone.

Sty'lobate (Gr. στυλος, *stu'los*, a pillar; βασις, *ba'sis*, a base). In *architecture*, generally, any basement on which columns are raised above the level of the ground; but especially applied to a continuous pedestal on which several columns are raised.

Stylohy'al (Gr. στυλος, *stu'los*, a style or pen; *hyoid* bone). A bone in the head of fishes, corresponding to the junction between the styloid process and hyoid bone.

Sty'loid (Gr. στυλος, *stu'los*, a style or pen; ειδος, *eidos*, shape). Like a style or pen: applied in *anatomy* to a process of the temporal bone.

Styp'tic (Gr. στυφω, *stu'phō*, I contract). Astringent: having the property of restraining bleeding.

Sub- (Lat. under). A preposition used in compound words, sometimes implying a lower position, sometimes a less or inferior degree.

Subac'id (Lat. *sub*, under; *acid*). Moderately acid.

Subal'tern (Lat. *sub*, under; *alter'nus*, alternating). In *logic*, applied to propositions which agree in quality but not in quantity.

Suba'queous (Lat. *sub*, under; *aq'ua*, water). Under water.

Subarach'noid (Lat. *sub*, under; *arach'noid*). Lying beneath the arachnoid membrane.

Subaxil'lary (Lat. *sub*, under; *axil'la*, an arm-pit). Placed under the axil or angle formed by a branch with the stem or by a leaf with the branch.

Subcar'bonate (Lat. *sub*, under; *car'bonate*). A salt containing less carbonic acid than a carbonate.

Subcar'buretted (Lat. *sub*, under; *carbon*). Containing less carbon than a carburet.

Sub'class (Lat. *sub*, under; *class*). A

subordinate class, consisting of orders allied to a certain extent.

Subcla'vian (Lat. *sub*, under ; *clavis*, a key). Lying under the clavicle or collar-bone.

Subcon'trary (Lat. *sub*, under ; *contra'rius*, contrary). Contrary in an inferior degree : in *geometry*, applied to similar triangles which have a common angle at the vertex, while the bases do not coincide ; in *logic*, applied to propositions which agree in quantity but differ in quality.

Subcor'date (Lat. *sub*, under ; *cor*, a heart). Somewhat like a heart in shape.

Subcos'tal (Lat. *sub*, under ; *cos'ta*, a rib). Under or within the rib.

Subcuta'neous (Lat. *sub*, under ; *cu'tis*, the skin). Under the skin.

Subcutic'ular (Lat. *sub*, under ; *cutic'ula*, the cuticle). Under the cuticle or scarf-skin.

Subcylin'drical (Lat. *sub*, under ; *cylin'drical*). Not perfectly cylindrical.

Subdu'plicate (Lat. *sub*, under ; *du'plex*, double). Having the ratio of the square roots : in *mathematics*, applied to the ratio which the square roots of two quantities have to each other.

Su'berate (Lat. *su'ber*, cork). A compound of suberic acid with a base.

Su'beric (Lat. *su'ber*, cork). Belonging to cork : applied to an acid produced by the action of nitric acid on cork and fatty bodies.

Sub'erose (Lat. *sub*, under ; *ero'do*, I gnaw). Appearing as if a little gnawed.

Sub'genus (Lat. *sub*, under ; *gen'us*). A subordinate genus, consisting of species allied to a certain extent.

Subglob'ular (Lat. *sub*, under ; *glob'ular*). Having a form approaching to globular.

Subgran'ular (Lat. *sub*, under ; *gran'ular*). Somewhat granular.

Subja'cent (Lat. *sub*, under ; *jac'eo*, I lie). Lying under or in a lower situation.

Subject (Lat. *subjic'io*, I place before). In *grammar* and *logic*, that regard-

ing which anything is affirmed or denied ; in *intellectual philosophy*, the personality of the thinker.

Subjec'tive (*Subject*). Relating to the subject ; applied in *philosophy* to the manner in which an object is conceived of by an individual subject ; in *medicine*, to symptoms observed by the patient himself.

Subjunc'tive (Lat. *sub*, under ; *jungo*, I join). Subjoined or added to something else ; in *grammar*, applied to a form of the verb expressing condition or supposition.

Sub'limate (Lat. *subli'mis*, exalted). To bring a solid substance by heat into the state of vapour, which condenses on cooling ; the substance produced by this process.

Sublima'tion (Lat. *subli'mo*, I raise up). The process of bringing solid substances by heat into the state of vapour which is condensed in cooling.

Sublime. *See* Sublimate.

Sublin'gual (Lat *sub*, under ; *lin'gua*, the tongue). Under the tongue.

Subluxa'tion (Lat. *sub*, under ; *luxa'tion*). An incomplete luxation or dislocation.

Submari'ne (Lat. *sub*, under ; *ma're*, the sea). Formed or lying beneath the sea.

Submaxil'lary (Lat. *sub*, under ; *maxil'la*, the jaw). Lying beneath the jaw.

Submen'tal (Lat. *sub*, under ; *mentum*, the chin). Under the chin.

Submu'cous (Lat. *sub*, under ; *mucous*). Lying beneath the mucous membrane.

Submul'tiple (Lat. *sub*, under ; *mul'tiple*). A quantity which is contained in another an exact number of times.

Subnas'cent (Lat. *sub*, under ; *nas'cor*, I am born). Growing underneath.

Subnor'mal (Lat. *sub*, under ; *norma*, a rule). In *conic sections*, the portion of a diameter intercepted between the ordinate and the normal.

Suboccip'ital (Lat. *sub*, under ; *oc'ciput*, the back of the head). Under or beneath the occiput.

Subœsophage'al (Lat. *sub*, under; *œsoph'agus*.) Beneath the œsophagus or gullet.

Suborbic'ular (Lat. *sub*, under; *orbic'ular*). Almost orbicular.

Subor'bital (Lat. *sub*, under; *or'bita*, the orbit). Applied to bones developed in the integument about the lower part of the orbit in fishes.

Sub'order (Lat. *sub*, under; *order*). A subdivision of an order, consisting of a number of allied genera.

Subor'dinate (Lat. *sub*, under; *ordo*, an order). In *geology*, inferior in the order of superposition.

Subo'val (Lat. *sub*, under; *oval*). Somewhat oval.

Subo'vate (Lat. *sub*, under; *o'vum*, an egg). Nearly in the shape of an egg.

Subox'ide (Lat. *sub*, under; *ox'ide*). An oxide containing a smaller proportion of oxygen than that in which the basic characters are most marked.

Subperitone'al (Lat. *sub*, under; *peritone'um*). Lying beneath the peritoneal membrane.

Sub'plinth (Lat. *sub*, under; *plinth*.) A plinth placed under the principal one.

Subro'tund (Lat. *sub*, under; *rotun'dus*, round). Nearly round.

Subsalt (Lat. *sub*, under; *salt*). A salt having an excess of the base.

Subscap'ular (Lat. *sub*, under; *scap'ula*, the shoulder-blade). Lying under the shoulder-blade, between it and the chest.

Subse'rous (Lat. *sub*, under; *serous*). Lying beneath a serous membrane.

Sub'soil (Lat. *sub*, under; *soil*). The bed or layer of earth which lies under the surface-soil, and on the base of rocks on which the whole rests.

Subspe'cies (Lat. *sub*, under; *spe'cies*). A subordinate species.

Substra'tum (Lat. *sub*, under; *stra'tum*). A stratum or layer lying under another.

Subsul'phate (Lat. *sub*, under; *sulphate*). A sulphate with excess of the base.

Subsul'tus (Lat. *sub*, under; *saltus*, a leaping). A twitching or convulsive motion.

Subtan'gent (Lat. *sub*, under; *tangent*). The segment of a produced or lengthened diameter or axis, intercepted between an ordinate and a tangent drawn from the same point in the curve.

Subtend' (Lat. *sub*, under; *tendo*, I stretch). To extend under or opposite to.

Subtrip'licate (Lat. *sub*, under; *triplex*, three-fold). In the ratio of the cube roots; in *mathematics*, the subtriplicate ratio of two quantities is the ratio which their cube roots have to each other.

Su'bulate (Lat. *su'bula*, an awl). Shaped like an awl.

Suc'cinate (Lat. *suc'cinum*, amber). A compound of succinic acid with a base.

Succin'ic (Lat. *suc'cinum*, amber). Belonging to amber; applied to an acid obtained from amber.

Suc'culent (Lat. *succus*, juice). Full of juice; applied to plants which have a juicy and soft stem or leaves.

Succus (Lat.) Juice.

Suc'tion (Lat. *sugo*, I suck). The act of sucking or drawing in fluid substances by removing the pressure of the air.

Sucto'rial (Lat. *sugo*, I suck). Fitted for sucking.

Sudorif'erous (Lat. *su'dor*, sweat; *fer'o*, I bear). Conducting perspiration.

Sudorif'ic (Lat. *su'dor*, sweat; *fac'io*, I make). Causing sweat or perspiration.

Sudorip'arous (Lat. *su'dor*, sweat; *par'io*, I produce). Producing or secreting perspiration.

Suffru'ticose (Lat. *sub*, under; *fru'tex*, a shrub). Partly shrubby: permanent or woody at the base, but decaying yearly above.

Sugil'lation (Lat. *sugil'lo*, I make black and blue). The mark left by a leech or cupping-glass; applied also to livid spots noticed on dead bodies.

Sul'cate (Lat. *sul'cus*, a furrow).

N

Furrowed ; deeply marked with longitudinal lines.

Sul'phate (*Sul'phur*). A compound of sulphuric acid with a base.

Sul'phide (*Sul'phur*). A compound of sulphur with another elementary substance, towards which it stands in the same relations as oxygen, so as to form a sulphur-acid or a sulphur-base.

Sul'phite (*Sul'phur*). A compound of sulphurous acid with a base.

Sulphocyan'ic (*Sul'phur* and *Cyan'ogen*). A name applied to an acid composed of sulphur, cyanogen, and hydrogen, found in the seeds and blossoms of cruciferous plants, and in human saliva.

Sulphovi'nic (*Sul'phur ;* Lat. *vi'num*, wine). A term applied to an acid produced by the action of sulphuric acid on alcohol.

Sulphur-acid. An acid in which the oxygen is represented by sulphur.

Sulphur-base. A base in which oxygen is represented by sulphur.

Sul'phuret (*Sul'phur*). A compound of sulphur with hydrogen or a metal, or other electro-positive body.

Sulph'uretted (*Sul'phur*). Combined with sulphur.

Sulphu'ric (*Sul'phur*). Belonging to sulphur : applied to an acid containing one equivalent of sulphur with three of oxygen commonly known as oil of vitriol.

Sul'phurous (*Sul'phur*). Containing sulphur ; applied to an acid containing one equivalent of sulphur and two of oxygen.

Sulphur-salt. A salt arising from the combination of a sulphur acid with a sulphur base, in each of which sulphur takes the place of oxygen.

Super- (Lat. above). A preposition used in compound words, signifying above or in excess.

Supercil'iary (Lat. *su'per*, above ; *ci'lium*, the eyebrow). Above the eyebrow.

Superfic'ial (Lat. *su'per*, above ; *fac'ies*, a face). On the face or outer surface ; superficial measure is the extent of any surface.

Superfic'ies (Lat. *su'per*, on ; *fac'ies*, a face). The surface of a body, capable of measurement in length and breadth.

Superimpo'se (Lat. *su'per*, above ; *impo'no*, I lay on). To lay on something else.

Superincum'bent (Lat. *su'per*, above; *incum'bo*, I lie on). Resting or lying on something.

Supe'rior (Lat. above). In *botany*, applied to the ovary when it is not adherent to the calyx, and to the calyx when it is adherent to the ovary ; also to the part of a flower nearest the axis or growing point.

Superja'cent (Lat. *su'per*, above ; *ja'ceo*, I lie). Lying above.

Superna'tant (Lat. *su'per*, above ; *na'to*, I swim). Floating or swiming on the surface.

Superposit'ion (Lat. *su'per*, above ; *po'no*, I place). A placing above ; in *geology*, the order in which rocks are placed over each other.

Su'persalt (Lat. *su'per*, above; *salt*). A salt with a greater number of equivalents of acid than of base.

Supersat'urate (Lat. *su'per*, above ; *sa'tur*, full). To add beyond saturation.

Superstra'tum (Lat. *su'per*, above ; *stra'tum*, a layer). A layer above another.

Supersul'phate (Lat. *su'per*, above ; *sulphate*). A sulphate containing more equivalents of acid than of base.

Supertem'poral (Lat. *su'per*, over ; *temporal-bone*). Applied to bones sometimes overarching the temporal fossæ in fishes.

Supervolu'te (Lat. *su'per*, above ; *volvo*, I roll). In *botany*, applied to leaves rolled on themselves in the leaf-bud.

Supina'tion (Lat. *supi'nus*, lying on the back). The act of turning the face or anterior part upwards.

Supina'tor (Lat. *supi'nus*, lying on the back). A name given to those muscles which turn the palm of the hand forwards or upwards.

Sup'plement (Lat. *sub*, under ; *pleo*, I fill). That which fills up the defects of any thing ; in *geometry*, the quantity by which an arc or angle falls short of 180 degrees or a semicircle.

Suppura'tion (Lat. *suppu'ro*, I turn into pus). The process of the formation of pus as a result of inflammation.

Su'pra- (Lat. *su'pra*, over). A preposition used in compound words, signifying over.

Supra-acro'mial (Lat. *su'pra*, above ; *acro'mion*). Lying above the acromion process of the scapula.

Supracreta'ceous (Lat. *su'pra*, over ; *cre'ta*, chalk). Applied to deposits lying over the chalk formation.

Supradecom'pound (Lat. *su'pra*, above ; *decom'pound*). In *botany*, applied to minutely divided or very compound leaves.

Suprafolia'ceous (Lat. *su'pra*, over ; *fo'lium*, a leaf). Inserted above a leaf or petiole.

Supracesophage'al (Lat. *su'pra*, over ; *æsopha'gus*). Above the œsophagus.

Supraoccip'ital (Lat. *su'pra*, above ; *oc'ciput*, the back of the head). A bone in the head of fishes, corresponding to the upper part of the occipital bone.

Supraor'bital (Lat. *su'pra*, over ; *or'bit*). Above the orbit or eye-socket.

Suprare'nal (Lat. *su'pra*, over ; *ren*, a kidney). Above the kidneys.

Suprascap'ular (Lat. *su'pra*, over ; *scap'ula*, the shoulder-blade). Above the shoulder-blade.

Supraspina'tus (Lat. *su'pra*, above ; *spina*, a spine). Above the spine : a name given to a muscle lying above the spine of the shoulder-blade.

Su'ral (Lat. *su'ra*, the calf of the leg). Belonging to the calf of the leg.

Surd (Lat. *sur'dus*, deaf). In *arithmetic* and *algebra*, a root which cannot be expressed in integral or rational numbers.

Suspen'sion (Lat. *suspen'do*, I hang up). In *chemistry*, the state in which bodies are held, but not in solution, in a fluid, so that they may be separated from it by filtration.

Suspen'sor (Lat. *suspend'o*, I hang). In *botany*, the cord which suspends the embryo, and is attached to the young radicle.

Sutu'ral (Lat. *sutu'ra*, a suture). Belonging to sutures ; in *botany*, applied to that form of dehiscence or separation of fruits which takes place at the sutures.

Su'ture (Lat. *suo*, I sew). A sewing : in *surgery*, the drawing together of a wound by sewing ; in *anatomy*, a seam or joint uniting the bones of the skull ; in *botany*, the part where separate organs unite, or where the edges of a folded organ adhere : the dental suture of the ovary is that next the centre, formed by the edges of the carpels : the dorsal suture is at the back, corresponding to the midribs.

Syco'sis (Gr. συκον, *su'kon*, a fig). A form of eruptive disease, affecting the skin of the chin, lower jaw, or upper lip, characterised by the formation of patches of tubercles.

Syllab'ic (Gr. συλλαβη, *sul'labē*, a syllable). In *grammar*, applied to the augment in the past tense of Greek verbs, which is formed by the addition of the vowel ε, so as to produce a new syllable.

Syl'lable (Gr. συν, *sun*, together ; λαμβανω, *lam'banō*, I take). A letter or combination of letters that can be uttered by a single effort of the voice.

Syl'logism (Gr. συν, *sun*, with ; λογι-ζομαι, *logi'zomai*, I think). In *logic*, an argument consisting of three terms, of which the first two are premises, and the last the conclusion.

Syllogis'tic (Gr. συν, *sun*, with ; λογιζομαι, *logi'zomai*, I think). Belonging to or in the form of syllogisms.

Symbleph'aron (Gr. συν, *sun*, with ; βλεφαρον, *bleph'aron*, an eyelid). A growing of the eyelids to the eyeball.

Sym'bol (Gr. συμβαλλω, *sumbal'lō*, I

compare). A visible object or character representing something.

Sym'metry (Gr. συν, *sun*, with ; μετρον, *met'ron*, a measure). The due proportion of one thing, as part, to another with respect to the whole ; in *botany*, applied in reference to the parts being of the same number, or multiples of each other.

Sympathet'ic (Gr. συν, *sun*, with ; παθος, *path'os*, suffering). Having common feeling ; in *anatomy*, applied to a system of nerves which are specially supplied to the viscera, and blood-vessels.

Sym'pathy (Gr. συν, *sun*, with ; παθος, *path'os*, suffering). Fellow-feeling : in *medicine*, applied to the production of a modified or diseased condition in an organ or part through action or a disease of some other organ or part.

Sym'phony (Gr. συν, *sun*, with ; φωνη, *phōnē*, voice). A consonance or harmony of sounds : a musical composition for a full band of instruments.

Sym'physis (Gr. συν, *sun*, together ; φυω, *phu'ō*, I grow). In *anatomy*, the union of bones by means of an intervening cartilage, so as to form an immovable joint ; applied also to the junction of the two halves of the lower jaw.

Sympesiom'eter (Gr. συμπιεζω, *sumpie'zō*, I press together ; μετρον, *met'ron*, a measure). An instrument for measuring the weight of the atmosphere by the compression of a column of gas.

Symp'tom (Gr. συν, *sun*, with ; πιπτω, *pip'tō*, I fall). Something that happens concurrently with another ; in *medicine*, a disordered function, or assemblage of disordered functions, becoming obvious in the course of a disease.

Symptomat'ic (*Symptom*). Belonging or according to symptoms ; produced from some apparent prior disorder or injury.

Symptomatol'ogy (Gr. συμπτωμα, *sump'tōma*, a symptom ; λογος, *log'os*, a discourse). The part of medicine which treats of symptoms.

Syn- or **Sym-** (Gr. συν, *sun*, with). A prefix in compound words signifying with.

Synæ'resis (Gr. συν, *sun*, with ; αιρεω, *hai'reō*, I take). A combination of two vowels into one.

Synalœ'pha (Gr. συν, *sun*, with ; αλειφω, *alei'phō*, I oil or anoint). In *prosody*, the process by which, when one word ends and the next begins with a vowel, the vowel of the first word is cut off, or absorbed in that of the second.

Synan'therous (Gr. συν, *sun*, with ; *anther*). Having the anthers widest in a tube round the style ; applied to some composite plants.

Synarthro'sis (Gr. συν, *sun*, together ; αρθρον, *arthron*, a joint). An immovable joint.

Syncar'pous (Gr. συν, *sun*, with ; καρπος, *karpos*, fruit). Having the carpels of a compound fruit completely united.

Synchondro'sis (Gr. συν, *sun*, with ; χονδρος, *chon'dros*, a cartilage). An articulation by cartilage ; applied especially to the joint formed by the sacrum with the ilium on each side.

Synchron'ic (Gr. συν, *sun*, with ; χρονος, *chron'os*, time). Happening at the same time ; performed in the same time.

Syn'chronous. *See* Synchron'ic.

Syncli'nal (Gr. συν, *sun*, with ; κλινω, *kli'nō*, I lean). In *geology*, applied to strata that dip from opposite directions downwards, or which incline to a common centre.

Syn'cope' (Gr. συν, *sun*, with ; κοπτω, *kop'tō*, I cut). A cutting off ; in *medicine*, fainting ; interruption of the action of the heart.

Syndesmol'ogy (Gr. συνδεσμος, *sundes'mos*, a ligament ; λογος, *log'os*, discourse). A treatise on ligaments.

Syndesmo'sis (Gr. συν, *sun*, with ; δεσμος, *des'mos*, a binding). The union of bones by ligaments.

Synec'doche (Gr. συν, *sun*, with ; εκδεχομαι, *ekdech'omai*, I take out). A figure in speech by which the whole is put for a part, or a part for the whole.

Syne'chia (Gr. συν, *sun*, with ; ἐχω, *ech'ō*, I hold). In *surgery*, an adhesion of the iris of the eye to the cornea or to the capsule of the crystalline lens.

Syngene'sia (Gr. συν, *sun*, with ; γενεσις, *gen'esis*, production). A term applied to a class of plants in the Linnæan system, in which the anthers are united, the filaments being mostly separate.

Syn'ocha (Gr. συνοχος, *sun'ochos*, holding together). A name formerly given to inflammatory fever.

Syno'chreate (Gr. συν, *sun*, together; Lat. *o'chrea*, a boot). In *botany*, applied to stipules which unite round the stem, on the opposite side from the leaf.

Syn'ochus (Gr. συνοχος, *sun'ochos*, holding together). A name formerly given to a mixed form of fever, intermediate between synochus and typhus.

Synod'ic (Gr. συν, *sun*, with ; ὁδος, *hod'os*, a way). In *astronomy*, applied to the common lunar month, or the period of time which the moon takes in returning to any given phase ; also to the motion of a planet considered merely in relation to that of the earth, without reference to its actual position in its orbit.

Syn'onym (Gr. συν, *sun*, with ; ὀνομα, *on'oma*, a name). A word having the same signification as another.

Synop'sis (Gr. συν, *sun*, with ; ὀψις, *opsis*, sight). A general view.

Synop'tic (Gr. συν, *sun*, with ; ὀψις, *opsis*, sight). Taking in at one view.

Syno'via (Gr. συν, *sun*, with ; Lat. *o'vum*, an egg). A fluid resembling the white of egg, secreted in the cavity of joints for the purpose of moistening them and facilitating motion.

Synovi'tis (*Syno'via ; itis*, denoting inflammation). Inflammation of a synovial membrane.

Syn'tax (Gr. συν, *sun*, together ; τασσω, *tassō*, I arrange). A connected system or order ; in *grammar*, the part which teaches the arrangement and connection of words.

Syn'thesis (Gr. συν, *sun*, together ; τιθημι, *tithēmi*, I place). A putting together ; the uniting of separate elements or constituents into a compound.

Synthet'ic (Gr. συν, *sun*, together ; τιθημι, *tithēmi*, I place). Relating to synthesis or composition.

Syn'tonin (Gr. συντονος, *sun'tonos*, stretched). Fibrin of muscle or flesh.

Sys'tem (Gr. συν, *sun*, together ; ἱστημι, *histēmi*, I place). A combination of things taken together ; a classification, real or theoretical, of parts or objects.

Systemat'ic (*System*). Formed according to a regular connection.

System'ic (*System*). Belonging to a system ; in *physiology*, relating to the system, or assemblage of organs of the body in general.

Sys'tole (Gr. συστελλω, *sustel'lō*, I contract). In *grammar*, the shortening of a long syllable ; in *physiology*, the contraction of the heart for carrying on the circulation.

Syz'ygy (Gr. συν, *sun*, with ; ζυγον, *zu'gon*, a yoke). A conjunction or coupling ; in *astronomy*, the line of syzygies is the diameter of the moon's orbit which is directed to the sun, its extremes being the points of conjunction and of opposition.

T.

Tabes (Lat.) A wasting.

Tab'ular (Lat. *tab'ula*, a table). In the form of a table ; arranged in laminæ or plates.

Tac'tile (Lat. *tactus*, touch). Relating to, or employed for, touch.

Tœ'nia (Gr. ταινια, *tainia*, a ribbon). The tape-worm.

Tænioid (Gr. ταινια, *tainia*, a ribbon ; ειδος, *eidos*, a form). Shaped like a ribbon, as the tapeworm.

Talc. A mineral consisting of magnesia, potash, and silica, closely resembling mica, arranged in broad, flat, smooth plates, translucent and often transparent.

Taliaco'tian Operation. In *surgery*, the operation of forming a new nose from the skin of the forehead or other part of the face.

Ta'lipes (Lat. *talus*, an ankle ; *pes*, a foot). A deformity known as club-foot.

Tan'gent (Lat. *tango*, I touch). In *geometry*, a straight line which touches a circle or curve in one point, and which, being produced, does not cut it.

Tan'nic (*Tan*). Applied to an acid existing in oak-bark, and in which depends its efficacy in tanning.

Tape'tum (Lat. a carpet). The coloured layer of the choroid coat of the eye.

Taphren'chyma (Gr. ταφρος, *taph'ros*, a ditch ; εγχυμα, *en'chuma*, tissue). A name for pitted vessels in vegetables.

Tar'digrade (Lat. *tar'dus*, slow ; *grad'ior*, I step). Advancing slowly.

Tarsal (*Tarsus*). Belonging to the instep, or to the cartilage of the eyelid.

Tar'sus (Gr. ταρσος, *tarsos*, a flat surface). The instep ; the cartilage supporting each eyelid ; also the last segment of the legs of insects.

Tartar'ic (*Tartar*, a deposit from wines). Applied to an organic acid which exists in tartar, and which is found in the juice of grapes and other fruits.

Tartari'sed (*Tartar*). Impregnated with tartar.

Tar'trate (*Tartar*). A neutral compound of tartaric acid with a base.

Tau'rine (Lat. *taurus*, a bull). Relating to a bull.

Taxider'my (Gr. τασσω, *tasso*, I put in order ; δερμα, *derma*, skin).

The art of preparing and preserving the skins of animals in their natural appearance.

Taxis (Gr. τασσω, *tasso*, I put in order). In *surgery*, a process by which parts that have left their proper situation are replaced by the hand without the aid of instruments.

Taxon'omy (Gr. ταξις, *taxis*, order ; νομος, *nom'os*, law). The department of natural history which treats of the laws and principles of classification.

Tech'nical (Gr. τεχνη, *techne*, art). Relating or belonging to a science or art.

Technol'ogy (Gr. τεχνη, *techne*, art ; λογος, *log'os*, discourse). A description of arts or of the terms used in arts.

Tectibran'chiate (Lat. *tectus*, covered; Gr. βραγχια, *bran'chia*, gills). Having covered gills ; applied to mollusca in which the gills are covered by the mantle.

Teg'men (Lat. *teg'o*, I cover). *See* Tegument.

Tegmen'tum (Lat. *teg'o*, I cover). The scaly coat covering the leaf-buds of deciduous trees.

Teg'ument (Lat. *teg'o*, I cover). A covering ; in *anatomy*, the skin ; in *botany*, see Tegmentum ; in *entomology*, the covering of the wings of the orthoptera, or straight-winged insects.

Tegumen'tary (*Tegument*). Belonging to or consisting of teguments or coverings.

Telangiec'tasis (Gr. τελος, *tel'os*, an end ; αγγειον, *angei'on*, a vessel ; εκτεινω, *ektein'o*, I stretch out). Distension of the vessels.

Tel'egram (Gr. τηλε, *tele*, at a distance ; γραφω, *graph'o*, I write). A message communicated by a telegraph.

Tel'egraph (Gr. τηλε, *tele*, at a distance ; γραφω, *graph'o*, I write). An instrument for communicating messages or news from a distance by means of signals representing letters or words : to transmit by means of a telegraph.

Telegraph'ic (*Tel'egraph*). Belonging to, or communicated by, a telegraph.

Telen'giscope (Gr. τηλε, *tēle*, far off; ἐγγυς, *en'gus*, near; σκοπεω, *skop'eō*, I look). An instrument combining the powers of the telescope and microscope.

Tel'escope (Gr. τηλε, *tēle*, at a distance; σκοπεω, *skop'eō*, I view). An optical instrument for viewing objects at a distance.

Telescop'ic (*Telescope*). Belonging to or seen by a telescope.

Tellu'ric (Lat. *tellus*, the earth). Belonging to or proceeding from the earth.

Tellu'ric (*Tellu'rium*, a kind of metal). Belonging to tellurium; applied to an acid consisting of tellurium and oxygen.

Tem'perament (Lat. *tem'pero*, I mix). Constitution; in *physiology*, a term applied to peculiar characters of the human body in health, each of which is specially liable to certain forms of disease.

Tem'perature (Lat. *tem'pero*, I mix or moderate). The state of a body with regard to heat and cold, especially as compared with another substance.

Tem'poral (Lat. *tem'pora*, the temples). In *anatomy*, belonging to the temples.

Tem'poral (Lat. *tempus*, time). In *grammar*, applied to a form of augment in the past tense of verbs, by which a short vowel is changed into a long one.

Tenac'ity (Lat. *tenax*, holding). The property which makes bodies adhere; in *physics*, the property by which a body resists the separation of its parts by extension in the direction of its length.

Tenac'ulum (Lat. *ten'eo*, I hold). An instrument used in surgery for laying hold of arteries or other parts in operating.

Ten'don (Gr. τενων, *ten'ōn*). The dense fibrous structure in which a muscle ends, and by which it is attached to bone.

Ten'on (Fr. from Lat. *ten'eo*, I hold).

In *architecture*, the end of a piece of wood cut into a rectangular prism, and received into a cavity in another piece called a mortise.

Tenot'omy (Gr. τενων, *tenōn*, a tendon; τεμνω, *temnō*, I cut). The operation of dividing a tendon.

Ten'sion (Lat. *tendo*, I stretch). The art of stretching, or the state of being stretched or strained.

Ten'tacle or **Tentac'ulum** (Lat. *ten'to*, I feel or try). A feeler: a thread-like organ, simple or branched, seated about the mouth or other part of the body of many invertebrate animals.

Tentaculif'erous (Lat. *tentac'ulum*, a feeler; *fer'o*, I bear). Producing or having tentacles.

Ten'tative (Lat. *ten'to*, I try). Experimental.

Tento'rium (Lat. *ten'do*, I stretch). In *anatomy*, a projecting of the dura mater, separating the cerebrum from the cerebellum.

Tenuiros'tral (Lat. *ten'uis*, thin; *ros'trum*, a beak). Having a slender beak, as the humming-bird, &c.

Tenu'ity (Lat. *ten'uis*, thin.) Thinness.

Tepida'rium (Lat. *tep'eo*, I am hot). The part of the ancient bath in which the garments were removed, before the sweating process commenced.

Teratol'ogy (Gr. τερας, *ter'as*, a monster: λογος, *log'os*, discourse). The study of monstrosities, or departures from the normal forms of beings.

Ter'cine (Lat. *ter'tius*, third). In *botany*, the innermost coat of an ovule.

Terebin'thinate (Gr. τερεβινθος, *terebin'thos*, turpentine). Belonging to or having the properties of turpentine.

Ter'es (Lat. round). In *anatomy*, applied to certain muscles, from their shape.

Te'rete (Lat. *te'res*, round). Cylindrical and tapering.

Ter'gal (Lat. *ter'gum*, the back). Belonging to the back.

Tergem'inal (Lat. *ter*, three times ; *gem'inus*, double). Thrice double.

Tergif'erous (Lat. *ter'gum*, the back ; *fer'o*, I bear). Bearing on the back ; applied to plants which bear their seeds on the back of the leaves, as ferns.

Ter'minal (Lat. *ter'minus*, a limit). Belonging to or placed at the end of an object.

Terminol'ogy (Lat. *ter'minus*, a term ; Gr. λογος, *log'os*, a discourse). The branch of a science or art which defines and explains the words and phrases used therein.

Ter'nary (Lat. *ter'ni*, three and three). Arranged in threes.

Ter'nate (Lat. *ter'ni*, three and three). In *botany*, applied to leaves having three leaflets on one stem.

Ter'ra (Lat.) The earth ; an earth, or earthy substance.

Terra'queous (Lat. *ter'ra*, earth ; *aq'ua*, water). Consisting of land and water.

Ter'reous (Lat. *ter'ra*, earth). Earthy.

Ter'tian (Lat. *ter'tius*, third) Occurring every third day.

Ter'tiary (Lat. *ter'tius*, third). Of the third order : in *geology*, a term applied to the formations above the chalk.

Tes'selated (Lat. *tes'sela*, a cube, or die). Formed in little squares, like a chess-board.

Test (Lat. *tes'tis*, a witness). In *chemistry*, a substance employed to detect the presence of any ingredient in a compound.

Tes'ta (Lat.) A shell; in *botany*, the outer covering of the seed ; sometimes applied to the coverings taken together.

Testa'ceous (Lat. *tes'ta*, a shell). Belonging to or having shells.

Testu'dinate (Lat. *testu'do*, a tortoise). Arched ; like the back of a tortoise.

Tetan'ic (*Tet'anus*). Belonging to or denoting tetanus.

Tet'anoid (*Tetanus* ; Gr. ειδος, *ei'dos*, shape). Resembling tetanus.

Tet'anus (Gr. τεινω, *tei'no*, I stretch). A disease characterised by violent and continued contraction of the muscles.

Tet'ra- (Gr. τεσσαρες, *tes'sares*, or τετταρες, *tet'tares*, four). A prefix in compound words, signifying four.

Tetrabran'chiate (Gr. τετρα, *tet'ra*, four ; βραγχια, *bran'chia*, gills). Having four gills ; applied to an order of cephalopods.

Tetracan'thous (Gr. τετρα, *tet'ra*, four ; ἀκανθα, *akan'tha*, a spine). Having four spines or thorns.

Tetrachot'omous (Gr. τετραχως, *tet'rachōs*, fourfold ; τεμνω, *tem'nō*, I cut). Branching in fours.

Tetradac'tylous (Gr. τετρα, *tet'ra*, four ; δακτυλος, *dak'tulos*, a finger, or toe). Having four toes.

Tetradynam'ia (Gr. τετρα, *tet'ra*, four ; δυναμις, *du'namis*, strength). A class of plants in the Linnæan system, having six stamens, of which four are longer than the other two.

Tet'ragon (Gr. τετρα, *tet'ra*, four ; γωνια, *gō'nia*, an angle). A figure having four angles; especially a square.

Tetrag'onal (Gr. τετρα, *tet'ra*, four ; γωνια, *gō'nia*, an angle). Belonging to a tetragon ; in *botany*, having four angles, the faces being convex.

Tetragyn'ia (Gr. τετρα, *tet'ra*, four ; γυνη, *gunē*, a female). An order of plants in the Linnæan system, having four pistils.

Tetrahed'ron (Gr. τετρα, *tet'ra*, four ; ἑδρα, *hed'ra*, a base). A figure bounded by four equilateral and equal triangles ; a triangular pyramid, with four equal and equilateral faces.

Tetrahexahed'ron (Gr. τετρα, *tet'ra*, four ; ἑξ, *hex*, six ; ἑδοα, *hed'ra*, a base). A solid bounded by twenty-four equal faces.

Tetram'erous (Gr. τετρα, *tet'ra*, four ; μερος, *mer'os*, a part). Consisting of four parts.

Tetran'dia (Gr. τετρα, *tet'ra*, four; ἀνηρ, *anēr*, a male). A class of plants in the Linnæan system, having four stamens.

Tetrapet'alous (Gr. τετρα, *tet'ra*, four ; πεταλον, *pet'alon*, a petal). Having four petals.

Tetraphyl'lous (Gr. τετρα, *tet'ra*, four; φυλλον, *phullon*, a leaf). Having four leaves.

Tetrap'odous (Gr. τετρα, *tet'ra*, four ; πους, *pous*, a foot). Having four feet.

Tetrap'terous (Gr. τετρα, *tet'ra*, four ; πτερον, *pter'on*, a wing). Having four wings.

Tetrap'tote (Gr. τετρα, *tet'ra*, four ; πτωσις, *ptōsis*, case). In *grammar*, a noun having four cases.

Tetraquet'rous (Gr. τετρα, *tet'ra*, four ; Lat *quad'ra*, a square). In *botany*, having four angles, the faces being concave.

Tetrasep'alous (Gr. τετρα, *tet'ra*, four ; *sepal*). Having four sepals.

Tetrasper'mous (Gr. τετρα, *tet'ra*, four ; σπερμα, *sper'ma*, seed). Having four seeds.

Tetrasyllab'ic (Gr. τετρα, *tet'ra*, four; συλλαβη, *sul'labē*, a syllable). Having four syllables.

Tetrathe'cal (Gr. τετρα, *tet'ra*, four ; θηκη, *thēkē*, a case). Having four thecæ, or loculaments.

Tet'rodon (Gr. τετρα, *tet'ra*, four; οδους, *od'ous*, a tooth). A genus of fishes having four large teeth.

Textile (Lat. *texo*, I weave). Woven, or capable of being woven.

Texture (Lat. *texo*, I weave). In *anatomy*, a name applied to the solid constituents of the body.

Thalamiflo'ral (Gr. θαλαμος, *thal'amos*, a bed : Lat. *flos*, a flower). A subclass of exogenous plants, in which the parts of the flower are inserted separately into the thalamus or receptacle.

Thal'amus (Gr. θαλαμος, *thal'amos*, a bed). In *anatomy*, a name given to a part of the brain from which the optic nerve is partly derived ; in *botany*, the receptacle of the flower, or part of the stem from which the flower grows.

Thal'logen (*Thallus* ; Gr. γενναω, *genna'ō*, I produce). Producing a thallus.

Thal'lophyte (*Thallus* ; Gr. φυτον,

phu'ton, a plant). A plant producing a thallus.

Thal'lus (Gr. θαλλος, *thal'lus*, a bough). In *botany*, the cellular expansion in cryptogamic plants bearing the analogues of fruit.

Thau'matrope (Gr. θαυμα, *thauma*, a wonder ; τρεπω, *trep'ō*, I turn). An optical toy, consisting of a disc having on successive divisions of its circumference pictures representing figures in a succession of different positions in performing some action, so that, when the disc is caused to revolve, the impressions made by figures on the eye remain and are combined, and the figure appears to pirouette before the eye.

The'ca (Gr. θηκη, *thēkē*, a sheath or case). In *botany*, the case containing the reproductive matter in some flowerless plants : in *anatomy*, a strong fibrous sheath, enclosing certain soft parts, as the spinal cord.

The'caphore (Gr. θηκη, *thēkē*, a sheath ; φερω, *pher'ō*, I bear). The roundish stalk on which the ovary of some plants is elevated.

Thecas'porous (Gr. θηκη, *thēkē*, a sheath ; σπορα, *spor'a*, a seed). Applied to fungi which have the spores in thecæ or cases

The'codonts (Gr. θηκη, *thēkē*, a sheath ; οδους, *od'ous*, a tooth). A tribe of extinct lizard-like reptiles having the teeth implanted in sockets.

Theod'olite (Perhaps Gr. θεαομαι, *thea'omai*, I view ; δολος, *dol'os*, stratagem). A surveying instrument for measuring horizontal angles, or the angular distance between objects projected on the plane of the horizon.

Theog'ony (Gr. Θεος, *Theos*, God ; γινομαι, *gin'omai*, I am born). The part of mythology which treats of the genealogy of heathen deities.

Theol'ogy (Gr. Θεος, *The'os*, God ; λογος, *log'os*, a discourse). Divinity ; the science of God and divine things.

The'orem (Gr. θεωρεω, *theūreō*, I see). In *mathematics*, a proposition to be proved by a chain of reasoning, the conclusion being stated ; in *arithmetic* and *algebra*, sometimes used to denote a rule.

Theoret'ical (*The'ory*). Pertaining to, or depending on, theory or speculation ; not practical.

The'orize (*The'ory*).   To form a theory ; to speculate.

The'ory (Gr. θεωρεω, *theōreō*, I see). A doctrine or scheme of things which terminates in speculation without a view to practice. An exposition of the general principles of a science, and the rules derived therefrom, as distinguished from an art : as the *theory* and practice of medicine : as distinguished from hypothesis, it means an explanation of phenomena founded on principles established on independent evidence, while an hypothesis is a proposition assumed to account for certain phenomena, and having no other evidence of truth than in giving a satisfactory explanation of the phenomena.

Therapeu'tic (Gr. θεραπευω, *therapeu'ō*, I heal). Healing; pertaining to the art of healing.

Therapeu'tics (Gr. θεραπευω, *therapeu'ō*, I heal). The part of medical science which describes the properties of medicines and their modes of administration.

Thermal (Gr. θερμος, *thermos*, warm). Belonging to heat ; warm ; applied to springs of which the temperature is above 60° Fahr.

Thermo - electricity (Gr. θερμη, *thermē*, heat ; *electricity*). Electricity developed by heat.

Ther'mo-electrom'eter. An instrument for ascertaining the deflagrating or heating power of an electric current.

Thermom'eter (Gr. θερμη, *thermē*, heat ; μετρον, *met'ron*, a measure). An instrument for measuring the heat or temperature of bodies, by the regular expansion of mercury or some other substance. The thermometers usually employed are Fahrenheit's, the Centigrade, and Reaumur's. In Fahrenheit's ther-

mometer, the space between the freezing and boiling points of water is divided into 180 degrees, the freezing point being marked as 32 degrees, and the boiling as 212. In the Centigrade thermometer the space is divided into 100 degrees ; and in Reaumur's into 80. Hence 5 degrees of the Centigrade, or 4 of Reaumur's thermometer, are equal to 9 of Fahrenheit.

Thermomet'ric (*Thermometer*). Belonging to the thermometer.

Ther'mo-mul'tiplier. A thermo-electric pile, used for detecting changes of temperature.

Ther'mophone (Gr. θερμη, *thermē*, heat ; φωνη, *phōnē*, sound). An apparatus for producing sound from heated bodies.

Ther'moscope (Gr. θερμη, *thermē*, heat ; σκοπεω, *skop'eō*, I view). An instrument for measuring minute differences of heat and cold.

Ther'mostat (Gr. θερμη, *thermē*, heat ; ιστημι, *histēmi*, I make to stand). An apparatus for regulating temperature in distilleries, baths, furnaces, &c.

Thermot'ics (Gr. θερμη, *thermē*, heat; The science of heat.

The'sis (Gr. τιθημι, *tithē'mi*, I place). A proposition to be maintained by argument.

Thorac'ic (*Thorax*).   Belonging to or contained in the chest.

Thorac'ic Duct. The vessel which conveys the chyle into the subclavian vein.

Thorax (Gr. θωραξ, *thōrax*, a breastplate). The chest, or the part of the body between the neck and the abdomen ; in *entomology*, the second segment of insects, or the part between the head and the abdomen.

Thrombus (Gr. θρομβος, *throm'bos*, a clot of blood). A small tumour of clotted blood that has escaped under the skin.

Thymus. A temporary organ, which exists at the lower part of the neck in children, disappearing gradually after the second year.

Thy'ro- or **Thy'reo-** (Gr. θυρεος, *thu'reos*, a shield). In *anatomy*, a prefix in compound words, implying connection with the thyroid cartilage.

Thy'roid (Gr. θυρεύς, *thu'reos*, a shield; εἶδος, *eidos*, form). Like a shield; in *anatomy*, applied to one of the cartilages of the larynx from its shape; also to a glandular body lying in front of this cartilage; and to arteries supplying this part.

Thyrsus (Gr. θύρσος, *thur'sos*, a light straight shaft). In *botany*, a kind of inflorescence resembling a bunch of grapes.

Thysanou'ra (Gr. θύσανος, *thu'sanos*, a tassel; οὐρά, *oura*, a tail). A family of wingless insects with fringed tails.

Tib'ia (Lat. a pipe or flute). The largest bone of the leg; so called from its supposed resemblance to an ancient flute.

Tib'ial (*Tib'ia*). Belonging to, or situated near, the tibia or large bone of the leg.

Timbre (French). An acoustic property, not yet explained, by which sounds of the same note and loudness are distinguished from each other.

Tinc'ture (Lat. *tin'go*, I tinge). In *medicine*, a solution, generally in spirit, of the active principles of any substance.

Tinni'tus Au'rium (Lat.). A ringing in the ears.

Tissue (French, *tissu*, woven). In *anatomy* and *botany*, the minute elementary structures of which organs are composed.

Titho'nic (Gr. Τιθωνός, *Tithō'nus*). Pertaining to those rays of light which produce chemical effects.

Tme'sis (Gr. τέμνω, *temnō*, I cut). In *grammar*, the division of a compound word into two parts, a word or words being inserted between them.

Tomen'tose (Lat. *tomen'tum*, down). Downy; covered with a down-like wool.

Tomen'tum (Lat. down). In *anatomy*, a term applied to the minutely divided vessels on the surface of the brain; in *botany*, a species of longish, soft, entangled hairs.

Ton'ic (Gr. τόνος, *ton'os*, that which tightens, or may be tightened). In *medicine*, increasing strength; applied also to spasmodic contractions which last steadily for a comparatively long time.

Tonic'ity (Gr. τόνος, *ton'os*, that which tightens). The property of muscles, by which they remain in a state of contraction, being at the same time counterbalanced by other muscles in a similar state.

Ton'sil (Lat. *tonsil'læ*). An oblong gland situated on each side of the fauces.

Tonsilli'tis (Lat. *ton'sillæ*, the tonsils; *itis*, denoting inflammation). Inflammation of the tonsils; a form of sore throat.

Topha'ceous (Lat. *toph'us*, a sand or gravel stone). Consisting of deposited calcareous matter.

Toph'us (Lat. a sand or gravel stone). A deposit of porous calcareous matter; in *medicine*, a chalky deposit on the joints from gout.

Topograph'ical (*Topog'raphy*). Descriptive of a place or country.

Topo'graphy (Gr. τόπος, *top'os*, a place; γράφω, *graph'ō*, I write). A description of a particular place, giving a notion of everything connected therewith.

Tor'mina (Lat. *tor'queo*, I twist). Griping pains.

Torna'do (Spanish, *tornar'*, to turn). A hurricane; especially applied to the whirlwind hurricanes prevalent in some tropical regions.

Tor'ose (Lat. *torus*, a protuberance). Swelling in protuberances or knobs.

Torrefac'tion (Lat. *torrefac'io*, I roast). The operation of drying or roasting.

Torricel'lian Vac'uum (*Torricel'li*, the inventor of the mercurial barometer). The space left in the upper part of a long tube closed at one end and filled with mercury, when it is inverted in this fluid, which still remains in the tube to the height of thirty inches.

Tor'rid (Lat. *tor'reo*, I roast). Dried with heat; extremely hot.

Tor'sion (Lat. *tor'queo*, I twist). A

twisting : force of torsion, a term employed to denote the effort made by a thread which has been twisted to untwist itself.

Torticol'lis (Lat. *tor'queo*, I twist; *collum*, the neck). Wry-neck.

Tor'tuous (Lat. *tor'queo*, I twist). Twisted ; winding.

Tor'ulose (Lat. *toru'lus*, a kind of ringlet). In *botany*, having successive rounded swellings, as the pods of some cruciferous plants.

Tor'us (Lat. a rope; also a bed). In *architecture*, a large moulding, with a semicircular section, used in the bases of columns; in *botany*, the receptacle or part of the flower on which the carpels are seated.

Tour'niquet (French). An instrument used in surgery for producing pressure on a blood-vessel so as to restrain hæmorrhage.

Toxæ'mia (Gr. τοξικον, *tox'icon*, a poison ; αιμα, *haima*, blood). A poisoned state of the blood.

Tox'ical (Gr. τοξικον, *tox'icon*, a poison). Poisonous.

Toxicohæ'mia (Gr. τοξικον, *tox'icon*, a poison ; αιμα, *hai'ma*, blood). See Toxæmia.

Toxicolog'ical (*Toxicol'ogy*). Relating to the branch of medicine which describes poisons.

Toxicol'ogy (Gr. τοξικον, *tox'icon*, a poison ; λογος, *log'os*, discourse). The branch of medical science which describes poisons, their effects, and antidotes.

Tox'odon (Gr. τοξον, *tox'on*, a bow ; ὁδους, *od'ous*, a tooth). An extinct genus of pachydermatous or thick-skinned animals, having teeth bent like a bow.

Tra'chea (Gr. τραχυς, *trachus*, rough ; ἀρτηρια τραχεια, *arte'ria tracheï'a*, the rough artery or air-tube). The windpipe, a cartilaginous and membranous tube, which conveys the air into and out of the lungs.

Tra'cheæ (Plural of *Tra'chea*). In *botany*, the spiral vessels of plants ; in *entomology*, the vessels by which air is carried to every part of the body in insects.

Tra'cheal (*Trachea*). Belonging to the windpipe.

Trachea'ria (*Trachea*). An order of arachnidan invertebrata, whose organs of breathing consist of tracheæ.

Trachei'tis (*Trachea* ; *itis*, denoting inflammation). Inflammation of the trachea ; croup.

Trachelip'odous (Gr. τραχηλος, *trache'los*, a neck ; πους, *pous*, a foot). Having the feet united to the head.

Trachen'chyma (*Trachea;* Gr. ἐγχυμα, *en'chuma*, a tissue). Vegetable tissue consisting of spiral vessels.

Tracheot'omy (Gr. τραχεια, *trachei'a*, the windpipe ; τεμνω, *temnō*, I cut). The operation of making an opening into the windpipe.

Tra'chyte (Gr. τραχυς, *trachus*, rough). A rock of volcanic origin, consisting of felspar, and having a harsh feel.

Trac'tile (Lat. *traho*, I draw). Capable of being drawn out in length.

Trac'tion (Lat. *traho*, I draw). Drawing ; the act of being drawn ; in *mechanics*, the act of drawing a body along a plane.

Trac'tor (Lat. *traho*, I draw). That which draws.

Trade-winds (*Trade* and *wind;* because favourable to navigation and trade). The constant winds which occur in the open seas to the distance of about thirty degrees north and south of the equator ; those on the north of the equator being from the north-east, and those on the south from the south-east.

Tra'gus (Gr. τραγος, *tra'gos*, a goat). In *anatomy*, a conical prominence projecting backwards from the front of the ear.

Trajec'tory (Lat. *trans*, across ; *jac'io*, I cast). The path of a moving body which is acted on by given forces.

Transcenden'tal (Lat. *trans*, beyond ; *scando*, I climb). Surpassing ; in *philosophy*, relating to that which goes beyond the limits of actual experience.

Tran'sept (Lat. *trans*, across ; *sep-*

tum, a partition). The transverse portion of a church built in the form of a cross.

Transfu'se (Lat. *trans*, across ; *fun'do*, I pour). To pour, as from one vessel into another.

Transfu'sion (*Transfuse*). A pouring from one vessel into another ; in *medicine*, the introduction of the blood of one person or animal into the vessels of another.

Trans'it (Lat. *trans*, across ; *e'o*, I go). In *astronomy*, the passage of a planet between the earth and the sun, so that it appears as a black round spot on the surface of the sun's disc ; the passage of a celestial body across the meridian.

Trans'it Circle. An apparatus for making astronomical observations, combining the functions of the mural circle and the transit instrument.

Trans'it Instrument. An instrument for determining the time at which an object passes the meridian, consisting of a telescope so arranged as to be capable of being directed to all points of the meridian.

Transit'ion (Lat. *trans*, across ; *e'o*, I go). A passage from one state to another ; in *geology*, a term applied to strata between the primary and secondary, containing remains of the lower invertebrate animals.

Trans'itive (Lat. *trans*, across ; *e'o*, I go). Passing ; in *grammar*, applied to verbs of which the action passes to an object.

Translu'cence (Lat. *trans*, through ; *lux*, light). The property of transmitting light, but not the images of objects.

Translu'cent (Lat. *trans*, through ; *lux*, light). Transmitting light, but not in such a way as to render objects distinct.

Transmuta'tion (Lat. *trans*, across ; *mu'to*, I change). The change of one substance or form into another.

Transpa'rency (Lat. *trans*, through ; *par'eo*, I appear). The property of allowing light to pass so that objects can be distinctly seen.

Transpa'rent (Lat. *trans*, through ; *par'eo*, I appear). Allowing the passage of light, so as to form distinct images of objects.

Transpira'tion (*Transpire*). The act of passing off in vapour from the surfaces of animals, or vegetables.

Transpire (Lat. *trans*, over ; *spi'ro*, I breathe). To pass off in vapour from the surfaces of animals or vegetables.

Transpose (Lat. *trans*, across ; *po'no*, I put). To change the order by putting one thing in the place of another ; in *algebra*, to bring a term of an equation to the other side.

Transuda'tion (Lat. *trans*, across ; *sudo*, I sweat). An oozing of fluid through membranes.

Transver'sal (*Transverse*). Lying across several lines so as to cut them all.

Transver'se (Lat. *trans*, across ; *verto*, I turn). Lying across ; in *geometry*, applied to the diagonals of a square or parallelogram.

Trap (Swedish *truppa*, a stair). In *geology*, originally applied to basaltic and greenstone rocks rising in masses like stairs ; but now denoting all granitic rocks which are not igneous or strictly volcanic.

Trape'zium (Gr. τραπεζα, *trapez'a*, a table). In *geometry*, a plane four-sided figure, with none of the sides parallel ; in *anatomy*, one of the small bones of the wrist.

Trape'zius (Gr. τραπεζα, *trapez'a*, a table). A somewhat square muscle attached to the shoulder and the spine in the neck.

Trap'ezoid (Gr. τραπεζα, *trapez'a*, a table ; ειδος, *eidos*, shape). In *geometry*, a plane four-sided figure having two of the opposite sides parallel : in *anatomy*, one of the bones of the wrist, somewhat resembling but smaller than the trapezium.

Traumat'ic (Gr. τραυμα, *trauma*, a wound). Relating to, or arising from, wounds.

Tra'vertin (Italian, *traverti'no*). A whitish limestone deposited from the waters of springs holding lime in solution.

Trem'atode (Gr. τρημα, tre'ma, a pore). An order of parasitic animals having suctorial pores.

Trepa'n (Gr. τρυπανον, tru'panon, a wimble). A circular saw for removing a portion of the skull.

Trephi'ne (Gr. τρεπω, trep'ō, I turn). A surgical instrument used for the same purpose as the trepan, of which it is a modification.

Tri- (Lat. tres, or Gr. τρεις, treis, three). A prefix in compound words, signifying three.

Triadel'phous (Gr. τρεις, treis, three ; ἀδελφος, adel'phos, a brother). Having the stamens united in three bundles.

Trian'dria (Gr. τρεις, treis, three ; ἀνηρ, anēr, a male). A class of plants in the Linnæan system having three stamens.

Trian'gle (Lat. tres, three ; an'gulus, an angle). A plane figure, having three sides and three angles.

Trian'gular (Triangle). Having the form of a triangle ; relating to a triangle ; applied to a series of numbers, such as 1, 3, 6, 10, 15, 21, &c., because the number of points expressed by any one may be arranged in an equilateral triangle; in botany, having three angles, the faces being flat.

Trias'sic (Gr. τριας, trias, a triad). In geology, a name given to the upper new red sandstone, from its consisting of three divisions in Germany, whence the term was introduced.

Triba'sic (Gr. τρεις, treis, three ; βασις, ba'sis, a base). In chemistry, applied to a class of salts which contain three atoms of base to one of acid.

Tribe (Lat. tri'bus). A division or class of people, sometimes originating from one forefather ; a number of animals or vegetables having certain characters in common.

Tri'brach (Gr. τρεις, treis, three ; βραχυς, brach'us, short). A foot in verse, consisting of three short syllables.

Tricap'sular (Lat. tres, three ; cap'-sula, a little chest). Having three capsules.

Tri'ceps (Lat. tres, three ; cap'ut, a head). Having three heads ; applied to muscles which arise by three heads.

Trichi'asis (Gr. θριξ, thrix, hair). A turning inwards of the eyelashes, so that they irritate the ball of the eye.

Trichop'terous (Gr. θριξ, thrix, hair ; πτερον, pter'on, a wing). An order of insects having hairy membranous wings.

Trichot'omous (Gr. τριχα, trich'a, thrice ; τεμνω, temnō, I cut). Divided into three parts.

Tri'chroism (Gr. τρεις, treis, three ; χροα, chroa, colour). An appearance which some bodies present of having three different colours, according to the way in which the rays of light traverse them.

Tricoc'cous (Gr. τρεις, treis, three ; κοκκος, kok'kos, a berry). Applied to a fruit consisting of a capsule with three cells, each containing one seed.

Tricos'tate (Lat. tres, three ; costa, a rib). Three-ribbed.

Tricus'pid (Lat. tres, three ; cus'pis, a point). Having three points : applied to a valve situated between the right auricle and ventricle of the heart.

Tricus'pidate (Lat. tres, three ; cus'-pis, a point). In botany, having three long points.

Tridac'tylous (Gr. τρεις, treis, three ; δακτυλος, dak'tulos, a finger, or toe). Having three fingers or toes.

Triden'tate (Lat. tres, three ; dens, a tooth). Having three teeth.

Trien'nial (Lat. tres, three ; an'nus, a year). Containing three years ; happening every three years.

Trifa'cial (Lat. tres, three ; fac'ies, a face). A term applied to one of the cranial nerves, from its division into three large branches, and distribution to the face and adjoining parts.

Trifa'rious (Lat. trifa'riam, in three ways). In three rows.

Tri'fid (Lat. *tres*, three; *findo*, I cleave). Cleft into three : in *botany*, divided half way into three parts.

Triflo'rous (Lat. *tres*, three; *flos*, a flower). Having three flowers.

Trifo'liate (Lat. *tres*, three; *fo'lium*, a leaf). Having three leaves.

Trifur'cate (Lat. *tres*, three; *furca*, a fork). Having three forks.

Trig'amous (Gr. τρεις, three; γαμος, *gam'os*, marriage). Having male, female, and neutral flowers in one head.

Trigem'ini (Lat. *tres*, three; *gem'ini*, double). Three-double; a name given to the fifth pair of cranial nerves, which are divided into three branches; otherwise called trifacial.

Tri'glyph (Gr. τρεις, *treis*, three; γλυφη, *glu'phē*, sculpture). In *architecture*, an ornament repeated at intervals in the Doric frieze, consisting of two gutters or channels cut to a right angle, and separated by their interstices from each other, and from half-channels at the sides.

Tri'gon (Gr. τρεις, *treis*, three; γωνια, *gō'nia*, an angle). A triangle.

Tri'gonal (Gr. τριγων, *trigōn*, a triangle). Belonging to a trigon or triangle.

Trigonomet'rical (*Trigonom'etry*). Relating to, or performed according to the rules of, trigonometry.

Trigonom'etry (Gr. τριγων, *trigōn*, a triangle; μετρον, *met'ron*, a measure). Literally, the art of measuring triangles; but now including all theorems and formulæ relating to angles and circular arcs, and the lines connected with them.

Tri'gonous (Gr. τρεις, *treis*, three; γωνια, *gō'nia*, an angle.) In *botany*, having three angles, the faces being convex.

Trigyn'ia (Gr. τρεις, *treis*, three; γυνη, *gunē*, a female). An order of plants in the Linnæan system, having three pistils.

Trihed'ral (Gr. τρεις, *treis*, three; ἑδρα, *hed'ra*, a base). Having three equal sides.

Tri'jugate (Lat. *tres*, three; *jugum*, a yoke) In *botany*, having three pairs of leaflets.

Trilat'eral (Lat. *tres*, three; *la'tus*, a side). Having three sides.

Trilin'gual (Lat. *tres*, three; *lin'gua*, a tongue). Written in three languages.

Trilit'eral (Lat. *tres*, three; *lit'era*, a letter). Having three letters.

Tri'lobate (Gr. τρεις, *treis*, three; λοβος, *lob'os*, a lobe). Having three lobes.

Tri'lobites (Gr. τρεις, *treis*, three; λοβος, *lob'os*, a lobe). A genus of fossil crustaceous animals, having the upper surface of the body divided into three lobes.

Triloc'ular (Lat. *tres*, three; *loc'ulus*, a little place). Having three cells.

Trim'erous (Gr. τρεις, *treis*, three; μερος, *mer'os*, a part.) Having three parts; applied to flowers which have three parts in the calyx, three in the corolla, and three stamens.

Trim'eter (Gr. τρεις, *treis*, three; μετρον, *met'ron*, a measure). A verse consisting of three measures.

Trinerv'ate (Lat. *tres*, three; *nervus*, a nerve). In *botany*, applied to leaves having three unbranched nerves extending from the base to the point.

Trino'mial (Lat. *tres*, three; *no'men*, a name). In *algebra*, a quantity consisting of three terms.

Trice'cia (Gr. τρεις, *treis*, three; οἰκος, *oi'kos*, a house). An order of plants in the Linnæan system, having male, female, and bisexual flowers on three separate plants.

Tripar'tite (Lat. *tres*, three; *par'tio*, I divide). Divided into three parts; applied to leaves divided into three parts down to the base.

Tripet'alous (Gr. τρεις, *treis*, three; πεταλον, *pet'alon*, a petal). Having three petals

Triph'thong (Gr. τρεις, *treis*, three; φθογγη, *phthon'gē*, sound). A combination of three vowels in one sound.

Triphyl'lous (Gr. τρεις, *treis*, three;

φυλλον, *phul'lon*, a leaf). Having three leaves.

Tripin'nate (Lat. *tres*, three; *pin'na*, a feather). In *botany*, applied to leaves in which there are three series of pinnation; bipinnate leaves being again divided down to the base of each division.

Trip'licate (Lat. *tres*, three; *plic'o*, I fold). Three-fold: applied to the ratio which the cubes of two quantities bear to each other as compared with the ratio which the two numbers bear to each other.

Triplicos'tate (Lat. *tri'plex*, threefold; *cos'ta*, a rib). In *botany*, applied to leaves which have three ribs proceeding from above the base.

Trip'tote (Gr. τρεις, *treis*, three; πτωσις, *ptō'sis*, case). A noun having three cases only.

Triquet'rous (Lat. *triquet'ra*, a triangle). Having three sides; in *botany*, having three angles, the faces being concave.

Trira'diate (Lat. *tres*, three; *ra'dius*, a ray). Having three rays.

Trisect' (Lat. *tres*, three; *sec'o*, I cut). To divide into three equal parts.

Trisec'tion (*Trisect*). Division into three parts.

Trisep'alous (Lat. *tres*, three; *sep'al*). Having three sepals.

Tris'mus (Gr. τριζω, *tri'zō*, I gnash). Lock-jaw; a kind of tetanus affecting the muscles of the jaw.

Trisoctahed'ron (Gr. τρις, *tris*, three times; ὀκτω, *ok'tō*, eight: ἑδρα, *hed'ra*, a base). A figure having twenty-four equal faces.

Trisper'mous (Gr. τρεις, *treis*, three; σπερμα, *sper'ma*, seed). Having three seeds.

Tris'tichous (Gr. τρεις, *treis*, three; στιχος, *stich'os*, a row). In three rows.

Trisul'cate (Lat. *tres*, three; *sul'cus*, a furrow). Having three forrows.

Tri'syllabic (Gr. τρεις, *treis*, three; συλλαβη, *sul'labē*, a syllable). Having three syllables.

Tri'syllable (Gr. τρεις, *treis*, three;

συλλαβη, *sul'labē*, a syllable). A word of three syllables.

Tritern'ate (Lat. *tres*, three; *ternate*). Divided three times in a ternate manner.

Tritox'ide (Gr. τριτος, *tri'tos*, third; *oxide*). The third degree of oxidation of a body.

Trit'urate (Lat. *tritu'ra*, a threshing or grinding). To rub or grind to a very fine powder.

Tritura'tion (*Trit'urate*). The act of reducing to a very fine powder.

Trival'vular (Lat. *tres*, three; *valvæ*, folding-doors). Having three valves.

Triv'ial (Lat. *triv'ium*, a highway). Common; trifling; in *botany*, applied to the name of the species, which, added to the generic name, forms the name of the plant.

Tro'car (Fr. *trois quart*, three-quarters, from its triangular point). A surgical instrument used in tapping.

Trocha'ic (*Troch'ee*). Consisting of trochees.

Trochan'ter (Gr. τροχαζω, *trocha'zō*, I run along). In *anatomy*, a name given to two prominences at the upper part of the thigh-bone, in which are inserted several of the muscles used in motion.

Tro'che (Gr. τροχη, *troch'ē*, a wheel)). A form of medicine in a circular cake for dissolving in the mouth.

Tro'chee (Gr. τρεχω, *trech'ō*, I run). A foot in verse consisting of two syllables, the first long, the next short.

Tro'chiform (Gr. τροχος, *troch'os*, a wheel; *for'ma*, shape). Resembling a wheel.

Troch'lea (Gr. τρεχω, *trech'ō*, I run). A pulley; applied in *anatomy*, to projections of bones over which parts turn as over pulleys.

Troch'oid (Gr. τροχος, *troch'os*, a wheel; εἰδος, *ei'dos*, shape). In *geometry*, a curve produced by the motion of a wheel.

Trochom'eter (Gr. τροχος, *troch'os*, a wheel; μετρον, *met'ron*, a measure). An instrument for computing the revolutions of a wheel.

Trope (Gr. τρεπω, *trep'ō*, I turn). In

rhetoric, a change in the signification of a word from a primary to a derived sense.

Troph'i (Gr. τρεφω, *treph'o*, I nourish). The parts of the mouth in insects employed in acquiring and preparing food.

Troph'osperm (Gr. τροφος, *troph'os*, one who feeds ; σπερμα, *sper'ma*, a seed). In *botany*, the part of the ovary from which the ovules arise.

Trop'ic (Gr. τρεπω, *trep'o*, I turn). A name applied to each of the two circles lying parallel to the equator at the distance of 23½ degrees north and south.

Trop'ical (*Trop'ic*). Belonging to the tropics.

Trun'cate (Lat. *trun'co*, I cut off). To cut or lop off.

Trun'cated (*Trun'cate*). Cut off ; applied to figures the angles or edges of which have been cut off.

Tu'ber (Lat. a mushroom or bunch). In *botany*, a thick underground stem, as the potato; in *anatomy*, a rounded projection of a bone.

Tu'bercle (Lat. *tuber'culum*, a little swelling). A little knob; in *medicine*, a peculiar diseased deposit in the lungs and various parts of the body, frequently attended by the symptoms known as those of consumption.

Tuber'cula Quadrigem'ina (Lat. Four-double tubercles). A name given to four rounded projections at the base of the brain.

Tuber'cular or **Tuber'culous** (Lat. *tuber'culum*, a little knob). Having knobs or tubercles.

Tuberculo'sis (Lat. *tuber'culum*, tubercle). In *medicine*, the name applied to the condition under which tubercle is deposited in the organs of the body.

Tuberif'erous (Lat. *tu'ber*, a knob ; *fer'o*, I bear). Bearing tubers, as the potato.

Tu'berose (Lat. *tu'ber*, a knob.) Having knobs or tubers.

Tuberos'ity (Lat. *tu'ber*, a knob). In *anatomy*, a kind of projection or elevation.

Tu'berous (Lat. *tu'ber*, a knob).

Knobbed ; consisting of tubers connected together.

Tubic'ola (Lat. *tu'bus*, a tube ; *col'o*, I inhabit). An inhabitant of a tube ; applied to an order of animals which live in calcareous tubes.

Tu'bifer (Lat. *tu'bus*, a tube ; *fer'o*, I bear). Bearing tubes.

Tu'biform (Lat. *tu'bus*, a tube ; *for'ma*, shape). Like a tube in shape.

Tu'bular (Lat. *tu'bus*, a tube). Having the form of a tube ; consisting of a tube or pipe.

Tu'bulated (Lat. *tu'bus*, a tube). In the form of a small tube ; furnished with a small tube.

Tu'bule (Lat. *tu'bus*, a tube). A small tube.

Tu'bulibran'chiate (Lat. *tu'bulus*, a little tube ; Gr. βραγχια, *bran'chia*, gills). Having the shell, which contains the branchiæ, in the form of a more or less regularly spiral tube.

Tufa (Italian, *tufo*). In *geology*, any porous vesicular compound.

Tumefac'tion (Lat. *tu'meo*, I swell ; *fac'io*, I make). In *medicine*, a temporary swelling or enlargement.

Tu'mour (Lat. *tu'mor*, a swelling). In *medicine*, a permanent swelling or enlargement.

Tu'mulus (Lat.). An artificial mound of earth.

Tung'state (*Tung'sten*). A compound of tungstic acid with a base.

Tu'nica (Lat.). A coat or covering.

Tu'nicated (Lat. *tu'nica*, a kind of garment). In *botany*, applied to a bulb covered by thin scales, as the onion ; in *geology*, to a class of mollusca, enveloped in an elastic tunic not covered by a shell.

Tur'binated (Lat. *tur'bo*, a top). Shaped like a top; in *conchology* and *botany*, conically spiral, large at one end and narrow at the other.

Turges'cent (Lat. *turges'co*, I swell). Growing large ; swelling.

Tu'rio (Lat. a tendril). A young shoot covered with scales sent up from an underground stem ; as the asparagus.

Tympan'ic (*Tym'panum*). Belonging to the tympanum or drum of the ear.

Tym'panum (Gr. τυμπανον, *tum'-panon*, a drum). In *anatomy*, the middle cavity of the ear; in *architecture*, the space in a pediment between the cornice of the inclined sides and the fillet of the corona; also the die of a pedestal and the panel of a door.

Tympani'tes (Gr. τυμπανον, *tum'-panon*, a drum). A distension of the abdomen by gas.

Type (Gr. τυπος, *tu'pos*, a figure or model). The perfect normal representation or idea of anything.

Ty'phoid (*Typhus* ; Gr. ειδος, *eidos*, shape). In *medicine*, a term applied to an asthenic or low form of fever : a fever characterised by general depression, and by an eruption of the skin with disturbance and morbid changes in the intestinal canal.

Typhoma'nia (*Typhus* ; Gr. μανια, *ma'nia*, madness). The low muttering delirium which accompanies typhoid fever.

Typh'oon (Gr. τυφων, *tu'phōn*, a storm). A furious whistling wind or hurricane.

Ty'phous (*Typhus*). Relating to typhus.

Ty'phus (Gr. τυφος, *tu'phos*, smoke or stupor). In *medicine*, a form of fever characterised by much depression, and by the appearance of an eruption on the skin.

Typ'ical (Gr. τυπος, *tu'pos*, a type). Having the characters of a type ; characteristic.

Typograph'ic (Gr. τυπος, *tu'pos*, a type ; γραφω, *graph'ō*, I write). Relating to printing.

Typog'raphy (Gr. τυπος, *tu'pos*, a type ; γραφω, *graph'ō*, I write). The art of printing.

U.

Udom'eter (Gr. ὑδωρ, *hudōr*, water ; μετρον, *met'ron*, a measure). A rain-gauge.

Ul'cer (Gr. ἑλκος, *helkos*, a sore). A loss of substance on the surface of parts, produced by some action going on in the part itself, or by the application of destructive agents.

Ul'cerate (*Ul'cer*). To form an ulcer ; to become ulcerous.

Ul'na (Gr. ὠλενη, *ō'lenē*, the elbow). The inner bone of the forearm, which forms part of the elbow joint.

Ul'nar (*Ul'na*). Belonging to or situated near the ulna.

Umbel (Lat. *umbel'la*, a little fan). In *botany*, a form of inflorescence in which numerous stalked flowers arise from one point, as in the carrot and hemlock.

Umbellif'erous (*Um'bel* ; Lat. *fer'o*, I bear). Producing umbels ; applied to an order of plants characterised by having the flowers arranged in umbels.

Umbel'lule (*Um'bel* ; Lat. *ule*, denoting smallness). A small or partial umbel.

Umbili'cus (Lat.) The navel ; in *botany*, the part of the seed by which it is attached to the pericarp.

Um'bonate (Lat. *um'bo*, the boss of a shield). Round, with a projecting point in the centre.

Um'bra (Lat. a shadow). In *astronomy*, the shadow of the earth or moon in an eclipse, or the dark cone projected from a planet or satellite on the side opposite to the sun.

Uncial. A term applied to a form of letters used in ancient manuscripts.

Un'ciform (Lat. *un'cus*, a hook ; *for'ma*, shape). Resembling a hook.

Un'cinate (Lat. *un'cus*, a hook). Having a hooked process.

Unc'tuous (Lat. *un'go*, I anoint). Oily ; having an oily feel.

Un'dulate (Lat. *unda*, a wave). To vibrate or move like a wave.

Un'dulated (Lat. *un'da*, a wave). Wavy ; in *botany*, applied to leaves with wavy or crisp margins.

Undula'tion (Lat. *un'da*, a wave). A waving motion, or formation of waves ; in *physics*, the vibration of a substance in the manner of waves.

Un'dulatory (Lat. *un'da*, a wave). Moving like waves.

Un'dulatory The'ory. In *optics*, the theory which supposes light to be produced by the undulation of a subtle fluid, as sound is produced by undulations of the air.

Unguic'ulate (Lat. *un'guis*, a nail or claw). Having claws.

Un'guiform (Lat *un'guis*, a nail or claw ; *for'ma*, shape). Like a claw.

Unguis (Lat). A nail or claw ; in *anatomy*, the name of a small bone of the face ; in *botany*, the lower part of a petal.

Un'gula (Lat). A hoof ; in *geometry*, a part cut off from a cylinder, cone, &c., by a plane passing obliquely through the base and part of the curved surface.

Un'gulate (Lat. *un'gula*, a hoof). Hoof-shaped ; having hoofs.

Uni- (Lat. *u'nus*, one). A prefix in compound words, signifying one.

Uniax'ial (Lat. *u'nus*, one ; *axis*). Having but one axis.

Unicel'lular (Lat. *u'nus*, one ; *cel'-lula*, a cell). Composed of one cell.

Unicos'tate (Lat. *u'nus*, one ; *cos'ta*, a rib). Having one rib.

Unifa'cial (Lat. *u'nus*, one ; *fac'ies*, a face). Having but one front surface.

Uniflo'rous (Lat. *u'nus*, one ; *flos*, a flower). Having but one flower.

Unig'enous (Lat. *u'nus*, one ; *gen'us*, a kind). Of one kind.

Unij'ugate (Lat. *u'nus*, one ; *ju'go*, I yoke). In *botany*, applied to a penninerved compound leaf, with only one pair of leaflets.

Unila'biate (Lat. *u'nus*, one ; *la'bium*, a lip.) Having one lip only.

Unilat'eral (Lat. *u'nus*, one ; *la'tus*, a side). Being on one side only ; having one side.

Unilit'eral (Lat. *u'nus*, one ; *lit'era*, a letter). Having one letter

Uniloc'ular (Lat. *u'nus*, one ; *loc'ulus*, a little place). Having one cavity.

Unip'arous (Lat. *u'nus*, one ; *par'io*, I bring forth). Bringing forth only one.

Uniper'sonal (Lat. *u'nus*, one ; *per-so'na*, a person). Having only one person.

Unipet'alous (Lat. *u'nus*, one ; *pet'al*). Having one petal only.

Unisex'ual (Lat. *u'nus*, one ; *sex'us*, a sex). Having one sex only ; applied to plants having separate male and female flowers.

U'nison (Lat. *u'nus*, one ; *so'nus*, a sound). A coincidence in sounds arising from an equality in the number of vibrations.

U'nivalve (Lat. *u'nus*, one ; *valve*). Having one valve only.

U'niverse (Lat. *u'nus*, one ; *ver'sus*, turned). The collective term for all the bodies which are the objects of astronomical observation.

Univ'ocal (Lat. *u'nus*, one ; *vox*, voice). Having only one meaning.

Unstrat'ified (*Un*, implying not ; Lat. *stra'tum*, a layer ; *fac'io*, I make). Not stratified ; in *geology*, applied to rocks which do not occur in strata or layers, but in shapeless masses.

Uranog'raphy (Gr. οὐρανος, *ou'ranos*, heaven ; γραφω, *graph'o*, I write). A definition of a heavenly body, as of a planet.

U'rate (*Uric*). A compound of uric acid with a base.

Ur'ceolate (Lat. *ur'ceola*, a pitcher). Shaped like a pitcher.

Ure'a. An organic compound formed in the animal body.

U'tricle (Lat. *utric'ulus*, a little bag). A little bag or cell ; in *botany*, a thin-walled cell, or a bladder-like covering.

Utric'ular (*U'tricle*). Containing utricles or vessels like small bags.

U'vea (Lat. *u'va*, a grape). The covering of dark pigment which lines the posterior surface of the iris in the eye.

U'vula (Lat. *u'va*, a grape). The small fleshy part which hangs down at the back of the soft palate.

V.

Vac'cinate (Lat. *vacca*, a cow). To introduce the cowpox into the human being, as a preventive of small-pox.

Vac'uum (Lat. *vacuus*, empty). Space devoid of all matter or substance.

Vagi'na (Lat. a sheath). In *botany*, the sheath formed by a petiole round a stem, as in grasses.

Vagi'nate (*Vagi'na*). Sheathed.

Vaginipen'nous (Lat. *vagi'na*, a sheath ; *penna*, a wing). Having the wings enclosed in a sheath.

Vallec'ula (Lat. *val'lis*, a valley ; *u'la*, denoting smallness). In *botany*, the interval between the ribs in the fruit of umbelliferous plants.

Val'vate (Lat. *val'væ*, folding doors). Having valves ; opening by valves : applied to æstivation and vernation, when the leaves in the flower-bud or leaf-bud are applied to each other by their margins only.

Valve (Lat. *val'væ*, folding doors). In *anatomy*, a fold of membrane in a tube or vessel preventing the backward flow of fluids.

Val'vule (*Valve*). A little valve.

Van'ishing Point. In *perspective*, the point at which an imaginary line, passing through the eye of the observer parallel to any original line, cuts the horizon.

Vaporiza'tion (*Va'por*). The rapid conversion of a fluid into a vapour by heat.

Va'riable (Lat. *va'rius*, changing). In the *differential calculus*, applied to quantities which are subject to continual increase or diminution.

Va'riable Elements. In *astronomy*, a method of viewing the effects of disturbing forces acting on a body moving in an elliptic orbit, which is supposed from time to time to change its position, form and magnitude in a minute degree.

Varia'tion (Lat. *va'rius*, changing). An alteration or partial changes ; in *arithmetic* and *algebra*, applied to

the different arrangements that can be made of any number of things, a certain number being taken together ; in *astronomy*, the inequality in the moon's apparent motion, which is greatest at conjunction and opposition, and least at the quadratures.

Varicel'la. The chicken-pox.

Var'icose (Lat. *va'rix*, a swollen vein). Enlarged ; applied to the veins when they are distended and present a knotty appearance.

Vari'ety (Lat. *va'rius*, changing). In *natural history*, a plant or animal differing from the rest of its species in some accidental circumstances, which are not permanent or constant, and are produced by the operation of such causes as climate, food, cultivation, &c.

Vari'ola (Lat. *va'rius*, spotted). The small-pox.

Vari'olous (*Vari'ola*). Relating to the small-pox.

Varix (Lat.). An uneven dilatation of a vein.

Vas'cular (Lat. *vas'culum*, a little vessel). Belonging to vessels ; consisting of, or containing vessels.

Vas'cular System. The collective name for the blood-vessels.

Vasculif'erous (Lat. *vas'culum*, a little vessel ; *fer'o*, I bear). In *botany*, applied to plants which have the seed-vessels divided into cells.

Va'siform (Lat. *vas*, a vessel ; *forma*, shape). Resembling vessels ; applied to a vegetable tissue called dotted vessels.

Veg'etable (Lat. *veg'eo*, I grow). A body having life, but without sensation or voluntary motion.

Veg'etate (Lat. *veg'eo*, I flourish). To grow, like plants.

Vegeta'tion (*Veg'etate*). The process of growing like plants.

Veg'etative (*Veg'etate*). Having the power of growing, or of producing growth in plants.

Vein (Lat. *vena*). In *anatomy*, a vessel which carries the blood towards the heart; in *botany*, applied to the midrib and its branches in a leaf; in *geology*, a fissure or rent filled with mineral or metallic matter, differing from the rock in which it occurs.

Veloc'ity (Lat. *velox*, swift). Swiftness; in *physics*, the measure of the rate at which a body moves.

Ve'na (Lat.). A vein.

Ve'na Portæ (Lat. the vein of the gate). The large vein which conveys the blood from the intestines into the liver.

Ve'næ Ca'væ (Lat. the hollow veins). The large veins which pour the blood collected from the body into the heart.

Vena'tion (Lat. *vena*, a vein). In *botany*, the arrangement of the veins in leaves.

Venesec'tion (Lat. *vena*, a vein; *sec'o*, I cut). The operation of letting blood by opening a vein.

Ve'nous (Lat. *vena*, a vein). Belonging to, or contained in the veins.

Venous System. In *anatomy*, the collective name for the veins.

Ventral (Lat. *venter*, the belly). Belonging to the belly; in *botany*, applied to that part of the carpel which is nearest the axis, or in front.

Ven'tricle (Lat. *venter*, the belly). A small cavity in an animal body; applied to two cavities of the heart, which propel the blood into the arteries, also to certain cavities in the brain.

Ven'tricose (Lat. *venter*, a belly). Distended; swelling out in the middle or unequally on one side.

Ve'nules (Lat. *ve'nula*, a little vein). In *botany*, the last branchings of the veins of a leaf.

Verbal (Lat. *verbum*, a word or verb). In *grammar*, derived from a verb.

Vermes (Lat. *ver'mis*, a worm). Worms; applied by Linnæus to all animals which could not be ranged under the heads of vertebrates and insects; but now restricted to the annelids and entozoa, or parasitic worms.

Vermic'ular (Lat. *ver'mis*, a worm). Pertaining to a worm; resembling the motion of a worm; shaped like a worm.

Vermicula'tion (Lat. *ver'mis*, a worm). The act of moving like a worm.

Ver'miform (Lat. *ver'mis*, a worm; *for'ma*, shape). Shaped like a worm.

Ver'mifuge (Lat. *ver'mis*, a worm; *fu'go*, I put to flight). Destroying or expelling worms.

Vermiv'orous (Lat. *ver'mis*, a worm; *vo'ro*, I devour). Eating worms.

Vernac'ular (Lat. *ver'na*, a bond-slave). Native; belonging to the country where one is born.

Ver'nal (Lat. *ver*, the spring). Belonging to the spring.

Verna'tion (Lat. *verno*, I bud or spring out). The arrangement of the young leaves within the bud.

Ver'nier. A small portable scale, running parallel with the fixed scale of a graduated instrument, for the purpose of subdividing the divisions of the instrument into more minute parts.

Verru'ca (Lat.). A wart.

Verru'cose (Lat. *verru'ca*, a wart). Warty; full of warts; having elevations resembling warts.

Ver'satile (Lat. *verso*, I turn). In *botany*, applied to anthers which are attached to the filament by a point at the back.

Ver'tebra (Lat. *verto*, I turn). A division or separate bone of the spinal column.

Ver'tebral (*Ver'tebra*). Belonging to a vertebra, or to the vertebræ; consisting of vertebræ.

Ver'tebrate (*Ver'tebra*). Having a vertebral column, or spine composed of a number of bones jointed together.

Ver'tebra'ta (*Ver'tebra*). Animals with a spine; including mammals, birds, reptiles, and fishes.

Vertex (Lat. *verto*, I turn). The top or summit.

Ver'tical (Lat. *vertex*, a top). Perpendicularly over-head, or to the plane of the horizon; standing upright; in *geometry*, applied to the opposite angles made by the intersection of two straight lines; in *astronomy*, to a circle passing through the zenith and the nadir, at right angles to the meridian.

Ver'ticil (Lat. *verticillus*, a pin or peg). In *botany*, a whorl, or form of inflorescence, in which the flowers surround the stem in a kind of ring, on the same plane.

Verticil'late (*Ver'ticil*). Having parts arranged in a whorl, or verticil.

Vertig'inous (*Verti'go*). Turning round; giddy.

Verti'go (Lat. *verto*, I turn). Giddiness.

Ves'icant (Lat. *vesi'ca*, a bladder). Producing a blister.

Ves'icate (Lat. *vesi'ca*, a bladder). To produce a blister.

Ves'icatory (Lat. *vesi'ca*, a bladder). Having the property of raising blisters.

Ves'icle (Lat. *vesic'ula*, a small bladder). A small blister; any small membranous cavity in plants or animals.

Vesic'ular (Lat. *vesic'ula*, a little bladder). Belonging to or having vesicles or little bladders.

Vessel (Lat. *vas*). In *anatomy*, any tube in which the blood or other fluid is formed or conveyed; in *botany*, a tube with closed ends.

Vexil'lary (Lat. *vexil'lum*, a standard). In *botany*, a form of æstivation in which the vexillum, or upper petal, is folded over the other.

Vexil'lum (Lat. a standard). In *botany*, the upper petal of a papilionaceous flower.

Via Lac'tea (Lat. the milky way). In *astronomy*, the galaxy or Milky Way, a region of the heavens presenting a whitish nebulous light, but consisting of innumerable stars crowded together.

Vi'able (Fr. *vie*, life; from Lat. *vivo*, I live). Capable of living.

Vi'aduct (Lat. *via*, a way; *duco*, I lead). An extensive bridge or series of arches for the purpose of conducting a road above the level of a ground in crossing a valley, or wherever it may be necessary to raise the road above the natural surface of the ground.

Viatec'ture (Lat. *via*, a way; Gr. τεκτων, *tektōn*, a builder). The art of constructing roads, &c.; civil engineering.

Vi'brate (Lat. *vi'bro*, I brandish). To swing or move to and fro.

Vi'bratile (*vi'brate*). Used for the motion of swinging to and fro.

Vibra'tion (Lat. *vi'bro*, I brandish). The act of moving to and fro quickly; in *mechanics*, the regular swinging motion of a suspended body, as a pendulum; in *physics*, the tremulous motion produced in a body when it is struck or disturbed by any impulse, by which waves or undulations are produced.

Vi'bratory (*Vibrate*). Having a vibratory motion.

Vib'rio (Lat. *vibro*, I shake). A name given to certain minute thread-like animalcules sometimes found in fluids.

Vibris'sæ. The stiff hairs which grow within the nostrils.

Vil'li (Lat. *villus*, wool or hair). In *anatomy*, minute projections from the surface of a mucous membrane, giving the appearance of the nap of cloth; in *botany*, long, straight, soft hairs on the surface of a plant.

Villos'ity (Lat. *villus*, wool or hair). The condition of being covered with villi.

Vil'lous (Lat. *villus*, wool or hair). Having a covering resembling hair or wool, or the nap of velvet or cloth.

Vina'ceous (Lat. *vinum*, wine). Pertaining to wine or grapes.

Vin'culum (Lat. from *vin'cio*, I bind). A bond or tie; in *algebra*, a line drawn over an expression consisting of several terms, to show that they are to be taken together.

Vi'nous (Lat. *vi'num*, wine). Belonging to, or having the quality of wine; applied to the process of fermentation which produces alcohol.

Vir'gate (Lat. *vir'ga*, a rod). Shaped like a rod.

Vir'tual (Lat. *vir'tus*, power or force). Being or acting in effect, not in fact ; in *optics*, applied to the focus from which rays, that have been rendered divergent, appear to issue ; in *mechanics*, to the velocity which a body in equilibrium would acquire in the first instant of its motion, if the equilibrium were disturbed.

Vir'ulent (Lat. *virus*, a poison). Very poisonous.

Vi'rus (Lat.). A poison ; in *medicine*, applied to the essential matter of a disease, which is capable of communicating the disease from one person to another.

Vis a Fronte. A force acting from the front or in advance.

Vis Iner'tiæ (Lat. the force of inaction). A term used to denote the power by which matter resists changes endeavoured to be made in its state.

Vis a Tergo (Lat. force from the back). A moving power acting from behind.

Vis In'sita (Lat. inherent force). The property by which a muscle, when irritated, contracts independently of the will of the animal, and without sensation.

Vis Medica'trix Natu'ræ (Lat. the healing power of nature). A term applied to denote the power by which a living body is able to throw off disease or recover from injury.

Vis Nervo'sa (Lat. nervous force). The property of nerves by which they convey stimuli to muscles.

Vis Plas'tica (Lat. plastic force). The formative power of plants and animals.

Vis Vi'tæ (Lat. force of life). Vital power or energy.

Vis'cera (Plural of Lat. *vis'cus*, an entrail). The organs contained in any of the great cavities of the body, especially the chest and abdomen.

Vis'coral (*Viscera*). Belonging to the viscera or internal organs.

Vis'cid or **Vis'cous** (Lat. *vis'cum*, bird-lime). Glutinous ; sticky.

Vis'cus (Lat.). An entrail, or organ contained in one of the great cavities of the body.

Vis'ible (Lat. *vid'eo*, I see). In *optics*, emitting or reflecting a sufficient number of rays of light to produce an impression on the eye.

Vis'ual (Lat. *vid'eo*, I see). Relating to sight.

Vi'tal (Lat. *vita*, life). Pertaining or contributing to life.

Vital'ity (*Vital*). The principle of life : the act of living,

Vitel'lary (Lat. *vitellus*, a yolk). Belonging to the yolk of an egg.

Vit'reous (Lat. *vit'rum*, glass). Belonging to, or consisting of glass : resembling glass.

Vit'reous Body. A large globular transparent structure occupying the centre of the eyeball, being the largest of the transparent media of the eye.

Vit'reous Electricity. A name sometimes given to positive electricity, because developed by rubbing glass.

Vitreous Humour. *See* Vitreous Body.

Vitres'cence (Lat. *vit'rum*, glass). Glassiness ; capability of being formed into glass.

Vitrifac'tion (Lat. *vi'trum*, glass ; *fac'io*, I make). The process of converting into glass by heat.

Vitrifi'able (Lat. *vit'rum*, glass ; *facio*, I make). Capable of being converted into glass by heat.

Vit'rify (Lat. *vit'rum*, glass ; *fac'io*, I make). To convert or be converted into glass by heat.

Vit'riol (Lat. *vit'rum*, glass). A name given to sulphuric acid and several of its compounds, probably from the glassy appearance of the crystals : oil of vitriol is sulphuric acid : blue vitriol, sulphate of copper : green vitriol, green sulphate of iron : red vitriol, red sulphate of iron : white vitriol, sulphate of zinc.

Vitriol'ic (*Vit'riol*). Belonging to or containing vitriol.

Vitt'a (Lat. a fillet or head-band). In *architecture*, the ornament of a capital, &c. ; in *botany*, (plural *vittæ*,) the receptacles of oil in the fruits of umbelliferous plants, as anise, carraway, fennel, &c.

Vit'tate (Lat. *vitta*, a band). In *botany*, applied to leaves which are striped.

Vivip'arous (Lat. *vivus*, alive; *par'io*, I bring forth). Bringing forth young alive ; in *botany*, applied to stems that produce leaf buds or bulbs in place of fruit.

Vocab'ulary (Lat. *vocab'ulum*, a word). A list of the words of a language.

Vo'cative (Lat. *vo'co*, I call). Calling.

Vol'atile (Lat. *volo*, I fly). Having the power of flying; capable of easily passing into an aëriform state.

Volatil'ity (*Vol'atile*). Capability of rising in an aëriform state.

Volat'ilize (*Vol'atile*). To cause to pass off in vapour, or in an aëriform state.

Volcan'ic (*Volca'no*). Belonging to or produced by volcanoes ; thrown out by volcanic eruptions.

Volca'no (Italian, from Latin *Vulca'nus*, the god of fire). An opening in the surface of the globe, generally in a mountainous elevation, giving issue from time to time to eruptions of melted matter.

Volit'ion (Lat. *volo*, I will). The act of willing.

Volta'ic (*Volta*). Relating to voltaism.

Volta'ic Bat'tery. An apparatus consisting of a series of pairs of plates of different metals—as zinc and copper—immersed in fluid, and connected by wires, for the development of voltaic electricity.

Volta'ic Electricity. The form of electrical action discovered by Galvani, but first correctly described by Volta, in which, any two conductors of electricity being brought into contact, an electric action is set up.

Vol'taism (*Volta*). A term for galvanism as produced by Volta's apparatus.

Voltam'eter (*Volta ;* Gr. μετρον, *met'ron*, a measure). An instrument for measuring the amount of a current of voltaic electricity by means of the quantity of water decomposed in a given time.

Vol'ume (Lat. *volvo*, I roll). Originally something rolled ; as much as is included in a roll ; dimension ; in *chemistry*, the relative or comparative measure of the combining atoms of gases.

Vol'untary (Lat. *volun'tas*, will). In *physiology*, acting under the direction of the will; produced by the will.

Volu'te (Lat. *vol'vo*, I roll). In *architecture*, a kind of spiral scroll used in capitals.

Vo'mer (Lat. a ploughshare). In *anatomy*, the small flat bone which separates the nostrils from each other.

Vor'tex (Lat. from *verto*, I turn). A whirlpool.

Vul'canist (Lat. *Vulca'nus*, the god of fire). In *geology*, a term applied to the supporters of an hypothesis which supposed that the older rock formations were of volcanic or igneous origin.

Vulcaniza'tion. A process of preparing india-rubber by impregnating it with sulphur.

Vul'nerary (Lat. *vulnus*, a wound). Useful in healing wounds.

Vulsellum (Lat. *vello*, I pull or pluck). A surgical instrument for seizing parts and drawing them into a convenient position for operation.

W.

Wacke. In *geology*, a German term for a soft earthy variety of trap-rock.

Weald-clay. In *geology*, the blue clay which forms part of the Wealden group.

Wealden (Sax. *wold*). In *geology*, a deposit prevailing in Kent and Sussex, consisting chiefly of clays and shales, with beds of indurated sand, sandstone, and shelly limestone.

Weight (Sax. *wiht*). The pressure

which a body exerts vertically downwards in consequence of the action of gravity.

Weld (Germ. *wellen*, to join). To unite two or more pieces, generally of iron, by hammering them together when heated.

Whirlpool (*Whirl* and *pool*). A body of water running round in a circle.

Whirlwind (*Whirl* and *wind*). A body of air moving in a circular or spiral form, as if round an axis, at the same time having a progressive motion.

Woulfe's Appara'tus. In *chemistry*, a bottle with two or more openings, used for generating gases.

Wormian Bones. The small triangular pieces of bone sometimes found lying between the other bones of the skull.

X

Xan'thic (Gr. ξανθος, *xan'thos*, yellow). Of, or belonging to yellow : yellowish ; having yellow as the type.

Xan'thogen (Gr. ξανθος, *xan'thos*, yellow ; γενναω, *genna'ō*, I produce). Yellow colouring matter in vegetables.

Xan'thophylle (Gr. ξανθος, *xan'thos*, yellow ; φυλλον, *phullon*, a leaf). Yellow colouring matter in plants.

Xan'thous (Gr. ξανθος, *xan'thos*, yellow). A term applied by Dr. Prichard to the variety of mankind including individuals with brown, yellow, or red hair.

Xiph'oid (Gr. ξιφος, *xiph'os*, a sword; ειδος, *eidos*, shape). Shaped like a sword.

Xiphosu'ra (Gr. ξιφος, *xiph'os*, a sword ; ουρα, *ou'ra*, a tail). A family of crustaceous animals with sword-shaped tails.

Xylo- (Gr. ξυλον, *xulon*, wood). A prefix in compound words, denoting relation to wood, or that wood enters into the composition.

Xylocar'pous (Gr. ξυλον, *xulon*, wood ; καρπος, *karpos*, fruit). Bearing fruit which becomes hard and woody.

Xylog'raphy (Gr. ξυλον, *xulon*, wood ; γραφω, *graph'ō*, I write). Engraving on wood.

Z.

Zen'ith. The point in the arch of the heavens which is vertically above the head of the spectator.

Zen'ith Distance. The distance of a heavenly body from the zenith, measured on the vertical circle passing through the zenith and the body.

Zen'ith Sector. An instrument for measuring the zenith distances of stars which pass near the zenith.

Zenograph'ic (Gr. Ζηνος, *Zēnos*, a genitive case of Ζευς, *Zeus*, Jupiter ; γραφω, *graph'ō*, I write). Relating to a description of the planet Jupiter, or characteristic of the appearance of this planet.

Ze'olite (Gr. ζεω, I boil : λιθος, *lith'os*, a stone). A term applied in *chemistry* to certain compounds, from their frothing when heated before the blow-pipe.

Zero (Italian, nothing). The point of a thermometer from which it is graduated : in the Centigrade and Reaumur's, it is the freezing point of water ; in Fahrenheit's, thirty-two degrees below the freezing-point.

Zeug'ma (Gr. ζευγνυμι, *zeugnu'mi*, I join). In *grammar*, a figure by which an adjective or verb that agrees with a nearer word, is also referred to another more remote.

Zinciferous (*Zinc;* Lat. *fer'o*, I bear).
Producing zinc.

Zinc'ous (*Zinc*). Relating to zinc ;
applied to the positive pole of a
galvanic battery.

Zo'diac (Gr. ζωδιον, *zō'dion*, a little
animal). The zone of the heavens
included within a space of the
celestial sphere extending a few
degrees north and south of the
ecliptic, and within which the apparent motions of the planets are
included.

Zodi'acal (*Zo'diac*). Belonging to
the zodiac.

Zon'ule (*Zone*). A small zone or
girdle.

Zoo- (Gr. ζωον, *zō'on*, an animal). A
prefix in compound words, implying
relation to animals.

Zoochem'ical (Gr. ζωον, *zō'on*, an
animal ; *chem'ical*). Relating to
the chemistry of animal bodies.

Zo'oid (Gr. ζωον, *zō'on*, an animal ;
ειδος, *ei'dos*, form). Resembling
an animal.

Zo'olite (Gr. ζωον, *zō'on*, an animal ;
λιθος, *lith'os*, a stone). A petrified
or fossil animal substance.

Zoolog'ical (*Zool'ogy*). Belonging to
zoology, or the classification of
animals.

Zool'ogist (*Zool'ogy*). One who is
skilled in the natural history of
animals.

Zool'ogy (Gr. ζωον, *zō'on*, an animal :
λογος, *log'os*, a discourse). The
science or natural history of the
animal kingdom ; the description
of the structure, habits, &c., of all
animals.

Zooph'agous (Gr. ζωον, *zō'on*, an
animal ; φαγω, *phag'ō*, I eat).
Eating animals.

Zoophor'ic (Gr. ζωον, *zō'on*, an animal;
φερω, *pher'ō*, I bear). Supporting
the figure of an animal.

Zo'ophyte (Gr. ζωον, *zō'on*, an animal ;
φυτον, *phuton*, a plant). In *natural
history,* a name given to bodies
resembling both animals and vegetables, and once supposed to par-
take of the nature of both.

Zo'ospore (Gr. ζωον, *zō'on*, an animal;
spore). A moving spore, provided
with cilia or vibratile organs.

Zoot'omist (*Zoot'omy*). One who dissects animals.

Zoot'omy (Gr. ζωον, *zō'on*, an animal;
τεμνω, *temnō*, I cut). Anatomy of
the lower animals.

Zoster (Gr. ζωστηρ, *zōstēr*, a girdle).
An eruptive disease which extends
round the waist like a girdle ; commonly called shingles.

Zygodac'tylous (Gr. ζυγον, *zugon*, a
yoke ; δακτυλος, *dak'tulos*, a finger
or toe). Having the toes formed
as if yoked together.

Zygo'ma (Gr. ζυγοω, *zugo'ō*, I yoke
together). A bony arch at the
upper part of the side of the face,
formed by the union of a process
from the temporal with one from
the malar bone.

Zygomat'ic (*Zygo'ma*). Belonging to
the zygoma.

Zymo'sis (Gr. ζυμοω, *zumo'ō*, I leaven).
In *medicine,* applied to diseases
which are epidemic, endemic, and
contagious, including fever, small-
pox, cholera, &c., which are believed to be produced by the action
of certain specific poisons.

Zymot'ic (Gr. ζυμοω, *zumo'ō*, I leaven).
Arising from zymosis or fermentation.

POPULAR SCIENCE,

DR. LARDNER'S "MUSEUM OF SCIENCE AND
ART." Complete in Twelve Single Volumes, ornamental boards, 18s. ;
or in Six Double Volumes cloth lettered, 21s. ; also handsomely half-bound
morocco, Six volumes, 1l. 11s. 6d.

The Planets: are they Inhabited Worlds?—Weather Prognostics—Popular Fallacies
—Latitudes and Longitudes—Lunar Influences—Meteoric Stones and Shooting
Stars—Railway Accidents—Light—Common Things: Air—Locomotion in the
United States—Cometary Influences—Common Things : Water—The Potter's
Art—Common Things: Fire—Locomotion and Transport—The Moon—Common
Things : The Earth—The Electric Telegraph—Terrestrial Heat—The Sun—Earth-
quakes and Volcanoes—Barometer, Safety Lamp, and Whitworth's Micrometric
Apparatus—Steam—The Steam Engine—The Eye—The Atmosphere—Time—
Common Things : Pumps—Common Things : Spectacles, The Kaleidoscope—
Clocks and Watches—Microscopic Drawing and Engraving—The Locomotive—
Thermometer—New Planets : Leverrier and Adams's Planet—Magnitude and
Minuteness—Common Things : The Almanack—Optical Images—How to observe
the Heavens—Common Things: The Looking Glass—Stellar Universe—The Tides
—Colour—Common Things : Man—Magnifying Glasses—Instinct and Intelligence
—The Solar Microscope—The Camera Lucida—The Magic Lantern—The Camera
Obscura—The Microscope—The White Ants: their Manners and Habits—The
Surface of the Earth—Science and Poetry—The Bee—Steam Navigation—Electro-
Motive Power—Thunder, Lightning, and the Aurora Borealis—The Printing Press
—The Crust of the Earth—Comets—The Stereoscope—The Pre-Adamite Earth—
Eclipses—Sound.

"The 'Museum of Science and Art' is the most valuable contribution that has
ever been made to the Scientific Instruction of every class of Society."—*Sir David
Brewster in the North British Review.*

LARDNER'S HANDBOOK OF NATURAL PHILO-
SOPHY. Forming a Complete course of Natural Philosophy. In Four
Volumes, 12mo, with 1334 illustrations, price 20s.

Also sold separately as under :—

MECHANICS. With 357 Illustrations. One Volume, 5s.

HYDROSTATICS, PNEUMATICS, and HEAT. 292 Illustrations. One
Volume, 5s.

OPTICS. With 290 Illustrations. One Volume, 5s.

ELECTRICITY, MAGNETISM, and ACOUSTICS. 395 Illustrations. One
Volume, 5s.

HANDBOOK OF ASTRONOMY. By Dr. LARDNER.
Third Edition. Revised and completed to 1867. By EDWIN DUNKIN,
F.R.A.S., Superintendent of the Altazimuth Department, Royal Observa-
tory, Greenwich. 138 Illustrations. Small 8vo, 7s. 6d.

"It is not very long since a lecturer was explaining some astronomical facts to his
pupils ; and in order to set the matter clearly before them, he referred to more than
one large and important volume on the subject, but without a decidedly satisfactory
result. One of the pupils, however, produced from his pocket a small unpretending
work (Dr. Lardner's *Handbook*), and that which a lengthy paragraph in the large work
had failed to make clear, was completely elucidated in a short pithy sentence in the
small book. A Third Edition of the well-known *Handbook of Astronomy* is now before
us. We can cordially recommend it as most useful to all those who desire to possess a
complete manual of the science and practice of astronomy in a portable and inexpensive form."—*Astronomical Register.*

LIEBIG'S FAMILIAR LETTERS ON CHEMISTRY.
Fourth Edition, revised throughout, with additional letters. 1 Vol. small 8vo, 7s. 6d. cloth.

COMMON THINGS EXPLAINED. By Dionysius
Lardner, D.C.L. Containing :—Air—Earth—Fire—Water—Time—The Almanack—Clocks and Watches—Spectacles — Colour— Kaleidoscope—Pumps—Man—The Eye—The Printing Press—The Potter's Art—Locomotion and Transport—The Surface of the Earth, or First Notions of Geography. (From "The Museum of Science and Art.") 233 Illustrations, 5s. cloth.

* Sold also in Two Series, 2s. 6d. each.

THE ELECTRIC TELEGRAPH. By Dr. Lardner.
New Edition. Revised and re-written by E. B. Bright, F.R.A.S., Secretary of the British and Irish Magnetic Telegraph Company; containing full information, in a popular form, of the Telegraphs at home and abroad, brought up to the present time ; including Descriptions of Railway Signalling Apparatus, Clock Regulating by Electricity, Admiral Fitzroy's System of Storm Warning, Messages, &c. 140 Illustrations. Small 8vo, 5s.

THE MICROSCOPE. By Dionysius Lardner, D.C.L.
(From " The Museum of Science and Art.") 147 Illustrations, 2s. cloth.

POPULAR GEOLOGY. By Dionysius Lardner, D.C.L.
(From "The Museum of Science and Art.") 201 Illustrations, 2s. 6d.

STEAM AND ITS USES: including the Steam Engine,
the Locomotive, and Steam Navigation. By Dionysius Lardner, D.C.L. (From "The Museum of Science and Art.") 89 Illustrations, 2s. cloth.

POPULAR ASTRONOMY. By Dionysius Lardner,
D.C.L. (From "The Museum of Science and Art.") 182 Illustrations, 4s. 6d.

NATURAL PHILOSOPHY FOR SCHOOLS. By Dr.
Lardner. 328 Illustrations. 1 Vol. 3s. 6d. cloth.

ANIMAL PHYSIOLOGY FOR SCHOOLS. By Dr.
Lardner. 190 Illustrations. 1 Vol. 3s. 6d.

LONDON : JAMES WALTON, 137, GOWER STREET.

JAMES WALTON,

Bookseller and Publisher to University College,

137, GOWER STREET.

Natural Science.

. THE ELECTRIC TELEGRAPH. By DR.
LARDNER. New Edition. Revised and re-written by
E. B. BRIGHT, F.R.A.S., Secretary of the British and
Irish Magnetic Telegraph Company ; containing full
information, in a popular form, of Telegraphs at
home and abroad, brought up to the present time ; including Descriptions of Railway Signalling Apparatus,
Clock Regulating by Electricity, Admiral Fitzroy's System
of Storm Warning, &c. 140 Illustrations. Small 8vo, 5s.

" It is capitally edited by Mr. Bright, who has succeeded in making
this one of the most readable books extant on the Electric Telegraph.
On the ground it takes up it is quite exhaustive ; and he who will carefully read the work before us, and can retain its chief facts in remembrance, may well be considered thoroughly posted up in all that
appertains to the Electric Telegraph to date."—*English Mechanic.*

HANDBOOK OF ASTRONOMY. By DR.
LARDNER. Third Edition. Revised and completed to
1867. By EDWIN DUNKIN, F.R.A.S., Superintendent
of the Altazimuth Department, Royal Observatory, Greenwich. 138 Illustrations. Small 8vo., 7s. 6d.

" It is not very long since a lecturer was explaining some astronomical
facts to his pupils ; and in order to set the matter clearly before them, he
referred to more than one large and important volume on the subject,
but without a decidedly satisfactory result. One of the pupils, however,
produced from his pocket a small unpretending work (Dr. Lardner's
Handbook), and that which a lengthy paragraph in the large work had
failed to make clear, was completely elucidated in a short pithy sentence
in the small book. A Third Edition of the well-known *Handbook
of Astronomy* is now before us, edited by Mr. Dunkin of the Greenwich Observatory, who has added to the text all that has lately been
discovered, so as to bring the work down to the present time. We
can cordially recommend it as most useful to all those who desire to
possess a complete manual of the science and practice of astronomy in
a portable and inexpensive form."—*Astronomical Reporter.*

NATURAL PHILOSOPHY FOR SCHOOLS.
By DR. LARDNER. 328 Illustrations. 4th Edit. 1 Vol. 3s. 6d.

"This will be a very convenient class-book for junior students in
private schools. It is intended to convey, in clear and precise terms,
general notions of all the principal divisions of Physical Science, illustrated largely by diagrams. These diagrams exhibit the forms and
arrangement of apparatus, and the manner of performing the most important experiments."—*British Quarterly Review.*

Natural Science.

HANDBOOK OF ELECTRICITY, MAGNE-TISM, AND ACOUSTICS. By DR. LARDNER.
Seventh Thousand. Revised and completed to 1866 by GEORGE CAREY FOSTER, F.C.S., Professor of Experimental Physics in University College, London. 400 Illustrations. Small 8vo, 5s.

"The book could not have been entrusted to anyone better calculated to preserve the terse and lucid style of Lardner, while correcting his errors, and bringing up his work to the present state of scientific knowledge. All we can say of the Editor's actual labours, is, that he has added much new matter to the old text, and that he has modified some of Dr. Lardner's statements in accordance with currently accepted doctrines and recent investigations. The work addresses itself to those who, without a profound knowledge of Mathematics, desire to be familiar with experimental physics, and to such we especially recommend it."— *Popular Science Review.*

HANDBOOK OF MECHANICS. By DR. LARD-NER.
Seventh Thousand. 357 Illustrations. Sm. 8vo, 5s.

HANDBOOK OF HYDROSTATICS, PNEU-MATICS, AND HEAT. By DR. LARDNER.
Seventh Thousand. 292 Illustrations. Small 8vo, 5s.

HANDBOOK OF OPTICS. By DR. LARDNER.
Fifth Thousand. 290 Illustrations. Small 8vo, 5s.

** *The above 4 Volumes form a Complete Course of Natural Philosophy.*

A FIRST BOOK OF NATURAL PHILO-SOPHY;
an Introduction to the Study of Statics, Dynamics, Hydrostatics, Optics, and Acoustics, with numerous Examples. By SAMUEL NEWTH, M.A., Fellow of University College, London. Eighth Thousand, with large additions. 12mo, 3s. 6d.

** This work embraces all the subjects in Natural Philosophy required at the Matriculation Examination of the University of London. The present edition has been to a large extent re-written, the points found to present most difficulty have been simplified, fresh matter has been introduced and new examples have been added.

NEWTH'S ELEMENTS OF MECHANICS,
including Hydrostatics, with numerous Examples. Fourth Edition. Small 8vo, 8s. 6d.

** The First Part contains all the subjects in Mechanics and Hydrostatics required for the B.A. and B.Sc. Examinations of the University of London.

AN ELEMENTARY TREATISE ON ME-
CHANICS, for the Use of the Junior University Students. By RICHARD POTTER, A.M., late Professor of Natural Philosophy in University College. Fourth Edit. 8s. 6d.

POTTER'S TREATISE ON OPTICS. Part I.
All the requisite Propositions carried to First Approxima-tions, with the construction of Optical Instruments. Third Edition. 8vo, 9s. 6d.

Part II. The Higher Propositions, with their applica-tion to the more perfect forms of Instruments. 8vo, 12s. 6d.

POTTER'S PHYSICAL OPTICS; or the Nature
and Properties of Light. A Descriptive and Experimental Treatise. 100 Illustrations. 8vo, 6s. 6d.

A GUIDE TO THE STARS. In Eight Plani-
spheres, showing the Aspect of the Heavens for every Night in the Year. 8vo, 5s.

COMMON THINGS EXPLAINED. By DIONY-
SIUS LARDNER, D.C.L. Containing: Air—Earth—Fire —Water—Time—The Almanack—Clocks and Watches— Spectacles — Color —Kaleidoscope — Pumps — Man—The Eye—The Printing Press—The Potter's Art—Locomotion and Transport—The Surface of the Earth. 233 Cuts. 5s.

STEAM AND ITS USES. By DIONYSIUS LARD-
NER, D.C.L. 89 Illustrations. 2s.

POPULAR GEOLOGY. By DIONYSIUS LARDNER,
D.C.L. 201 Illustrations. 2s. 6d.

POPULAR ASTRONOMY. By DR. LARDNER.
Containing: How to Observe the Heavens—Latitudes and Longitudes—The Earth—The Sun—The Moon—The Planets : are they Inhabited ?—The New Planets—Lever-rier and Adams's Planet—The Tides—Lunar Influences— The Stellar Universe—Light—Comets—Cometary Influ-ences—Eclipses—Terrestrial Rotation—Lunar Rotation— Astronomical Instruments. 182 Illustrations. 4s. 6d.

POPULAR PHYSICS. By DR. LARDNER. Con-
taining : Magnitude and Minuteness — Atmosphere — Thunder and Lightning — Terrestrial Heat — Meteoric Stones—Popular Fallacies—Weather Prognostics—Ther-mometer—Barometer—Safety Lamp—Whitworth's Micro-metric Apparatus — Electro-Motive Power—Sound—Magic Lantern — Camera Obscura — Camera Lucida — Looking Glass—Stereoscope—Science and Poetry. 85 Illustns. 2s.6d.

THE MICROSCOPE. By DIONYSIUS LARDNER, D.C.L. 147 Engravings. 2s.

THE BEE AND WHITE ANTS. Their Manners and Habits. By DR. LARDNER. With Illustrations of Animal Instinct and Intelligence. 135 Illustrations. 2s.

*** *The Seven preceding Volumes are from the "Museum of Science and Art."*

HENRY'S GLOSSARY OF SCIENTIFIC TERMS for General Use. 12mo, 3s. 6d.

"To students of works on the various sciences, it can scarcely fail to be of much service. The definitions are brief, but are, nevertheless, sufficiently precise and sufficiently plain; and in all cases the etymologies of the terms are traced with care."—*National Society's Monthly Paper.*

Chemistry.

DR. HOFMANN'S MODERN CHEMISTRY. Experimental and Theoretic. Small 8vo, 4s. 6d.

"It is in the truest sense an introduction to chemistry; and as such it possesses the highest value—a value which is equally great to the student, new to the science, and to the lecturer who has spent years in teaching it."—*Reader.*

BUNSEN'S GASOMETRY: comprising the leading Physical and Chemical Properties of Gases, with the Methods of Gas Analysis. By DR. ROSCOE. 8vo, 3s. 6d.

Baron Liebig's Works.

"Side by side, as long as husbandry shall last, will these three names shine in co-equal glory—Antoine Lavoisier, Humphry Davy, Justus Liebig. To Lavoisier belongs the noble initiation of the work; to Davy, its splendid prosecution; to Liebig, its glorious consummation. Embracing in his masterly induction the results of all foregone and contemporary investigation, and supplying its large defects by his own incomparable researches, Liebig has built up on imperishable foundations, as a connected whole, the code of simple general laws on which regenerated agriculture must henceforth for all time repose."—*International Exhibition Report*, 1862.

THE NATURAL LAWS OF HUSBANDRY. 8vo, 10s. 6d.

FAMILIAR LETTERS ON CHEMISTRY, in its Relations to Physiology, Dietetics, Agriculture, Commerce, etc. Fourth Edition. Small 8vo, 7s. 6d.

LETTERS ON MODERN AGRICULTURE. Small 8vo, 6s.

In a New and Elegant Binding for a Present.

In 6 Double Volumes, handsomely bound in cloth, with gold
ornaments and red edges, Price £1 1s.

LARDNER'S MUSEUM OF SCIENCE AND ART,

containing :—The Planets ; are they inhabited
Worlds? — Weather Prognostics — Popular Fallacies in
Questions of Physical Science—Latitudes and Longitudes—
Lunar Influences—Meteoric Stones and Shooting Stars—
Railway Accidents—Light—Common Things: Air—Loco-
motion in the United States—Cometary Influences—Com-
mon Things: Water—The Potter's Art—Common Things:
Fire—Locomotion and Transport, their Influence and
Progress—The Moon—Common Things—The Earth—The
Electric Telegraph—Terrestrial Heat—The Sun—Earth-
quakes and Volcanoes—Barometer, Safety Lamp, and
Whitworth's Micrometric Apparatus—Steam—The Steam
Engine—The Eye—The Atmosphere—Time—Common
Things: Pumps—Common Things: Spectacles, the Kalei-
doscope—Clocks and Watches—Microscopic Drawing and
Engraving — Locomotive — Thermometer — New Planets :
Leverrier and Adams's Planet—Magnitude and Minuteness
—Common Things: the Almanack—Optical Images—
How to Observe the Heavens—Common Things : the
Looking-glass—Stellar Universe—The Tides—Colour—
Common Things: Man—Magnifying Glasses—Instinct
and Intelligence—The Solar Microscope—The Camera
Lucida—The Magic Lantern—The Camera Obscura—The
Microscope—The White Ants : their Manners and Habits
—The Surface of the Earth, or First Notions of Geography
—Science and Poetry—The Bee—Steam Navigation—
Electro-Motive Power—Thunder, Lightning, and the Au-
rora Borealis—The Printing Press—The Crust of the Earth
—Comets—The Stereoscope—The Pre-Adamite Earth—
Eclipses—Sound.

"The 'Museum of Science and Art' is the most valuable contribution
that has ever been made to the Scientific Instruction of every class of
society."—*Sir David Brewster in the North British Review.*

"The whole work, bound in six double volumes, costs but the price of
a Keepsake; and whether we consider the liberality and beauty of the
illustrations, the charm of the writing, or the durable interest of the
matter, we must express our belief that there is hardly to be found among
the new books, one that would be welcomed by people of so many ages
and classes as a valuable present."—*Examiner.*

*** *The Work may also be had in 12 single Volumes 18s. Ornamental
Boards, or handsomely half-bound morocco, 6 Volumes, £1 11s. 6d.*

Popular Physiology.

ANIMAL PHYSIOLOGY FOR SCHOOLS.
By Dr. LARDNER. With 190 Illustrations. Second Edition. 1 vol. 3s. 6d.

Professor De Morgan's Works.

ELEMENTS OF ARITHMETIC.
By AUGUSTUS DE MORGAN, late Professor of Mathematics in University College, London. Eighteenth Thousand. Small 8vo, 5s.

THE BOOK OF ALMANACKS.
With an Index of Reference by which the Almanack may be found for every year, up to A. D. 2000. 5s.

FORMAL LOGIC ;
or, the Calculus of Inference, Necessary and Probable. 8vo, 6s. 6d.

SYLLABUS OF A PROPOSED SYSTEM OF LOGIC.
8vo, 1s.

Mathematical Works and Tables.

NEWTH'S MATHEMATICAL EXAMPLES ;
a Graduated Series of Elementary Examples in Arithmetic, Algebra, Logarithms, Trigonometry, and Mechanics. Crown 8vo, 8s. 6d.

TABLES OF LOGARITHMS COMMON AND TRIGONOMETRICAL TO FIVE PLACES.
Under the Superintendence of the Society for the Diffusion of Useful Knowledge. Fcap. 8vo, 1s. 6d.

FOUR-FIGURE LOGARITHMS AND ANTI-LOGARITHMS.
On a Card, 1s.

BARLOW'S TABLES OF SQUARES, CUBES, SQUARE ROOTS, CUBE ROOTS AND RECIPRO-CALS of all Integer Numbers up to 10,000. Royal 12mo, 8s.

LESSONS ON FORM ;
an Introduction to Geometry, as given in a Pestalozzian School, Cheam, Surrey. By CHARLES REINER. 12mo, 3s. 6d.

"It has been found in the actual use of these lessons, for a considerable period, that a larger average number of pupils are brought to study the Mathematics with decided success, and that all pursue them in a superior manner."—*Rev. Dr. Mayo.*

LESSONS ON NUMBER; as given in a Pestalozzian School, Cheam, Surrey. By CHARLES REINER. The Master's Manual. New Edition. 12mo, cloth, 5s.

A COURSE OF ARITHMETIC, as Taught in the Pestalozzian School, Worksop. By J. L ELLENBERGER. Post 8vo, 5s.

Logic.

ART OF REASONING: a Popular Exposition of the Principles of Logic, Inductive and Deductive. By SAMUEL NEIL. Crown 8vo, 4s. 6d.

AN INVESTIGATION OF THE LAWS OF THOUGHT, on which are founded the Mathematical Theories of Logic and Probabilities. By the Late GEORGE BOOLE, LL.D., Professor of Mathematics in Queen's College, Cork. 8vo, 14s.

Monetary History of England.

THE MYSTERY OF MONEY Explained and Illustrated by the Monetary History of England from the Norman Conquest to the present time. 8vo, 7s. 6d.

Economy in Food.

DR. EDWARD SMITH'S PRACTICAL DIETARY FOR FAMILIES, SCHOOLS, AND THE LABOURING CLASSES. Fourth Thousand. Small 8vo, 3s. 6d.

CHAP. I. Elements of Food which the Body requires.—II. What Elements can be supplied by Food.—III. Qualities of Foods: Dry Farinaceous Foods—Fresh Vegetables—Sugars—Fats, Oil, &c.—Meats—Fish—Gelatin—Eggs—Cow's Milk—Cheese—Tea and Coffee—Alcohols—Condiments.—IV. Dietary for Families—Infancy—Childhood—Youth—Adult and Middle Life—Old Age.—V. Dietary in Schools.—VI. Dietary of the Labouring Classes—Present Mode of Living—Most suitable Dietary—Best arrangement of Meals—Specimens and Proposed Dietaries.—VII. Dietary of the Labouring Classes. Cooking Depôts and Soup Kitchens.

"Dr. Smith's book is by far the most useful we have seen upon all the practical questions connected with the regulation of food, whether for individuals or families."—*Saturday Review.*

Sir George Ramsay's Works.

THE MORALIST AND POLITICIAN; or,
Many Things in Few Words. Fcap 8vo, cloth, 5s.

"A book which reminds us in its style of some parts of Coleridge's
'Aids to Reflection,' without affecting to emulate its power. Without
being profound, it is thoughtful and sensible."—*Notes and Queries.*

INSTINCT AND REASON; or, The First Principles of Human Knowledge. Crown 8vo, 5s.

PRINCIPLES OF PSYCHOLOGY. 8vo, 10s. 6d.

Drawing.

LINEAL DRAWING COPIES for the Earliest
Instruction. 200 subjects on 24 sheets mounted on 12
pieces of thick pasteboard. By the Author of "DRAWING
FOR YOUNG CHILDREN." 5s. 6d.

By the same Author:

EASY DRAWING COPIES FOR ELEMEN-
TARY INSTRUCTION. Simple Outlines without
Perspective. 67 Subjects. Price 6s. 6d.

Sold also in Two Sets:

Set I.   Price 3s. 6d.   Set II.   Price 3s. 6d.

PERSPECTIVE :   ITS   PRINCIPLES   AND
PRACTICE. By G. B. MOORE, Teacher of Drawing in
University College.   In Two Parts.   Text and Plates.
8vo, cloth, 8s. 6d.

Singing.

THE SINGING MASTER COMPLETE.   1 vol.
8vo, 6s. cloth.
   I. First Lessons in Singing, and the Notation of Music.
        8vo, 1s.
  II. Rudiments of the Science of Harmony. 8vo, 1s.
 III. The First Class Tune Book. 8vo, 1s.
  IV. The Second Class Tune Book. 8vo, 1s. 6d.
   V. The Hymn Tune Book.   70 Popular Psalm and
        Hymn Tunes. 8vo, 1s. 6d.

⁎⁎ Any Part may be purchased separately.

WERTHEIMER, LEA AND CO., PRINTERS, FINSBURY CIRCUS.

www.ingramcontent.com/pod-product-compliance
Lightning Source LLC
Chambersburg PA
CBHW030819270326
41928CB00007B/803